## DATE DUE

| | | | |
|---|---|---|---|
| | | | |
| | | | |
| | | | |
| | | | |
| | | | |
| | | | |
| | | | |
| | | | |
| | | | |
| | | | |
| | | | |
| | | | |
| | | | |
| | | | |
| | | | |
| | | | |
| | | | |
| | | | |

# Biology and Revolution
# in Twentieth-Century China

# Asia/Pacific/Perspectives
## Series Editor: Mark Selden

# Biology and Revolution in Twentieth-Century China

LAURENCE SCHNEIDER

ROWMAN & LITTLEFIELD PUBLISHERS, INC.
*Lanham • Boulder • New York • Toronto • Oxford*

ROWMAN & LITTLEFIELD PUBLISHERS, INC.

Published in the United States of America
by Rowman & Littlefield Publishers, Inc.
A wholly owned subsidiary of The Rowman & Littlefield Publishing Group, Inc.
4501 Forbes Boulevard, Suite 200, Lanham, MD 20706
www.rowmanlittlefield.com

P.O. Box 317, Oxford OX2 9RU, United Kingdom

British Library Cataloguing in Publication Information Available

**Library of Congress Cataloging-in-Publication Data**

Schneider, Laurence A., 1937–
    Biology and revolution in twentieth-century China / Laurence Schneider.
        p. cm.—(Asia/Pacific/Perspectives)
    Includes bibliographical references and index.
    ISBN 0-7425-2696-8 (alk. paper)
        1. Biology—China—History—20th century. 2. Genetics—China—History—20th century. 3. Communism and science—China—History—20th century. I. Title. II. Series.

QH305.2.C6S36 2003
570'.951'0904—dc21

                                                                        2003046651

Printed in the United States of America

♾™ The paper used in this publication meets the minimum requirements of American National Standard for Information Sciences—Permanence of Paper for Printed Library Materials, ANSI/NISO Z39.48-1992.

**For Lilly**
**Number one grandchild**

# Contents

## Contents

# Abbreviations

| | |
|---|---|
| BDCC | Biographic Dictionary of Chinese Communism |
| BDRC | Biographical Dictionary of Republican China |
| BIC | Biotechnology in China |
| CASS | Zhongguo kexueyuan shengwuxue ge yanjiuso |
| CMB | China Medical Board |
| DCB | Directory of Chinese Biologists |
| DSSI | Directory of Selected Science Institutions Mainland China |
| FBIS | Foreign Broadcast Information Service |
| GMRB | Guangming ribao |
| JFRB | Jiefang ribao |
| JPRS | Joint Publications Research Service |
| KRZZ | Kangri zhanzheng shiji jiehfang qu kexue ishu fazhan shiliao |
| KTGT | Keji tizhi gaige di tansuo |
| KXTB | Kexue tongbao |
| NCNA | New China News Agency |
| RAC | Rockefeller Archive Center |
| RMRB | Renmin ribao |
| SCMP | Survey of the China Mainland Press |
| SXTB | Shiwuxue tongbao |
| YTJ | Yichuanxue taolun ji |
| ZBT | Zeran bianzhengfa tongxun |
| ZKJ | Zhongguo kexuejia zedian |
| ZXN | Zhongguo xiandai nongxue jiachuan |
| ZXS | Zhongguo xiandai shengwuxue jiachuan |

# Acknowledgments

My deepest appreciation and thanks to James F. Crow and Richard P. Suttmeier for carefully reading and critiquing the entire manuscript for this book. Encouragement in the form of expert advice and valuable research materials came graciously from Li Peishan and Fan Dainian when this project was in its infancy. The faculty, staff, and libraries at Fudan University and the Chinese Academy of Sciences Institute of Genetics provided invaluable cooperation and guidance. I am especially grateful to the Chinese scientists and educators in the PRC, Taiwan, and the United States who generously agreed to be interviewed for this study.

The last stages of research and the time to put a manuscript together were made possible by a 1997-1998 grant from the National Science Foundation, Science and Technology Studies Program, in conjunction with a sabbatical leave from the history department of Washington University.

And finally, my enthusiastic thanks to Mark Selden for his perceptive and challenging high-octane editorial advice and encouragement.

# Introduction:
# Controlling Nature, Science, and Scientists

This is a study of how modern science was transferred to China, how it was established there and diffused throughout culture and institutions. It explores the far-reaching and contentious political and developmental consequences of these processes. The history of genetics and evolutionary theory in China is used here—as a case study—to exemplify the centrality of the scientific enterprise to the fierce political conflicts of revolutionary and reform agendas, and to the development process itself.

Genetics and evolutionary theory are followed from their first appearance in China in the 1920s to their much transformed condition in the People's Republic at the turn of the century. The narrative demonstrates that by 1949 a small, intellectually robust and—by international standards—sophisticated community of geneticists was created in China as a cumulative result of Chinese training in the United States, American philanthropy and missionary schools in China, and the educational policies of the Nationalist government. After 1949, that community was subjected to harshly oscillating state and Party policies that virtually destroyed it through a combination of Lysenkoism and repression, before its restoration and flourishing in the final decades of the twentieth century. Altogether, this eighty-year-long history is rooted in the selected, exemplary careers that are followed from the story's beginning to its conclusion. It is, in effect, local history. The narrative not only provides a continuous examination of the professional lives of practicing scientists, it deals as well with the creation and operation of scientific institutions in which these people worked, and the product of

1

their work in research and education.

A wide spectrum of concerns faced by all the sciences was confronted with particular sharpness by genetics and evolutionary theory over the course of their development: dependency on foreigners; applied versus basic science; the role of the state and the Party; the conflict of science and revolutionary ideology; the relationship between science and development. All of these came into play quite dramatically in the 1950s when the Chinese Communist Party adopted Soviet Lysenkoism, banned classical genetics education, research, and publication, and sought to develop a "people's science" to energize economic development. This was the defining moment for genetics and evolutionary theory in twentieth-century China: It determined the fate of all effort and accomplishment that came before; and it became a closely watched microcosm of the troubled relations between science and the state after 1949. It also sharply framed the issues that would subsequently reverse Lysenkoism's hegemony over biology in China. Because of Lysenkoism's moral and intellectual importance to the entire science community in China, a detailed description and analysis of it takes place at the narrative's midpoint where it becomes clear that the perception and treatment of geneticists by Lysenkoism anticipated the later treatment of the science community during the Great Leap Forward (1957-1958) and the Cultural Revolution (1966-1976).

A detailed analysis of the post-Cultural Revolution efflorescence of genetics concludes this study. Here, there is an opportunity to see in detail, at the local level, how complex national politics and a profusion of science policy actually affected the growth and performance of a discrete area of science with important implications for Chinese development.

The narrative and analytical approach taken here assumes that while one sorts out the messiness of individual careers and institutional development, the actual research conducted by the scientists must be kept in plain view. It is assumed that one cannot speak meaningfully about scientists and a science community without describing and evaluating their research. That research provides the most reliable insight into the quality of their training, their preparation for teaching, and their ability to mediate the transfer of science from the international science community to China. It also permits us to gauge the stakes in high-profile political campaigns that sought to challenge, control, and restructure scientific enterprise in the service of state goals. For in China as elsewhere, the course of modern science has been embedded in the political and economic priorities of states and social revolutions.

In brief, the narrative details the shifting combination of forces acting on the development of the scientific enterprise, on particular scientists' careers, and on the development of a variety of complex educational or research institutions. It describes the influences on science of unanticipated interactions of government science policy and ubiquitous foreign influences.

## Analyses and Arguments

Among the organizing structures of this study are themes of *control*: the control of nature with science, the control of science by nation and state, the control of scientists by state and party. In China, among the central questions raised repeatedly throughout the twentieth century by politicians and scientists alike are how to modernize rapidly without becoming dependent on other nations and without compromising egalitarian social values. Some leaders of the Chinese Communist Party found answers to these questions in populist conceptions of science that were offered as alternatives to modern/Western, professional science. In their aggregate, these alternatives are designated as "mass science," wherein, it was argued, the control of nature, science, and scientists could be achieved.

## Controlling Nature

> For the purpose of attaining freedom in society, man must use social science to understand and change society and carry out social revolution. For the purpose of attaining freedom in the world of nature, man must use natural science to understand, conquer, and change nature and then attain freedom from nature.

> —Mao Zedong, 1940[1]

> We cannot wait for kindnesses from nature; our task is to wrest them from her.

> —Motto of Lysenkoism, oft-quoted in China during 1950s

Understanding nature has seldom been entertained as an end in itself in modern China. Science has been thought of as an instrumentality, and controlling and transforming nature its assigned goal. Often, this has created a bias for applied science and an impatient, even denigrating attitude toward basic ("pure") science. Furthermore, some influential scientists and policymakers have adhered to a transformist (or voluntarist) understanding of science—a belief that nature (like society) is infinitely malleable, and that once the proper methods are employed, humanity will not have "to beg favors from nature" but will be able to command it and shape it at will.

Modern Chinese scientists (before the 1980s) rarely expressed any awareness of, let alone respect for, earlier Chinese ecological traditions (daoism and fengshui for example) that emphasized natural balances and harmonious coexistence with nature. Before and after 1949, governments—with encouragement from the United States or the Soviet Union—promoted mammoth projects to control rivers, mine resources, expand railroads, and increase industrial production. And like the United States, the Soviet Union, and Japan among many others, little regard was given to environmental consequences. In its rhetorical extreme, during Mao's Great Leap Forward in 1958, the pursuit of these goals was characterized as a militant and rapacious "war on nature."[2] Beginning in the 1980s, notwithstanding Chinese environmentalists' efforts, government sponsored megaprojects (particularly in hydropower and water diversion) have been launched, transforming Mao's war on nature into a battle on an unprecedented scale in China.

Not until the very recent advent of biotechnology in China, did the field of genetics even remotely come to fit the above profile of science. For most of the period surveyed in this study, the field of genetics displayed no impatience with basic science and no utopian transformism—perhaps because of the relative youth of the field, perhaps because of the training of its first Chinese practitioners in the cautious, highly disciplined, experimental research tradition of T. H. Morgan. Even in the field of agricultural genetics, explicitly mandated to increase the food supply, these same constraints and attitudes prevailed.

The first generations of these American- (and some European-) trained geneticists were all but overwhelmed with the responsibilities of institution building, teaching, and writing or translating textbooks. Nevertheless, many geneticists were actively engaged in research, and much of that research—especially in plant genetics—had direct and successful applications. After 1949 and under the influence of Soviet Lysenkoism, however, the practical, applied contributions of earlier research were ignored or denied, and the entire genetics community was accused of self-indulgent play at puzzles and irrelevant speculation. Genetics was banned in education, research, and publication on the grounds of its "bourgeois idealism" which showed itself most clearly, it was charged, in the geneticists' "passive" attitude toward nature. (They were in good company: Darwin, as we shall see, was accused of the same malfeasance.)

Lysenkoites and the CCP charged that the scientific attitude of geneticists and the content of their science were totally useless for achieving the only appropriate goal of science—the control of nature and its manipulation for the benefit of the nation and the masses. The reality, however, was that Chinese agricultural scientists, using advanced genetics made demonstrable contributions to the improvement of food grains, industrial plants, and even strains of silk worms. On the other hand, Lysenkoist biology could neither support its grandiose claims to producing new species of plants nor even its more modest claims to improving common food plants.

It may not be easy in the early twenty-first century to think of genetics, now transforming the world of nature before our eyes, as a "passive" science. We might therefore get a better feel for the powerful objections to genetics by looking a bit further at the transformists' particular concerns with some of its fundamental concepts and principles.

The transformist tradition that developed in China adhered to a belief that organic life was fundamentally shaped by environment, not by any fixed, continuous substance or structure within organisms (like chromosomes or genes). In a scheme very close to that of Jean Baptiste Lamarck's (1744-1829) ideas, the transformist believed environmental pressures to be the source of organic change and evolution. Like Lamarck, it was further believed that organic changes brought about by the environment could be inherited by the progeny of the changed organism. Modern genetics, while taking full cognizance of the role of environment in biological reproduction and evolution, has of course taken as one of its principles the noninheritability of "acquired characteristics." This position, combined with the concept of the gene as the medium of inheritance was considered by the transformists effectively to obviate the prospect of control promised by their vision, a vision based on the assumption that if one controls the environment, one controls the organism, heredity, and the direction and pace of evolution and development.

The modern transformist tradition in China appears to have originated within the utopian socialist thought that was so attractive to Chinese intellectuals from the early twentieth century. Most prominent here were the formulations of Peter Kropotkin (1842-1921), the expatriate Russian geographer and anarchist social revolutionary. Kropotkin's works were translated into Chinese and widely circulated during the first two decades of the century.[3]

Kropotkin's transformist attitude toward nature underpinned his optimistic belief that humanity, through "direct action," would be able to control the natural environment and hence the direction of biological evolution.[4] His attitude toward nature, however, was tempered by his belief in its organic unity and the integral relation of humanity and nature—both subject to the same natural processes.[5]

Kropotkin's transformist ideas, imbedded in his anarchist social philosophy, initially came to China by way of Japan and France. Further contact with elements of other transformist traditions came from Chinese who studied agricultural science in Japan, and from Chinese biologists who acknowledged a strong affinity for Lamarckian principles of heredity and evolution, thanks to their advanced training in biology in France or in the Low Countries during the 1920s and 1930s.

When Soviet Lysenkoism introduced its version of biological transformism to China in 1949, its rapid and widespread acceptance was in part attributable to the existence of a diverse community of scientists and nonscientific intellectuals committed to Lamarckism and a transformist attitude toward nature. Lysenkoism's motto (guided by Marx's critique of philosophy) stated that where the "old biology" had devoted itself merely to understanding nature, Lysenkoism's new aim for biology was to change nature. And it claimed to be replacing Darwin's "passive" attitude toward evolution with its determined efforts to guide the evolutionary process using its alleged ability to create new species in short order.

## Controlling Science

We mustn't copy everything indiscriminately and transplant mechanically.
—Mao Zedong, "On the Ten Major Relationships," 1956

For those Chinese devoted to the control of nature with modern science, there was a persistent dilemma. The process of acquiring the necessary science ostensibly required dependency on foreigners—whether Americans during the Republic or Soviets in the first decade of the People's Republic—and raised uncomfortable questions for the Chinese about China's global cultural status. Sometimes, this generated various intellectual subterfuges to mitigate the sense of dependency. In the latter part of the nineteenth century, for example, government officials sponsored Western science and technology in educational institutions and industrial enterprises devoted primarily to enhancing Chinese self-defense. The officials defended these activities with the famous formula that declared that "Chinese learning" remained China's essential core, superior to "Western learning" which was a mere utilitarian device, adopted to protect and preserve the Chinese core.

By the early twentieth century, Chinese reformers and revolutionaries alike were abandoning this formula and openly advocating the adoption of scientific culture as a necessary means for China to energize its lagging social evolution and thereby to survive the challenges of foreign incursion and competition. And if it meant that—for the time being—the Chinese would have to acquire their science from the West, it was argued that science was not inherently "Western." Rather, it was "modern," that is, the product of a universal process of social and cultural evolution that all societies would experience unless they were blocked from following their natural proclivities. This argument was rarely helped along by scientists or nonscientists with references to China's own scientific or proto-scientific traditions. Certainly, in biology, such efforts were at best erratic and superficial.

In this formulation, science takes on a very powerful historical role. It is both a product of social evolution and a continual source of social progress. For some, it followed that the establishment of scientific culture did not in itself require that Chinese society reshape itself any further in the image of the West.

While these rationalizations may have been satisfying to many intellectuals and political leaders, there was nevertheless the persistent reality that during the first decades of the twentieth century the development of science in China was largely in the hands of foreigners, rested on Chinese researchers trained abroad, and was heavily subsidized by foreign funds. Alongside a small number of Chinese schools, missionary schools at the secondary and college levels pioneered science curricula and eventually created institutions that trained China's first generations of scientists. Missionaries likewise mediated the transfer of Western medicine and medical education.

In missionary and Chinese schools alike, foreign dominion over science was further evidenced in the use of foreign textbooks, some translated, but many in their original language—English, German, or French. In some instances those same languages were used in the classroom, depending on the science in question and the location of the classroom. Educators (foreign and Chinese) expressed growing concerns that the foreign textbooks never addressed the flora, fauna, or geology of China. And they lamented the lack of systematic efforts to create a standardized Chinese vocabulary for scientific terminology in all of the sciences, essential for useful translations of foreign textbooks as well as for writing original Chinese texts.[6]

In spite of missionary and Chinese efforts, a proper undergraduate education in many of the sciences was, until at least the mid-1920s, rare in China. Thousands flocked to the United States, Japan, and Europe for their scientific foundations and on their return to China made it possible to implement the first stage of the "nationalization" of science—replacing foreigners with Chinese science teachers.

Over the course of the 1920s and early 1930s, China's appropriation of science from foreigners was encouraged by virulent antiforeign, anti-Christian political movements. The Guomindang (Nationalist) government, on its accession to power in 1928, was markedly devoted to this appropriation. The new regime was committed, often well beyond its resources, to the rapid expansion of scientific and technical education in national higher education. This commitment was informed by a number of overlapping visions pursued by intellectual leaders of the Nationalists and by others outside the Party who saw it as the only available means to achieve their goals. For the former, science was seen as a necessary instrument for building a strong industrial state; for some of the latter, science was the source of an enlightened society and the achievement of progress in all areas of life. American Progressivism, especially John Dewey's pragmatism, strongly guided Chinese attitudes toward both of these approaches to modern science.

The Guomindang facilitated the development of science curricula, modeled largely on those of the United States, and it also established quotas for the percentage of science and technical majors in order to increase them rapidly. Additionally, the government created modern China's first academy of sciences, the Academia Sinica, conceived as the national center for planning and conducting scientific research. It was meant to be a symbol of China's cultural sophistication and independence.

The nationalistic intent of these efforts was explicit and expressed in policy goals like the "Sinicization of science." New or restructured public colleges and universities, like National Central and Qinghua, were proud to have an all-Chinese faculty, especially in the sciences. Chinese language textbooks, translations and originals, became more frequently used in the science classrooms; and national agencies were undertaking the endless task of creating standardized Chinese science terminology and publishing the results in lexicons.

By the early 1930s, the Guomindang regime exhibited significant ambivalence toward the missionary community. On one hand, the Party denounced imperialism and the missionary role in it; yet the missionaries were useful to national programs in science education or rural reform, and sometimes proved to be sympathetic and effective intermediaries with their home governments.

Guomindang ambivalence toward the foreign presence in China was attributable in part to the strong role played by foreign philanthropic institutions in promoting scientific development. Initially, the Party attempted to direct the use of foreign funds to its own, often technology-intensive, projects—like expanding the national railroad system. It was also critical of the continued flow of the majority of these funds into science programs at missionary schools rather than Chinese schools.

The Rockefeller Foundation, as we shall see, put large amounts of money into this effort, especially into biomedical education and research. Its work was complemented by the China Foundation, created and protected by treaties with the United States for the explicit purpose of developing China's science education and research. Since the Guomindang was perennially short of funds to carry out its ambitious science programs, it was never inclined—on nationalistic or anti-imperialist grounds, or otherwise—to prohibit the foundations' largesse to science in the public schools and in the Academia Sinica. The government's science policies and those of the foundations eventually worked together smoothly.

Hence, before 1949, the Chinese state—especially under the Guomindang—consistently sought foreign science and technology; and, over time, it exercised greater agency through de facto cooperation with foreigners who promoted and sustained the development of science in China. Likewise, within China's modern intelligentsia and political leadership that development was strongly desired and encouraged. Additionally, over time, significant foreign efforts became explicitly devoted to facilitating the creation of a self-perpetuating Chinese scientific community, independent of foreign control and attentive to the needs of China.

For all of these reasons, the transfer of modern science to China (from the West or from the Soviet Union) will not be viewed here as a conventional case of "cultural imperialism" in the simplistic sense of the unilateral imposition by an imperial power of its ideas or institutions upon a helpless colonized society. Nor will it be viewed from an understanding of cultural imperialism that holds that the imperialist nations encouraged the development of those scientific activities that served their own exploitative designs. (Thus, developing the fields of geology and botany, for example, could provide information of use to foreign commercial interests.)[7] Whatever merit these conceptions might have, neither confronts the complexity of motivations nor the degrees of agency on the part of the Chinese in their encouragement of scientific development.

It is nevertheless fruitful to consider the analyses of cultural imperialism put forward by leading Chinese parties, particularly as they referred to biology. The transfer of biology will be problematized here as cultural imperialism in two different instances—the first expressed by the Guomindang, the second by the

Chinese Communist Party. The Guomindang implied a notion of cultural imperialism when it occasionally criticized foreign controls of science that neglected issues appropriate to China's needs. Guomindang leadership argued that the Chinese state ought to be sovereign in deciding what sciences to develop and for what purposes.

From at least the late 1930s, important elements of the Chinese Communist Party used a more comprehensive "hegemonic" assessment of foreign methods and motives for promoting modern science and technology in China. Their argument (which would reappear in much refined form in the 1950s and 1960s) suggests that in addition to all of the selfish and crass motivations there is a much more important political and ideological one: The scientific and technical approach to modernization was promoted by Americans, Europeans, *and* the Guomindang as an alternative to social revolution. This approach would (hopefully) keep China out of the hands of the revolutionaries and open for business with the West; and it would create a professional, technocratic intelligentsia whose cultural umbilical is tied to the bourgeois, capitalist, counter-revolutionary West. This latter analysis does not deny to China the importance of science and technology development. Rather, it raises the profoundly difficult question of finding a form of development that is not dependent on the capitalist West and is compatible with social revolution.[8]

After 1949, even though the Western missionary schools and philanthropies were gone from China, the CCP's concerns about *who controls science* resonated with significant parts of the science community and the intellectuals more broadly. We shall see that the problem of dependency—this time however on the Soviet Union—was revisited; and many members of the science community (especially during the 1956 Hundred Flowers campaign) came to accuse the CCP of indiscriminate borrowing and a humiliating re-creation of the cultural problems they had worked so hard to resolve decades earlier. (For example, they charged that there was excessive reliance on the Russian language, Russian textbooks, and Soviet technical advisers, and the imposition of Soviet scientific biases.) Some Chinese scientists argued that—thanks to the policies of the CCP—Soviet science had become no less colonial than the Western science from which the CCP and the Soviet Union were supposed to have rescued China. Dependence, however, was only part of the problem, for many Chinese scientists trained in the United States, Europe, and Japan charged that China was borrowing *bad* science—Lysenkoism being the clearest example. In the latter half of the present study, we shall see genetics and evolutionary theory caught up in this rancorous debate that was fueled both by the problem of dependence and by conflicting scientific paradigms, particularly those of East and West, categories sharpened both by the Chinese revolution and the Cold War.

At the heart of all this controversy lay the third of my designated control problems—the control of scientists.

## Controlling Scientists

> It is . . . wrong to label a particular theory in medicine, biology or any other
> branch of natural science 'feudal,' 'capitalist,' 'socialist,' 'proletarian,' or
> bourgeois.
>
> —Lu Dingyi, "Let a Hundred Flowers Blossom," 1956

Before and after 1949, the Chinese state has periodically expressed reluctance to become dependent on foreign science in the various ways suggested
above. After 1949, it has likewise expressed a desire not to become dependent
on Chinese scientific professionals. Guomindang tensions with the professional
science community had been no different than its problems with the intellectual
community as a whole; however, they were fundamentally different from the
most serious problems that the science community had with the CCP.

The Guomindang had problems at one time or another with all elements of
the intellectual community, including scientists; but there was nothing inherent
in the nature of modern/Western science or the professional scientific community that antagonized it. There was, however, always the potential that any cor-
porate group in society might constitute a threat or challenge to the ultimate cor-
porate authority that the Guomindang explicitly sought for itself. Among its
most provocative political actions during the late 1920s and 1930s, for example,
were its efforts to impose party control (danghua) on education and the Academia Sinica. There were intellectuals of all kinds who resisted Guomindang politicization and who dissented from one or another policy; some of these just hap-
pened to be scientists. On the other hand, some of the Guomindang's staunchest
supporters were scientists.

At the core of the CCP's troubled relationship with the science community
was a set of overlapping concerns: The Chinese Communist Party's relentless
combination of mobilization and egalitarianism were seen by some to be jeop-
ardized by specialized scientific experts divorced from the revolution. Revolutionaries feared that scientists were becoming an elite concerned with enhancing
their status and committed to their occupational position rather than the nation
and the people. They were out of touch with China's needs because—isolated in
urban universities and laboratories—they were cut off from the masses, from
labor and production, from nature. In Franz Schurmann's characterization, "the
professional intellectuals [during the Great Leap Forward] were denounced for
their technological fetishism, for their arrogant conviction that modern scientific
and technological learning was only accessible to the educated."[9] During the
Great Leap Forward, the Party set out to overcome the deepening division of
labor and the "three big divisions"—between intellectual and manual labor, agriculture and industry, the city and the country. Consequently, intellectuals in
general and scientists in particular came under attack.[10]

The roots of this social criticism of science, like the transformist tradition
mentioned above, go back to the utopian socialist literature coming to China in
the early decades of the twentieth century—especially Kropotkin's critique of

the scientific enterprise. His major points resonate strongly throughout the early Chinese socialist movement as well as the CCP's Maoist-led discourse on science.

Kropotkin had decried the specialization demanded by the modern organization of science. Specialization, of any kind, was rejected by him because it fragments the whole person, whose natural freedom, autonomy, and self-fulfillment are a function of using the full range of human intellectual and physical abilities. Specialization, for example, leads to the separation of those who labor with their hands from those who labor with their minds; it leads to dependency where there ought to be mutual aid; and as specialists assert their importance, one over another, it leads to hierarchies where there ought to be equality. Just as Kropotkin did not want individuals to alienate their political autonomy and judgment by turning over power to representatives, he did not want them to alienate their spontaneous understanding of nature and their natural inventiveness to professionals, to specialists.

Kropotkin further argued that specialization not only threatened to produce undesirable consequences for the social order, it also threatened scientific creativity itself. The trend toward specialization, he said, betrays an ignorance of the sources of industrial society's inventiveness, and threatens to dry them up: "Far from being inferior to the 'specialized' young persons manufactured by our universities, the *complete* human being, trained to use his brain and his hands, excels them, on the contrary, in all respects, especially as an initiator and an inventor in both science and technics."[11]

The great inventors of the nineteenth century, Kropotkin claimed, were "not men of science nor trained engineers." Rather it was manual workers who "have invented, or brought to perfection, the prime motors and all that mass machinery which has revolutionised industry." "Is it not striking, indeed," he argued, "that the steam engine, the steamboat, the telephone, the telegraph, the weaving-machine [etc.] and thousands of less important things, have *not* been invented by professional men of science . . . ?" On the contrary,

> men who hardly had received any education at school, who had merely picked up the crumbs of knowledge from the tables of the rich, and who made their experiments with the most primitive means are the real makers of civilization. . . . [T]hese men . . . knew something which the savants do not know—they knew the use of their hands . . . they knew machines . . . they had breathed the atmosphere of the workshop and the building yard.[12]

But Kropotkin feared that, in his own day, unless the situations of the professionals were rectified, further inventiveness was in jeopardy:

> The flight of genius which has characterized the workers at the outset of modern industry has been missing in our professional men of science. And they will not recover it as long as they remain strangers to the world, amidst their dusty book shelves; as long as they are not workers themselves, amidst other workers.[13]

Kropotkin's notion of scientific or technical creativity, then, is one that operates from the bottom up: from those that labor with their hands to the society in general. It operates from the particular to the general, from the concrete and material to the abstract: The inventiveness and creativity of the worker do not flow from knowledge of general laws, but rather, the former lead to the articulation of those laws in the first place.[14]

It should be noted in passing that in China's New Culture discourse on the role of the masses one hears arguments compatible with Kropotkin's notion of technical inventiveness. Beginning around 1915, one of the New Culture movement's goals was to find indigenous (instead of foreign) traditions that would replace what was characterized as China's Confucian, classical, upper-class culture. Though the discourse was often led by intellectuals who were neither utopians nor socialists, the main lines of their argument bore striking resemblance to Kropotkin's—especially where it was claimed that all significant cultural creativity originated at the "bottom" of the social hierarchy. Indigenous traditions of mass culture were thus offered as the basis of China's authentic historical identity and the point of departure for China's new culture.[15]

The linkage of native, mass, and creativity is recapitulated in another critique of the professional science community, a critique based on social class. The premise here is that science, like all elements of culture, has a class nature and hence lacks universality. At least from the 1940s, in Yan'an a desultory Chinese Marxist discourse addressed the implications of the class-contingent nature of science and regularly concluded that science derived from capitalist, bourgeois society was not only useless but inimical to a socialist society. In this discourse, Western-derived and eventually Soviet-derived professional science were denied a place in the substructure as a productive force; nor were scientists considered workers.

During the 1950s and 1960s, the class-nature-of-science idea was persistently wielded against the professional science community, periodically with damaging force. This is evidenced in the repeated efforts of some CCP leaders (Zhou Enlai, Lu Dingyi, et al.) to counter the idea with arguments or fiats that bestowed legitimacy on the professional science community. Notwithstanding these efforts over more than two decades, highly contentious ideas of the "class-nature of science" and "mass science" persisted until Mao was laid to rest.

As we shall see in the chapters discussing Lysenkoism, the class-nature position argued that bourgeois science was idealist, formalistic, and metaphysical; that it had a penchant for excessive theorizing, fruitless experimentation, meaningless mathematization, and statistical abstraction. All of this supposedly reflected a distinct breakdown of bourgeois science's contact with nature and the material bases of society. The prominence of probabilistic analysis in bourgeois science was considered to be sufficient evidence that it had abandoned the idea of causality and hence surrendered its ability to control nature. The class nature of science guaranteed that the Chinese had imported this impotent bourgeois science as well as the social relations of bourgeois science. This would eventuate in the aforementioned division-of-labor problems, elitism, and a slavish imitation of foreign technical doctrines.[16]

The solution to these problems required the implementation of a properly dialectical-materialist science. It also demanded a broad sharing of scientific and technical power which would foster industrial and agricultural change "according to distinctively Chinese patterns."[17]

For some, the answer was to be found in mass science.

## Mass Science

> To a large extent, inventions come not from experts or scholars but from the working people at various times. Most of the important inventions of history, whether in China or in any other country, have come from the oppressed classes, from among those in lower social position, younger in age, less learned, in bad circumstances, and those who even suffered setbacks and discrimination.
>
> —"Science Is No Mystery," *People's Daily*, June 6, 1958

There are two prominent conceptions of mass science that have been employed by the CCP.[18] The first, and most radical in its implications, sees the common people as the ultimate source of scientific discovery and technical innovation (perhaps of all cultural creativity) and the site of scientific validification. This is primarily attributed to the common peoples' direct involvement through labor in the processes of production. The second conception has been called a "mass mobilization" approach to science innovation and application. Here, emphasis is on the interaction of the working masses and organizing agents (including the CCP and mass organizations) that bring the workers' unstructured experience and intuitions together with those of professional scientists (or with systematic principles and techniques like those advocated by Lysenkoism). The former notion of mass science emphasizes the spontaneous, natural creativity and inventiveness of workers and leaves no place for professional science specialists; the latter emphasizes the need for discipline and guidance to be brought to the masses from the outside.

Schematically, the "mass mobilization" approach to science stands between the radical conception and that of a professional Western-model of science. The latter informed the science that developed under the Guomindang, and—notwithstanding Lysenkoism—also did so in certain areas expanded under Soviet tutelage during the 1950s. In Yan'an Communism (1935-1949), and then in Mao's China (1949-1976), this professional approach to science sometimes coexisted uneasily with the other two, sometimes alternated with them. The result for scientists, plainly seen in the fields of genetics and evolutionary theory, was challenging, to put it mildly. It was terribly wasteful of human resources, expensive in opportunity costs, and it ultimately inhibited the achievement of scientific technical goals.

The two conceptions of mass science shared some common criticisms of professional specialized science (whether Western or Soviet inspired); they obviously are a variation on class-basis-of-science critique. Both approaches denied the validity of science derived from formal education in "ivory tower" universities isolated from field and factory and lacking connection with application and production. By the same token, critics of professional specialized science had little use for its devotion to research and experiment without specific applied goals (a characterization that simply ignored the highly applied goals in many fields, including plant genetics). Additionally there was criticism for its reliance on abstraction (especially mathematical abstraction) and theory, all of which purportedly distanced science from nature, material reality, and the ability to control and change them. We shall see that the field of genetics was under constant attack on these grounds during the 1950s.

Additionally, these critiques argued that the elitist proclivities of professional science, especially its emphasis on specialized expertise, have been a serious impediment to social equality, and an aid to the oppressive division of labor in capitalist/industrial society. In China, such expertise was characterized as lacking ideological correctness ("redness") and was periodically branded as a source of bourgeois values and even counterrevolution.

In the late 1950s, Great Leap Forward campaign rhetoric expressed great faith in the spontaneous creativity of the masses (as in the epigraph to this section); but almost invariably that faith was compromised with a felt need to mobilize, structure, and guide the masses in their creative endeavors. Why so? Because, one argument went, the masses must initially be liberated from the oppressive and hegemonic belief that they are helpless and must rely on the expertise of the formally educated. A second argument, well articulated in Lysenkoist transformism, is that the masses are burdened with the "superstitious" belief that the natural order is not to be tampered with and, at best, can only be altered slowly. The Great Leap Forward magnified Lysenkoism's contempt for natural laws and its unbounded optimism for the outcome of human intervention into natural processes. Thus, the mass mobilization approach to science was supposed to be, among other things, a Party-led effort to liberate the masses from their misunderstanding of nature, and to enable their natural creativity to be reinstated.

Evaluations of Mao's populism suggest that he was of two minds about all expressions of mass spontaneity: his faith in its creative importance was bracketed by his Leninist proclivity to guide and restrain.[19] This may well be at least a partial explanation of the apparent strain between Great Leap Forward populist rhetoric and the prevalence of mass mobilization practices in which the spontaneous creativity of the masses was scripted and directed by the CCP. It may also help in understanding the particular practice of restructuring the division of labor through a variety of "combinations," such as the "three unifications" which required the Party cadre, the technical expert, and the hand laborer to work together.

Schurmann's analysis of this practice indicates that the "three-unifications movement" of the late fifties was aimed at resolving the contradictions between leader, intellectual, and worker not just by putting the three together, but by creating little melting pots in which the division of labor disappeared. "[E]ach was expected in effect to become the other," he observes, "workers becoming technicians, technicians becoming workers, and both sharing leadership with the cadres."[20]

One hoped-for product of this process was the *"duomianshou,"* the jack-of-all-trades, the omnicompetent ideal person for the new society that had been introduced in Yan'an, eclipsed in the early 1950s by Soviet science, and reintroduced in the Great Leap Forward. Here was a utopian resolution to the loose ends, inflated rhetoric, and potential conflicts surrounding the notion that mass science was the answer to the problems of dealing with nature, depending on foreigners, and trusting science to scientists. In Meisner's summary:

> Now the masses of workers and peasants themselves were to become scientists and engineers, mastering modern technology and learning the necessary expertise in the course of their daily productive activities. They were to study while they worked, and apply their newly acquired knowledge to immediate productive needs. There was thus to be no separate stratum of experts and intellectuals . . . but only a new generation of politically conscious *duomianshou* emerging from the masses. Everyone, it was said, will be a mental laborer and at the same time a physical laborer; everyone can be a philosopher, scientist, writer, and artist.[21]

The *duomianshou* ideal would seem not so much to resolve the issues of mass science as to abandon them for a utopian vision that would have warmed Kropotkin's heart. And indeed, it seems even to resonate with the rare utopian abandon of Marx—often cited in Great Leap Forward literature—when he too envisioned a way out of division of labor:

> In a communist society, where nobody has one exclusive sphere of activity but each can become accomplished in any branch he wishes, society regulates the general production and thus makes it possible for me to do one thing today and another tomorrow, to hunt in the morning, fish in the afternoon, rear cattle in the evening, criticize after dinner, just as I have in mind, without ever becoming hunter, fisherman, shepherd or critic.[22]

## Genetics and Evolutionary Theory in Post-Mao China

The final chapters of this study chart the efforts of the Chinese state, the CCP, and the science community to transcend this Maoist approach, understood as destructive of science, and to reenter the international mainstream of scientific development from the late 1970s to the end of the century. Outside of China, genetics had gone through a sea change that had largely left Chinese biologists behind and struggling to catch up. These chapters will noticeably be more concerned than earlier ones with science policy and political economy because the Party and state had bet their future on the success of science in remaking the Chinese economy and restoring the legitimacy of the Chinese Communist party.

We shall see that a once-embattled genetics (now referred to as molecular biology, genetic engineering, or biotechnology) has been moved to the head of the line and become a vessel of great expectations. Post-Mao leadership continues to think of science as a means of controlling nature, but—unlike its political predecessor—genetics, in all of its ramifications, is held in the highest esteem and considered a formidable transformist tool. And as for the control of scientists, it is fair to say that state and Party have come to a tacit understanding with the science community: scientists stay out of politics and the Party stays out of science. In the last chapters of this study, however, it will be argued that state and Party have in effect imposed a new and different form of control on science—control by the market, the "bottom line." Whether the state can control the market, once having unleashed it, is another question. Meanwhile, genetics, like most fields of science, must justify its work not only in its ability to control and change the biological world, but ultimately also by showing that this ability is marketable and a source of wealth and power for the national economy.

# Notes

1. Mao, "Speech at the Inaugural Meeting of the Natural Science Research Society of the Border Region," February 5, 1940. Cited in Frederic Wakeman, Jr., *History and Will* (Berkeley, Calif.: University of California Press, 1973).

2. This theme is pursued in Judith Shapiro, *Mao's War against Nature: Politics and the Environment in Revolutionary China* (Cambridge: Cambridge University Press, 2001).

3. See Robert Scalapino and George T. Yu, *The Chinese Anarchist Movement* (Bekeley, Calif.: University of California Press, 1961); Peter Zarrow, *Anarchism and Chinese Political Culture* (New York: Columbia University Press, 1990); and Arif Dirlik, *Anarchism and the Chinese Revolution* (Berkeley, Calif.: University of California Press, 1991).

4. Kropotkin's transformist ideals are especially evident in the pieces collected in *Evolution and Environment: Collected Works of Peter Kropotkin, Vol. 11*, George Woodcock, ed. (Montreal: Black Rose, 1995). Pps. 111-243 comprise "Thoughts on Evolution," a series of essays published 1910-1915 in the journal *Nineteenth Century & After*. Also see analysis of Kropotkin's transformism as part of a larger Russian tradition of natural philosophy in Douglas Weiner, "The Roots of 'Michurinism': Transformist Biology and Acclimatization as Currents in the Russian Life Sciences," *Annals of Science*, 1985, no. 42:243-60.

5. Kropotkin on nature: Stephen Osofsky, *Peter Kropotkin* (Boston, Mass.: Twayne, 1979), chapter 4.

6. For the problem of scientific terminology as well as other related issues, see James Reardon-Anderson, *The Study of Change* (Cambridge: Cambridge University Press, 1991).

7. For example, see Lewis Pyenson, *Civilizing Mission: Exact Sciences and French Overseas Expansion, 1830-1940* (Baltimore, Md.: Johns Hopkins University Press, 1993).

8. For discussion of the technical and technocratic approaches to revolution under the Guomindang, see Wen-hsin Yeh, *The Alienated Academy: Culture and Politics in Republican China, 1919-1937* (Cambridge, Mass.: Harvard University Press, 1990), chapter 7.

9. Franz Schurmann, *Ideology and Organization in Communist China* (Berkeley, Calif.: University of California Press, 1967), 91.

10. In addition to Schurmann's analysis, Maurice Meisner's work has been invaluable for making sense of Mao's and the CCP's approaches to the problem of the division of labor in China. See Meisner's, "Marx, Mao, and Deng on the Division of Labor in History," in *Marxism and the Chinese Experience*, eds. Arif Dirlik and Maurice Meisner (Armonk, N.Y.: M. E. Sharpe, 1989), chapter 5. Also see Meisner's *Marxism, Maoism and Utopianism* (Madison, Wis.: University of Wisconsin Press, 1982).

11. Peter Kropotkin, *Fields, Factories, and Workshops: Or Industry Combined with Agriculture and Brain Work with Manual Work* (New York: Greenwood reprint, 1968), 215. Reprinted from the original 1898 anthology of essays Kropotkin published from 1888-1890 in the journals *Nineteenth Century* and *Forum*.

12. *Fields, Factories*, 206.

13. *Fields, Factories*, 210-11.

14. *Fields, Factories*, 209-12.

15. This discourse was devoted not to science, but to culture in the sense of "literature": vernacular fiction, poetry, folk song, and folktale. An important formulation of this position is in Hu Shi, *Baihua wenxue shi* [History of vernacular literature] (Shanghai: Commercial Press, 1929).

16. In addition to chapter 4 below, see Douglas R. Weiner, *Models of Nature: Ecology, Conservation and Cultural Revolution in Soviet Russia* (Pittsburgh, Pa.: University of Pittsburgh Press, 1988). His discussions of I. I. Prezent's role in Lysenkoism provides dramatic details of the class-based criticism of "bourgeois" science. See 221-23 for the attack on mathematicization.

17. See Rensselaer Lee, "Mass Innovation and Communist Culture: The Soviet and Chinese Cases," in *Technology and Communist Culture*, ed. Frederic J. Fleron, Jr. (New York: Praeger, 1977), 290.

18. This and related terminology and discussion below are taken from Richard P. Suttmeier's detailed schematization in *Research and Revolution: Science Policy and Societal Change in China* (Lexington, Mass.: D. C. Heath, 1971), 49, 89-92.

19. For example, Meisner, *Marxism, Maoism and Utopianism*.

20. *Ideology & Organization*, 76.

21. Meisner, "Division of Labor," 100.

22. Karl Marx, "Critique of the Gotha Program," in *Karl Marx and Friedrich Engels, Selected Works* (Moscow: Foreign Languages Publishing House, 1949), vol. 2: 23. Meisner cites the passage in "Division of Labor," and in *Marxism, Maoism and Utopianism* (7-8, 192). Meisner argues in the latter (192) that the Great Leap Forward "vision of the future was derived from classical Marxist sources, and Maoists drew on the more utopian strains in the Marxist-Leninist tradition. Nothing was more frequently quoted in the theoretical and popular literature of the Great Leap era than the famous passage [by Marx]" cited above.

I am suggesting that one can see in Great Leap Forward utopianism the legacy of Kropotkin's anarchism as well, though one would not expect Mao to cite an anarchist authority to legitimate his political idiosyncrasies. It is further worth noting that Marx's passage cited here virtually exhausts his utopian vision on the subject and that it was only available to Chinese readers years after the writings of Kropotkin and other anarchists were available in translation. Rather than say that Great Leap Forward utopianism was derived from classical Marxist sources, perhaps it would be more accurate to say that those canonical sources were cited to legitimize a position influenced much earlier by utopian socialist/anarchist literature in China. After all, Meisner, throughout *Marxism, Maoism and Utopianism*, convincingly indicates in detail the affinities between Maoism and utopian socialist thought, including that of Kropotkin.

# Part I

## Republican China, 1911-1949

# Prologue:
# Independence through Dependence

These first three chapters follow the development of education and research in genetics and evolutionary theory at three universities in China: National Central University, in Nanjing, which became the flagship campus for the aspiring national university system; Yanjing University, in Beijing, an American missionary school, linked to the Rockefeller Foundation's Peking Union Medical College; and Nanjing University, an American missionary school that developed a superior agricultural science program closely tied to Cornell University.

From the early 1920s and throughout the war years (1937-1945), China's earliest and best biology programs and training in genetics developed in these schools with concerted and substantial support from American philanthropic organizations. Together, they were responsible for implementing the transfer of genetics education and research, mainly from the United States, to China's developing national education system. Biology faculty from these schools were regularly drafted by new Chinese universities to develop new biology departments; students from these schools regularly became leading biology faculty at the new universities when they returned from advanced training abroad. Disparate as these three universities appear at first glance, their life science programs were collectively at the center of an expanding web of connections that closely linked Chinese life scientists to each other and to major American scientists and institutions.

This is not to say that there was no tension or rivalry between Chinese and foreign schools. Foreign philanthropy regularly privileged the latter with significant funding that just as easily could have gone to the former. In the case of ag-

21

ricultural biology, for example, we shall see how the decisions of foreign foundations decisively determined that agricultural biology should thrive, in a preconceived fashion, at a missionary school rather than at a rival national university. Such developments resulted in obvious competition between Chinese and foreign schools for the best students and faculty. The missionary schools examined here initially sent the bulk of their biology graduates to work in other missionary schools where the pay was higher and more dependable than in Chinese schools. In time, however, these tensions were lessened when many of the science faculty in Chinese universities had received their undergraduate training in missionary schools before getting their Ph.D. abroad.

Initially, at the three universities examined here the Chinese faculty who worked in genetics and related areas all received advanced training at Columbia University or Cornell at a time when these schools were sites for some of the world's most creative contributions to basic genetics research. T. H. Morgan's seminal group, for example, was based at Columbia in the 1920s and provided the graduate education for China's earliest geneticists. When Morgan's group moved to Caltech in 1928, China's first generation geneticists stayed in close touch with it through collaborative research and through their own students, whom they sent to Caltech for graduate training. In this way, China's geneticists stayed abreast of and participated in the momentous developments leading to the so-called "New Synthesis" in biology. This referred to a melding of experimental biology with the techniques and traditions of naturalist description and taxonomy, which enabled the fruitful application of contemporary genetics to Darwinian evolutionary theory. In time, Chinese geneticists used their initial connections with the American science community to place outstanding students in other programs that were doing significant genetics research, such as University of Texas at Austin, Washington University in St. Louis, and University of California, Berkeley.

By 1937 and the war with Japan, new biology departments emulating American examples were appearing in the gradually expanding national university system, and genetics courses were standard parts of their curriculum. Geneticists holding doctorates from American universities were regularly chairs of these and other biology departments (such as National Central, Qinghua, Amoy, and Yanjing Universities). Additionally, new government institutions were established to support basic and applied agricultural research; these agencies relied heavily on the plant geneticists trained at Nanjing and Cornell, and other schools. At this time in China, there were on the order of three dozen working geneticists holding American doctorates, and there were approximately twice that number of undergraduate and graduate students studying the subject in American universities. The scope of the term "genetics" was of course expanding to include molecular biology and biochemistry, and Chinese students abroad were beginning to be enticed into these new frontiers as well.

In short, less than two decades after China's first geneticist began teaching and doing research in China, a well-trained, sophisticated cohort of geneticists was available to play a central role in the establishment of the life sciences.

Judging on the basis of their education abroad, their ongoing research, and their performance as educators, they were admirably suited to teach the early generations of Chinese biology students, to mediate future international developments of genetics, and to prepare aspiring undergraduates for the rigors of advanced training abroad.

Collectively, these geneticists and other biologists at the three universities were instrumental in establishing some fundamental attitudes toward the life sciences. They contributed to the ongoing differentiation of science from scientism in China, particularly by wresting the discourse on heredity and evolution in China from the monopoly of Social Darwinism, utopian socialism, and other social philosophies. Within the life sciences, they established formal, professional, international standards for biological research and education. One can see this in their teaching curricula, textbooks, research methodology, and the adaptation of standard scientific terminology to the Chinese language.

Additionally, the new biology community in China made a conscious effort to follow mainstream biology (especially from the United States) toward the application of findings in genetics to Darwinian evolutionary theory. At the same time, they made explicit, reasoned rejections of anti-Darwinian ideas about evolution and heredity which continued to have a following within some areas of the life sciences as well as in nonscientific intellectual circles.

Much less self-conscious, and with considerably less open discussion, was an attitude toward the role that the scientific community should play in solving China's most fundamental social and economic problems. There was a common understanding that the application of science to national problems was appropriate and necessary. In the scientific community represented by the three universities examined here, however, there was tacit and explicit support for the belief that China's basic needs could be satisfied with technical solutions, and by implication, that it was not also necessary to engage in social economic reform or revolution. This will be seen most clearly in the development of agricultural biology.

Finally, this examination of the initial development of biology provides insight into the problem of foreign control of science in China. A kind of resolution is reached through a paradox—independence through dependency—a belief that the foreign schools and foundations were enabling China to acquire the science it desired, and that eventually this would provide China with the scientists and the institutions necessary for autonomy.[1] Two philanthropic foundations played ubiquitous roles in this process, especially where the life sciences were concerned.

In all aspects of its development in China, genetics depended heavily on the largesse and guidance of foreign-based philanthropy. Generally, the same could be said of virtually all areas of the sciences; but the small, fledgling field of genetics was especially well supported by philanthropic agencies. Such philanthropy was evident in the missionary universities, like Yanjing and Nanjing, that pioneered the development of science curricula in China. They relied on funding from their American religious constituencies as well as from tuition from their students. Philanthropy, from similar American sources, sometimes took the form of emergency relief for one or another of China's devastating famines, and in one notable case famine relief funds were used to endow a prominent plant genetics and improvement program.

## The Rockefeller Foundation's China Medical Board

Above all, two well-endowed institutions, which had overlapping boards of directors, played profound roles in shaping the development of modern science in China before 1949 and in directing the flow of foreign philanthropy into scientific education and research. These were the Rockefeller Foundation's China Medical Board and the China Foundation for the Promotion of Education and Culture. More than any other agencies in China, foreign or domestic, these two mediated the transmission of American—and to a lesser degree, European—science to China and enabled their domestication. Over the course of the 1920s, the foundations came to recognize that this required much more than the transplantation of whole scientific institutions, such as the Rockefeller's luxurious Peking Union Medical College, founded in 1921.

Between 1925-1949, the period when the two foundations worked together, they collectively spent about U.S.$15,000,000 directly for the development of natural science education and research in China. This is over and above the tens-of-millions the Rockefeller Foundation spent just on Peking Union Medical College before and during this period. These figures do not take into consideration the millions of U.S. dollars which the foundations, through matching grants, encouraged Chinese industry, commerce, and government to contribute. During these years, the U.S. dollar's value in China was such that U.S.$3,000 was a rare and outstanding salary and research budget for a senior scientist at a major university. And for U.S.$75,000 it was possible to build a three-story, 20,000 sq. ft., fireproof, university science building, completely equipped for teaching physics, chemistry, and biology. Altogether, the two foundations funded over a hundred different institutions—including hospitals, libraries, research institutes, schools, and professional societies. On the order of one thousand people received fellowships for graduate science training and/or research.

Eventually, their policies reflected an understanding that the practice and development of advanced scientific medicine and all areas of science research required an infrastructure of education, communication, physical plant, and equipment. From the early twentieth century, this infrastructure was being established piecemeal in China under the auspices of various central and provincial government agencies and various missionary organizations; but there was no coherent plan or model, nor was there a steady, dependable source of funding. After 1928, the Guomindang (Nationalist) government provided China relatively greater unity and stability, and it advocated the strengthening and expansion of all areas of science education and research. It established institutions and agencies among whose responsibilities were the planning and control of all science and technology development. The Academia Sinica (National Academy of Sciences, established in 1928), for example, was supposed to have been both a set of institutes devoted to research and the central planning agency for all science research. Often, however, the government reluctantly had to acknowledge that its ambitious plans for science were dependent on the financial prowess, the expertise, and the policies of the two foundations.

Underlying the initial efforts of the two foundations was a notion of science called the "Johns Hopkins model"—because of the strong influence that the Johns Hopkins Medical School had on the Rockefeller Foundation's work in China. This model was constructed around 1913, in the wake of important medical reforms in America. Where medicine was concerned, this was science thought of as clinical practice wed to laboratory research, a devotion to the expansion of knowledge, to the pursuit of basic explanations of nature and the discovery of laws. For many Chinese promoting modern science, the Hopkins model was taken generally to be an embodiment of America's growing emphasis on pure science or pure research. And since America's wealth and power were considered to be products of science, it was argued that the Chinese nation should emulate America's example.

In China as well as America, the pure science ideal was a controversial one. Influential Chinese and American critics argued that the goal of pure science was inappropriate for immediate Chinese cultural realities and social needs. For example, the Rockefeller Foundation made the following observation in 1924:

> Among those who are interested in the promotion of science in China there is a marked division into two groups, one believing that at present attention should be devoted almost exclusively to teaching the application of modern science in such fields as medicine, engineering, and agriculture, as they are now known in the west. This group holds that it would be a mistake to encourage scientific research for some time to come. Others believe that while the application of existing scientific knowledge is needed, teaching and research in pure science should not be slighted at this critical period in the educational development of China, and that the scientific workers there should be enabled to keep up with their literary colleagues in productive research if they are to have their due share of leadership.[2]

After 1928, the Guomindang government's concerns for the practical value of science was often a source of conflict with this ideal and was an important impetus for the shift of the Rockefeller Foundation's China policy toward practical and applied scientific concerns.[3] Nevertheless, even when the Hopkins model was not explicitly the dominant one for the Rockefeller and China foundations, it remained a distant goal toward which patient and calculated growth was aimed.

The China Medical Board's original China policy was formulated around 1914, and focused on two goals: upgrading the hospitals and medical colleges run by American missionaries, and building from scratch two world-class medical complexes—one in Beijing and one in Shanghai. For economic reasons, it only built one, the Peking Union Medical College, a sophisticated and very expensive complex made up of a medical school, a teaching hospital, and a biomedical research unit. It was called "the Johns Hopkins of China" and its extraordinary history has been well chronicled.[4]

China Medical Board policy shifted and broadened when it was learned that Chinese students were not sufficiently trained in the basic sciences to meet its medical school's rigorous standards. And so, Peking Union Medical College set up its own preparatory department. Here, from 1921 to 1925, top graduates of (mostly missionary) colleges and universities advanced their knowledge of physics, chemistry, and biology in order to prepare for the tough demands of the medical school curriculum. The preparatory department, however, was a stopgap measure. The China Medical Board determined that there was only one way to guarantee a steady flow of medical students with adequate science education, and that was to raise the quality of science education throughout the colleges and universities of China. This would provide students for Peking Union as well as for the missionary and the Chinese medical schools which the China Medical Board planned to upgrade.

The next shift in China Medical Board policy, during the early 1920s, saw it funding key public and private Chinese—instead of foreign missionary—universities and colleges in all areas of science education. Nankai University in northeastern China and National Southeastern (soon to become National Central) University in Nanjing were the initial beneficiaries of grants for science buildings, equipment, and operating costs. Then, in 1925 Peking Union Medical College disbanded its preparatory department and in effect turned that preparatory function over to nearby Yanjing University, for which the China Medical Board endowed it generously.

Over the next decade, the China Medical Board continued to shift away from funding foreign institutions in China and toward Chinese institutions, and away from medicine and hospital plants to basic science education and biomedical research. The China Medical Board hired its own science education coordinator to plan and implement a policy for enhancing science curricula in Chinese higher education; and American consultants were brought to China year after year to evaluate the science programs and make suggestions for their improvement.[5]

These policy shifts in part reflect a response by the China Medical Board to the antiforeign, antimissionary sentiment burgeoning in China during the mid-1920s. This is explicit in the writings of Roger Greene (1881-1947), the leading author of China Medical Board policy.[6] It also reflects a long-range policy of Greene and the Rockefeller Foundation to provide the Chinese with the means to generate their own, high-level medical and biomedical research community. By the mid-1920s, Greene and his colleagues were quite aware that Peking Union Medical College was in itself not the means, and that it had created more educational problems than it had solved.

The China Medical Board became actively devoted to creating a self-perpetuating science education community throughout Chinese colleges and universities. The bulk of its funding continued going to programs in key colleges and universities, programs which turned out excellent students and acted as models for new and expanding schools. In addition to providing funds for science buildings and for research and teaching equipment, the China Medical Board also provided large sums for endowments. No less important were its hundreds of advanced research fellowships that brought outstanding American scientists to Peking Union Medical College as visiting instructors, and brought Chinese science educators and researchers there for advanced training with them. Other fellowships sent Chinese scientists to the United States for postgraduate education. The field of genetics profited from China Medical Board financial support in all these areas, and could not have developed as quickly and well without it.

Clearly, the China Medical Board did not trust the development of science education to provincial or central ministries of education, or to the Guomindang. It used its financial and intellectual resources as leverage to elicit the kinds of reforms and policies it wanted in science education. Through a program of strategic seed money and matching grants, it was able to involve Chinese sources of funding, including the Guomindang government, in its enterprise.[7]

## The China Foundation

The China Medical Board's policies and projects were complemented by and usually coordinated with those of the China Foundation[8] for the Promotion of Education and Culture, established in 1924. This institution was a latter-day development of the Boxer Indemnity Fund, that peculiar agency established in 1908 with funds that the Chinese imperial government was initially required to pay to the United States for the costs of the 1900 Boxer Rebellion. The Boxer Fund was mandated to send Chinese men and women, selected by competitive examination, to the United States for higher education. It did so for thousands, including hundreds who were trained in biology, and just about everyone who was trained before 1928 as a specialist in genetics. The Boxer Fund also supported Qinghua College in Beijing, a preparatory school for those who wanted to go to America for a college education.

In 1924, the Boxer Fund was converted to new purposes. Most of it went to endow the new polytechnic Qinghua University (which replaced Qinghua College); the remainder endowed a program to promote scientific education and research. The latter program was administered by a self-perpetuating board of Chinese and American directors, independent of the Chinese and American governments. In accordance with its original constitution, the board was composed of a majority of Chinese, many of whom were prominent educators and scientists. Though it may have been coincidental, it was nonetheless helpful for the development of genetics that one of its practitioners sat on the board or presided over it throughout the 1930s, the foundation's most active period. Additionally, two and sometimes three China Medical Board administrators (including Roger Greene, the China Medical Board's director) sat on the China Foundation board of directors. This not only facilitated coordination between the two foundations, it often resulted in the experienced China Medical Board administrators using the China Foundation to carry out their own policies.[9]

The original goals of the China Foundation were ambitious if simply stated, and they remained steadfast. They were to provide China with the institutions and tools for modern science education and research. In the foundation board's first meeting, it interpreted its mandate from the United States' government to be "the development of scientific knowledge and . . . the application of such knowledge to the conditions of China through the promotion of technical training, of scientific research, experimentation and demonstration, and training in science teaching."[10]

Between 1925 and 1949, the foundation's largest expenditures were made in two areas. The first was for the complete costs of constructing the National Library of Peking and endowing it with funds to build its initial collection. The foundation's object was to provide the library with the largest and most complete collection of scientific books and periodicals in China, thus making it the hub of a national science library network. The second area of large grants was devoted to building up science teaching and research capabilities in Chinese universities and in the Academia Sinica.

The China Foundation provided advanced science training and research fellowships in all disciplines; they were determined by competitive application and peer review. Generous one- and two-year stipends made it possible for a total of four hundred men and women to carry out research projects or complete postgraduate degrees in China, the United States, and Europe. In addition, the foundation endowed scientific research professorships to support the research of established scientists and to bring them into contact with younger members of their discipline.

In science education, the China Foundation worked closely with the China Medical Board in its effort to build up college and university infrastructure; but it went beyond the China Medical Board into secondary school science education. It funded in-service training for secondary science teachers, sponsored the writing and publication of Chinese-language textbooks, and the translation of exemplary foreign textbooks. It also created and distributed science curriculum packages, complete with laboratory equipment, manufactured to its own specifications.[11]

China Foundation enterprise is also evident in its efforts to make foreign language scientific terminology more accessible to China. In the 1920s, it was common to find college and university science instructors lecturing in the language of the country of his/her training. It was the rule to find science textbooks used in their original English, German, or French. Toward the elimination of these linguistic burdens and culturally offensive dependencies, the foundation encouraged two kinds of effort: It sponsored the translation and publication of historically significant Western science texts and current advanced-level textbooks. Additionally, it sponsored the establishment of nationally standardized translation of scientific terminology, and then made sure that these were circulated by way of glossaries, lexicons, and dictionaries. Linguistic problems of this nature were first addressed in China by nineteenth-century Western missionaries. From 1916, formal government agencies undertook the task; and in the late 1920s, the new Ministry of Education and the Academia Sinica assumed responsibility for these perennial problems.[12]

In sum, the China Foundation and the China Medical Board used their financial and intellectual resources to facilitate the development of a natural science education and research community that included Chinese and foreign institutions, Chinese and foreign scientists, and financial support from the public and private sectors. At the same time, the foundations successfully carried out policies aimed at diminishing Chinese dependence on foreigners and foreign institutions for modern science. They enabled and sped up the process that made it possible for well-trained Chinese to replace foreign science faculty in the missionary colleges and universities; and they provided Chinese colleges and universities the wherewithal to develop their own science programs, with all-Chinese faculty. And the foundations even contributed substantially to the budgets of scientific institutions that the Guomindang government created explicitly to enable China's scientific independence—the national library and various Academia Sinica institutes, for example.

The latter contradiction was not lost on the Guomindang or the foundations, and it seems to have been accepted as an inescapable one as long as the Guomindang's ambitious expansion of the central government left it short of funds. The more devoted the government was to rapid scientific growth and independence, the more it would have to rely on the foundations and the generosity of foreigners.

The complex interactions of Chinese and foreign scientists, educators, and institutions is nowhere better seen than at National Central University, designated by the Guomindang as the keystone of a national university system, and model for the development of science education.

## Notes

1. Of course, this paradoxical resolution of the need for foreign dependency was neither limited to Chinese thinking about science nor to China itself: For example, see Suzanne Pepper, *Radicalism and Educational Reform in Twentieth-Century China: The Search for an Ideal Developmental Model* (Cambridge: Cambridge University Press, 1996), 113-17; and Richard J. Samuels, *"Rich Nation Strong Army": National Security and the Technological Transformation of Japan* (Ithaca, New York: Cornell University Press, 1994), 271-78.

2. Rockefeller Foundation, *Annual Report*, 1924:254. And see discussions on this issue by these important participants in the formation of China Medical Board science policy for China: Paul Monroe, "Address" (December 1921), *Chinese Social & Political Science Review*, 1923, no. 6: 143-48. Roger Greene, "Education in China and the Boxer Indemnities," *Chinese Social and Political Science Review*, 1923, no. 7: 199-207; and Greene's memo to George Vincent, January 15, 1925 (RAC: IV.2.B9.Box 62). For Paul Monroe see text and note below.

3. The most important policy statement on this issue is S. M. Gunn, "China and the Rockefeller Foundation," Shanghai, January 23, 1934 (RAC: Ser. 601.Box 12). A version of this report is discussed in J. C. Thomson Jr., *While China Faced West: American Reformers in Nationalist China, 1928-1937* (Cambridge, Mass.: Harvard University Press, 1969), 125-29. There is a summary of this policy in Rockefeller Foundation, *Annual Report*, 1935: 3, 329.

4. See Mary Brown Bullock, *An American Transplant: The Rockefeller Foundation and Peking Union Medical College* (Berkeley, Calif.: University of California Press, 1980).

5. These China Medical Board policy shifts are formally stated in the Rockefeller Foundation's *Annual Reports*: (1) Shift to Chinese institutions, 1922: 32-33, 248; (2) Shift away from hospitals to medical education, 1925: 374; (3) Shift to medical research, away from medical education, 1929: 175-76.

The China Medical Board's science coordinator was N. Gist Gee (1876-1937), an American biologist who taught biology at Soochow University from 1901 to 1920. See William J. Haas' biography of Gee, *China Voyager* (Armonk, New York: M. E. Sharpe, 1996. George Twiss (b. 1863), an associate of Paul Monroe, was hired by the Chinese National Association for the Advancement of Education (partially funded by the China Medical Board) in 1922-1924 to survey and give advice about science teaching in China. His report was published as *Science and Education in China* (Shanghai: Commercial Press, 1925). The single most detailed report is by the China Medical Board consultant, W. E. Tisdale, "Report of a Visit to Scientific Institutions in China," December 1933 (RAC: 601.D.Box 40).

6. See Roger Greene's communications to (1) George Vincent, June 15, 1925, and (2) N. Roosevelt of the *New York Times*, April 19, 1927 (RAC: IV.2.B9.Box 62, 124). Also see H. S. Houghton's confidential memo to Greene, March 4, 1927, and Greene's "Statement before Committee on Foreign Relations (House)," Washington, D.C., January 29, 1927" (both in RAC: IV.2.B9.Box 124).

7. For examples of matching grants, see Rockefeller Foundation *Annual Report*, 1923: 246 (for Southeastern University), and 1929: 226 (for Yanjing University).

8. My observations on the China Foundation are based largely on (1) published annual reports of the China Foundation, 1926-1940, 1947, 1948; (2) *Summary Report of the Activities of the China Foundation, 1925-1945*, n.p., December 1945; (3) "China Foundation—Report of the Director Presented to the Emergency Committee, 30 June 1943," (RAC: IV.B.9); (4) Roger Greene's correspondence and diary (RAC: China Medical Board/IV.2.B9, and Series 601); (5) Hoh Yam-tong, "The Boxer Indemnity Remissions and Education in China," Ph.D. diss. (Teachers College, Columbia University, 1933); (6) Yang Tsui-hua, *Zhongjihui dui kexue de zanzhu* (Patronage of Sciences: The China Foundation for the Promotion of Education and Culture) (Taibei: Institute of Modern History, Academia Sinica, 1991).

9. The science education liaison between the two boards was N. Gist Gee. Examples of the linkage between the two foundations can be seen in (1) "Record of a Conversation

between Roger Greene and Kuo Ping-wen [Guo Bingwen, the president of Southeastern University]," May 6, 1925 (RAC: China Medical Board/Ser.II.Box 63), and in (2) "Proposal for Aid to the Medical School of the Central National University." This is an evaluation by Greene, March 6, 1929 (RAC: China Medical Board/Ser.601/Box 24). Page 3 discusses cooperation between the two foundations with this aid.

10. "A Summary Report of the Activities of the China Foundation," n.p., December 1946, 3.

11. RAC: China Medical Board/Ser.II.Box 84, and China Foundation *Report,* 1929, no. 3: 7, 27-29.

12. For an extensive discussion of the problem of scientific terminology in nineteenth- and early-twentieth-century China, see James Reardon-Anderson, *The Study of Change,* and David Wright, *Translating Science: The Transmission of Western Chemistry into Late Imperial China, 1840-1900* (Leiden: Brill, 2000).

# 1

# Biology at National Central, the Model University

During the 1920s and 1930s, the National Central University in Nanjing embodied many of the plans, aspirations, and problems of China's educational and political reformers. Here, in 1921, the first biology department was established in a Chinese university. The department was under the direction of the American-trained biologist Bing Zhi (1886-1965), who received a classical education in his native Honan Province as well as an introduction to modern subjects in his original pursuit of a traditional civil service degree. The civil service examination system was abolished in 1905, however, and in 1909 he chose to be educated in the United States under the sponsorship of a Boxer Foundation fellowship. He studied biology at Cornell, specializing in entomology with J. G. Needham, and received a B.S.A. in 1913 and Ph.D. in 1918. It was at Cornell, in 1914, that Bing along with other Chinese students at the university was a founding member of the Science Society of China, the country's first organization devoted to the promotion of modern science. He was also a major contributor to the society's journal, *Kexue* (Science), modeled on the journal of the American Association for the Advancement of Science. From 1918-1920, he was a postdoctorate scholar in physiology at the Wistar Institute in Philadelphia before returning to China and his new university post.[1] In his first year, he arranged for China's first genetics course to be taught in his department by Chen Zhen (1894-1957) just returned from study at Columbia University.

33

National Central became the university's name only after a long series of developments begun in 1903, when it was founded in Nanjing as a regional college-preparatory school in the wake of late-imperial education reforms. Until the 1911 revolution, the school continued to train men for the civil service bureaucracy with a classical curriculum enhanced with "modern" subjects ranging from mathematics and science to foreign languages, world history, and geography. When education reforms resumed under the republic, this locally prestigious campus in 1915 became a provincial teacher's college, in the Western sense, training secondary schoolteachers for the new public system. Then, it set a course for national prominence.[2]

## The University's Political Economy, 1920-1925

Under the leadership of its new president, Guo Bingwen (1880-1969), the school was made a national university at the end of 1920 when the biology department was established. This meant that it would raise its entrance requirements, open its student enrollment to the entire country, and receive funding from the central government as well as from Jiangsu Province. It also meant that in addition to its core arts and sciences college, it could create or incorporate professional schools. It rapidly added five of them, in education, medicine, engineering, commerce, and agriculture, which was the initial home of the biology department.[3]

President Guo was also the dean of the school of commerce. As a young student in his native Jiangsu Province, he was trained at a "Western-style" school for a career with the customs service in Shanghai. From 1906 to 1914, he was a student in the United States, first as a science major at Wooster College in Ohio, then at Teachers College, Columbia, where he completed his doctorate in 1914. The following year he was back in Nanjing, as dean of the Teachers' College that he and his fellow provincials aimed to make into the rival of the great National Beijing University. They wanted to make their school the model campus for the future national university system, and "to dominate educational policies of the nation."[4]

During the five years of his presidency, Guo Bingwen used his local and foreign connections well to bring in the funds required for such rapid and extensive development. Locally, he himself was able to go to Shanghai bankers and borrow money to keep the university solvent when the government was typically late in disbursing funds. Guo was also able to take advantage of Jiangsu provincial pride and commercial-industrial needs. In 1921, for example, he brokered donations from the Chinese Flour Mill Association and the Chinese Cotton Mill Association for experimental stations respectively in wheat and cotton. The Jiangsu Bankers' Association contributed to entomological research (Bing Zhi's field); and the university received complete funding for a new library from the provincial warlord, with whom Guo enjoyed good relations.[5]

Guo's foreign connection proved to be vital to the development of natural science and medicine at the university. His Teachers' College mentor had been Paul Monroe (1869-1947), a professor of "international education," who had been an important adviser for the Rockefeller Foundation's work in China since 1913. The foundation's China Medical Board saw Guo's presidency of the new national university as a target of opportunity; and by 1922, it determined that National Central was the key Chinese school to invest in for the development of the natural sciences.[6]

Up to this time, the China Medical Board's general goal had been to develop "biomedical science" education in key universities, and thereby produce students trained well enough to enter the Rockefeller Foundation's spectacular new medical school, Peking Union Medical College. Now, the China Medical Board wanted to use National Central to create another world-class school of medicine, this time in Shanghai. Roger Greene, the China Medical Board's director, envisioned students training as undergraduate premeds at National Central and then going on to the university's publicly funded medical school in Shanghai—a process to be shaped by China Medical Board standards and strategic grants.[7]

Toward those long-range goals, in 1922 the China Medical Board awarded National Central the funds to build a huge, fully equipped natural sciences building. Later, the China Medical Board's continuing generosity to the sciences at National Central combined with that of the China Foundation and helped to develop one of the strongest overall science programs in the country—in spite of the fact that the China Medical Board never realized its plan for a new medical school in Shanghai.[8]

Guo Bingwen's educational vision was not limited to emulating American universities and their science programs. His administration occurred during the high tide of China's iconoclastic "New Culture Movement," centered at Beijing University, with which his values were partially in sympathy. He was, however, quite put off by what he considered to be the uncompromising extremism and disruptiveness of such movements, which seemed to preoccupy Beijing University students to the detriment of their studies. At least from his student days at Columbia, Guo was an ardent anticommunist and he regularly attributed China's student radicalism and political radicalism among university intellectuals to communist conspiracy.[9]

Nevertheless, Guo actually anticipated New Culture proponents in his 1914 Ph.D. dissertation, where he laid out his educational platform: He wanted once and for all to sever the old link between education and government office, and to provide universal public education open to women. Under Guo's leadership, in 1921 the university was the first in China to admit women and to have a woman dean. And like the New Culture intellectuals, Guo advocated the use of vernacular Chinese as the medium of education, and the elimination of the old classical curriculum.[10]

On these latter two points, President Guo was potentially in conflict with an important segment of his faculty who voiced some of the nation's most trenchant criticisms of New Culture iconoclasm. That group was for the most part based in the humanities, but it received strong support from some scientists as well. It was led by literature and philosophy specialists whose recent doctorates were from Harvard, where they had become disciples of Irving Babbitt's (1865-1933) "New Humanism." As such, they defined a culturally conservative position on the issues of China's classical heritage and the use of classical Chinese language. They attacked the New Culture "populism" that advocated democracy in aesthetic and social values as well as in politics. They argued that the salvation of China's cultural identity, its "national essence," was vested in a cultured elite of intellectuals like themselves. All of this they promoted in their long-lived journal, *Xue Heng (The Critical Review).*[11]

Fortunately for Guo Bingwen, his vice president for Arts and Sciences was Liu Boming (1884?-1923), a Northwestern Ph.D. in philosophy, who mediated the campus factions and brought them together under his ecumenical cultural philosophy. Liu argued that "classical values" were essential to guide and control the reforms necessary for China's survival. He equated Western classical values (he was a specialist in ancient Greece) with those of China; and he claimed that classical Western values still underlay those aspects of Western culture that the Chinese wanted to emulate. Therefore, the Chinese could continue to draw on their own classical heritage, maintain their distinct identity, and successfully carry out the necessary Western-inspired reforms. In particular, Liu acknowledged the importance of natural science in China's new society; but he also cautioned that natural science must be embedded in the new "moral education" of China. [12]

## The Development of Biology at National Central

No member of the faculty better embodied Liu Boming's balanced and critical approach to education and cultural reform than Hu Xiansu (1894-1968), Bing Zhi's partner in the development of National Central's biology department. Hu got his B.A. in botany from the University of California at Berkeley in 1916 and, like Bing Zhi, was a founding member of the Science Society of China. He was brought to National Central in 1922 as chair of the biology department and head of its taxonomy division. These responsibilities were given to him while Bing Zhi took charge of the division of zoology and busied himself with the establishment and management of the privately funded Biology Research Laboratory of the Science Society of China, located in Nanjing.[13]

During 1923-1925, Hu returned to the United States to get a Ph.D. in botany from Harvard, where he forged close and long-lasting relations with the Arnold Arboretum. He also grew close and sympathetic to Babbitt's New Humanism and its Harvard-trained Chinese disciples who had become his colleagues. Having joined the *Xue heng (Critical Review)* circle of cultural critics, Hu now became an outspoken and articulate foe of vernacular language reforms, and generally of what he and his group called the excessiveness of New Culture ideology.[14]

It is notable that Hu Xiansu did not feel obliged to link his promotion and practice of modern science with the radical iconoclasm he attributed to the foreign-trained faculty at Beijing University, a major source of contemporary student movements. Hu went on record with his bitter resentment of university faculty who encouraged students to extremism without having provided them first with the educational foundation to make intelligent judgments. In the case of the explosive antiforeign May Fourth Movement (1919), Hu rebuked those professors who incited students to violence and the destruction of property, with the goal of destroying a social system and culture for which no one had replacements. Most reprehensible of all, he charged, was the use of education as a political weapon and students as cat's-paws.[15]

Notwithstanding such criticism, Hu was in cautious sympathy with the anti-foreign/anti-Christian sentiments he found in the student movements of 1925, when he returned to China from Harvard. Perhaps provoked by his Harvard sojourn, Hu publicly expressed his contempt for the ideas that Euro-American peoples were superior to the "yellow races," that followers of Christianity were superior to "heathens," and that the new culture of America and Europe was superior to China's old culture. Christian education in China, he concluded, had been of use to imperialism and "cultural encroachment" and warranted the criticism it was currently receiving. Still, after taking such a position, Hu never entertained the consideration that Western science might be part of that cultural encroachment, or that the biology to which he was so devoted was promoted and supported by foreign institutions in China, including missionary schools. Nor did he express any sense of potential conflict between Western science and "China's old culture."[16]

Hu Xiansu's sentiments on these issues appear not in the least to inhibit his promotion of biology. He and Bing Zhi quickly came to dominate access to and control of China's major new institutions for funding and carrying out biological education and research. In 1925, Hu was made director of the new privately endowed Fan Memorial Research Institute based in Beijing. It was exclusively devoted to taxonomic studies, concerned essentially with the collection and classification of botanical and zoological specimens.

A few years later, both Bing and Hu were founding members of the new national academy of sciences, the Academia Sinica. In all of these endeavors, Bing and Hu garnered continuous funding support from the China Foundation. In this way, they steered the development of biology in the direction of taxonomy rather than experimental, reductionist sciences like physiology or genetics. Hu's work became preeminent in the taxonomic study of Chinese flora, the field that dominated not only modern botany in China, but all of modern biology, well into the 1950s.[17]

The relative roles of taxonomic and experimental sciences in China became a charged issue for the Chinese science community as well as for its foreign advisers and benefactors. Some advisers to the China Medical Board and China

Foundation, like Bing Zhi's mentor J. G. Needham, rather dramatically encouraged the development first of taxonomy and systematics so that Chinese biology could be based on China's own natural environment, and not on that of North America and Europe, which was the subject of the English language botany and zoology textbooks used in Chinese universities. Additionally, institutions like the Arnold Arboretum encouraged the growth of taxonomy in order to enrich their collections and expand knowledge of the world's flora.[18]

Other foreign advisers to the China Medical Board criticized and deprecated the development of taxonomics, an area that they at best considered "useful," but not calling "for the use of imagination."[19] These critics could not understand how the China Medical Board had hoped to accomplish its goal of enhancing "pre-medical" science education by encouraging the development of such fields. These tendentious observations assumed that the China Medical Board completely controlled the development of science research in China and that its board of directors was completely of one mind.[20] The China Foundation's significant contributions to taxonomic research, on the other hand, at least in part reflect the strength of taxonomy and its practitioners when the foundation began operating in 1925. And so part of the explanation to the dominant place of taxonomy is that, in the beginning, it was fortunate to have as its proponents consummate academic entrepreneurs, based in the country's most dynamic Chinese university.[21]

There was still another consideration, for the work of naturalists like Hu Xiansu and Bing Zhi spoke directly to the nativism and cultural nationalism that evolved in China over the 1920s, and which we have already seen expressed in Hu Xiansu's response to the anti-Christian movement of 1925. Chinese scientists coined the term "local science" to express a felt need to distance science from its Western sources and make it more relevant to immediate Chinese needs. Local sciences (like botany or geology) were contrasted to "universal science" (like physics) which dealt with abstract principles independent of earthly locale. And these scientists heeded those foreign advisers who cautioned them to beware of "borrowed biology"—resulting from the simple fact that virtually all biological research and teaching were carried on by people educated abroad and trained in Western flora and fauna and their problems. These scientists were admonished to develop the interest and the tools necessary to study China's indigenous natural world.[22] This same sentiment was expressed by some of China's leading geologists, for whom the importance of their science was the knowledge of their own land and the riches it might contribute to national wealth and power.[23] In 1928, all of this seemed to be summed up in the inaugural speech of Fu Sinian, one of the directors of the just-opened Academia Sinica when he called for the "nationalization" of science.[24]

The taxonomic interests of Bing Zhi and Hu Xiansu did not deter them from hiring Chen Zhen at National Central to create genetics courses and to develop experimental biology. Clearly, they wanted their department's curriculum to reflect those of the American schools that trained them; and they wanted a faculty representing a broad spectrum of American biology departments. With the addition of Chen Zhen, the department had representatives of Columbia, Cornell, and Harvard.

## Chen Zhen and the Introduction of Genetics

Chen Zhen, a Jiangsu native, got his undergraduate biology from the competent department at the missionary Nanjing University, from 1914 to 1918. Then, supported by the Boxer Foundation, he attended Qinghua College in Beijing, preparing for graduate study in America. During 1920-1921, he was submerged in an intensive nonthesis M.A. program in biology at Columbia. Genetics was his main interest and he was fortunate to have as his adviser and primary instructor E. B. Wilson (1856-1939), chair of the zoology department.[25]

Chen was of course very much in the right place at the right time for his interest in genetics. E. B. Wilson was by this time a major figure in the development of classical genetics. In 1905, he had laid a cornerstone of chromosome theory (simultaneously with Nettie Stevens at Bryn Mawr) by demonstrating that sex was determined by a specific chromosome. And from that same date, T. H. Morgan was based at Columbia.

The record does not reveal whether Chen actually studied with Morgan; however, Chen's early teaching and research do show that he was well trained in Morgan's work. Chen was especially devoted to Morgan's epochal *The Mechanism of Mendelian Heredity*, published in 1915, and then updated and expanded in 1919 as *The Physical Basis of Heredity*.[26] In this work, Garland Allen writes, the Morgan group made "the first major consolidation of gains in the revolution in genetics." Here they "laid out in a rigorous but non-specialized way the fundamentals of the Mendelian-chromosome theory."[27] Chen Zhen's teaching and research were guided by a number of key points made by these books. Among them were their fundamental "commitment to the actual physical reality of Mendelian factors," and their demonstration that "the Mendelian factor hypothesis could be related to cell research data about chromosomes." Chen's later research projects clearly were inspired by some of the avenues opened up by the Morgan group, especially regarding the "relation of Mendelian mutation and variation to the origin of species by natural selection."[28]

In his pioneering genetics course at National Central, Chen assigned Morgan's *The Physical Basis of Heredity* as one of the two textbooks. The other was the 1918 E. B. Babcock and R. E. Clausen, *Genetics in Relation to Agriculture*, a hefty and solid introduction to classical genetics and its practical applications. The use of this latter text is partially explained by the fact that National Central's biology department was lodged in the Agriculture School. Both texts were read in the original English, a feat typically required by natural science courses in China. In fact, National Central acknowledged its curriculum's reliance on English language texts and lectures by establishing an English competency test as part of its new entrance requirements in 1921. Chen's one semester course enrolled twelve students who were expected to attend two lectures and six laboratory hours per week.[29]

## Chen Zhen's General Biology Textbook

One of Chen's long-lasting contributions to the early development of biology in China, and its "nationalization," was his 1924 Chinese-language general biology textbook, *Putong shengwuxue*.[30] In retrospect, it is considered the most famous Chinese-authored, Chinese language biology textbook up to 1949, by which time it had gone through many editions and printings and had been used in middle-school as well as college courses. Geneticists trained during these decades remember the text with respect as their first introduction to genetics, or as the best book to study before a college entrance examination in biology.[31] More broadly, the book is celebrated as an exemplary effort in the development of science in China because of its author's broad command of the field and because it provided a reliable alternative to a foreign-authored, foreign language textbook. Further, the textbook was of special significance because it was the first in China to take an approach to biology that gave genetics and evolution a major role.

Chen's textbook was in part derived from his classroom lectures, and it comprises, topic by topic, succinct distillations of a score of major English language textbooks and monographs, all cited in a detailed bibliography. Profuse illustrations are likewise lifted from these acknowledged sources. It is written in a pellucid vernacular Chinese and pays careful attention to technical vocabulary by accompanying key terms with their English originals, and by providing an extensive English-Chinese glossary in the back matter. Chen prides himself in being able to supplant foreign, English language texts with his Chinese language textbook. Nevertheless, he pointedly cautions his readers that, even when the Chinese language has fully developed its modern biology vocabulary, it will be necessary to continue to use the Latin-based nomenclature that is the universal medium for classification—if China is to be part of the world science community.[32] Chen did not see his use of Chinese as a means of separating China's science from that community, but rather as a means of making it more accessible and less foreign.[33]

In his introduction to *Putong shengwuxue*, Chen argues the need for a textbook like his because the few Chinese language biology texts thus far published had specialized in either zoology or botany, or were general texts of a popular or "metaphysical" nature. No Chinese language text discussed the phenomena common to all life—from protoplasm, to reproduction, heredity, and evolution. Nor, previous to his book, had any introduced the various theoretical approaches to these phenomena, the evidence used to substantiate them, and the current states of major theoretical discourses.[34]

Loyally, and effectively, Chen's textbook follows these guidelines. Beginning with the material basis of life, his narrative then proceeds systematically through basic organic processes illustrated with organisms of increasing complexity. Each discussion in turn usually revolves around the scientists and theoretical controversies that were germane to the dominant contemporary understanding of the biological phenomenon at issue. And with each of his descriptions of how a process works, Chen repeatedly tells his readers that this is the conclusion of a particular scientist's research, based not on speculation but on hard evidence. Because of this approach, Chen's textbook sometimes reads like a history of biology, affording his readers the opportunity to see that current understanding has developed over time and is likely to change in the future. This historical sensibility is further carried out in a short appendix that is in effect an outline history of Western biology that is structured by minibiographies (and cameo pictures) of the great biologists, from the ancient Greeks to Mendel. In spite of Chen's national self-consciousness, however, there is virtually no reference to China's premodern traditions of biology, biological ideas, or natural philosophy.

## The Problem of Heredity

*Putong shenguwxue* devotes more than a third of its pages to heredity and evolution. And while the nature and sophistication of Chen's textbook presentation are unprecedented in China, the Chinese periodical press had reliably introduced these subjects piecemeal over the previous decade in plain language national journals like *Dongfang zhazhi* (Eastern Miscellany) and *Kexue* (Science). Mendel and his rules for inheritance patterns were first introduced to China in 1913, and from the outset were reverently equated with the importance of Darwin and evolution. A steady and growing stream of periodical literature about Mendel continued, and in 1920-1921 a first complete translation of his rules appeared in one of China's many new scholarly journals. The following year, the centenary of Mendel's birth was celebrated, and by 1923 college textbooks for general biology and for agriculture science were published with long, systematic, and well-illustrated chapters on Mendel's laws. [35]

The introduction of genetics to China followed the same path and timetable. Preceding Chen's textbook, major periodicals took their readers through the development of genetics from August Weismann in the nineteenth century to the contemporary work of T. H. Morgan. Further, journals outlined the basics of cell biology and the role of chromosomes, and explained various schools of thought on the relationship of Mendelism, chromosome theory, and natural selection in evolutionary theory. By 1923, Ph.D.'s who returned from Cornell's plant genetics program were publishing discussions of inheritance that deftly employed the concept of the gene. Some of this literature illustrated its arguments with material derived from the author's own research. Additionally, those interested in genetics were quick to establish standardized Chinese translations for basic genetics terminology. [36]

In Chen Zhen's textbook, the discussion of heredity is for the most part (both narrative and illustrations) derived from T. H. Morgan's *The Physical Basis of Heredity*. Chen Zhen's narrative begins with the device of a make-believe discourse between the biologist August Weismann (1834-1914) and Charles Darwin in order to establish two fundamentals: that the material basis of heredity is located in a discrete organic structure of the cell (germplasm), not suffused generally throughout an organism (somatoplasm); and second, that acquired characters cannot be inherited. This is followed by a detailed, well-illustrated exposition of Mendel's principles that introduces the basic concepts of a dominant and recessive character, backcrossing, segregation, and independent assortment.

These basics in place, the rest of Chen's discussion of heredity is an elaboration of the principle that chromosomes are the physical basis of Mendelian factors and hence the physical basis of heredity. Carefully following Morgan's 1919 summation, Chen economically lays out the key issues and concepts, beginning with the "well substantiated fact" that chromosomes do not equate with just one Mendelian factor, but are the basis for many. The concept of "linkage" is then introduced to explain the association of characters over generations— sex-linked characters being especially illustrative.[37]

Next, repeatedly citing the research of "Morgan's group," Chen gives the concept of the gene its most extensive introduction in Chinese. He explains that Morgan's work has led to an understanding that chromosomes are comprised of units called "genes" (yinji) that are the physical basis of Mendel's factors (yinzi). And then Chen follows Morgan through three basic aspects of heredity: First, that these genes sit on chromosomes in a "lineal order," having a fixed position and sequence. Second, that there occurs an association ("linkage") of certain genes owing to their position on the same chromosome (not to their being tied together, so potentially any two genes could be linked). And finally, there is the process of "crossing over," that pairing of chromosomes at meiosis which results in interchanges of their genes.[38] Chen concludes this section on heredity with examples of sex-linked inheritance to illustrate the mechanics of heredity in which all of these processes and concepts are at play. He apologizes to his readers for not having the space to provide (as Morgan did) more illustration and cite more evidence. Nevertheless, in the space of forty pages, he did provide an accurate and intelligible outline of contemporary genetics theory.

## The Problem of Evolution

Chen's discussion of evolution was certainly one of the best summaries to date in China. It was preceded in China by a body of literature and discourse far larger and more complex than the earlier writing on heredity. From the beginning of the twentieth century, the idea of evolution had great currency in China, but for decades it was deeply submerged in extrascientific discourse about the nature of historical progress and China's ability to compete with other nation-states. Two chapters of the *Origin of Species* were carefully translated into Chinese and published in 1903-1904, and the first complete translation appeared in 1920.[39] Typically, however, Darwin's ideas were conveyed darkly through social Darwinist literature aimed at social and cultural reform in China. Among the earliest and most celebrated of this genre was an interpretation of T. H. Huxley's *Evolution and Ethics* by the reformer Yan Fu (1853-1921).[40] This was required reading by generations of Chinese revolutionaries and reformers, and for China's post-1911 revolution "enlightenment intellectuals" whose cultural reform agenda was suffused by a belief in an "evolutionary cosmology" and a natural law of progress.[41]

During this same era in China, the writings of anarchist philosopher Peter Kropotkin also had significant influence on the idea of evolution. Decades before the first full Chinese translation of Darwin's *Origin*, virtually all of Kropotkin's writings had been translated into Chinese. If it is true, as recent scholarship convincingly argues—that anarchist philosophy and values came to suffuse Chinese discourse on revolution and modernization—then it follows that Kropotkin's thought was central to that process.[42]

The great appeal of Kropotkin's anarchism began where its evolutionary cosmology envisioned a world advanced by "mutual aid"—not by struggle—for humanity and all living things, and at the same time was deeply optimistic about the possibility for universal progress.[43] Characteristic of the Russian tradition, Kropotkin insisted on an evolutionary process without competition. Further, he preferred to think that variation, adaptation, and selection were fundamentally a function of the "direct action" of environment on organisms, and that environmentally induced variations could be inherited. Where humanity was concerned, Chinese anarchists understood Kropotkin's "environment" to mean society and institutions, education and labor. Restructuring this social environment according to anarchist principles was the way that humans could redirect their own evolution and guarantee their survival in a utopian future.[44]

While Kropotkin's legacy in China all the more strengthened the conceptual bond between evolution and progress, it also contributed to a set of principles that more broadly challenged biological thought well into the second half of the century. First and foremost was its distinctly Lamarckist emphasis on the role of an organism's environment (as opposed to some genetic material within the organism) in organic change and variation. Next, was the equally Lamarckist principle that such changes could be inherited.[45] Finally, there was Kropotkin's transformism, his belief that the above principles insured the infinite malleability of nature and the potential for humans to control and direct nature by controlling and manipulating environments.[46]

More modest, direct, and evenhanded efforts to introduce Darwinian evolution to China had actually begun in the nineteenth century, in treaty-port missionary publications. The tradition of science reportage begun at that time was continued by a number of Chinese periodicals, from the early years of the twentieth century,[47] and evolution became a routine subject in the science sections of China's growing periodical press aimed at general audiences. These journals seldom discussed this subject apart from its social-Darwinian and sometimes even its eugenic implications. Nevertheless, they displayed a comfortable grasp of the basic terminology, mechanics, and implications of evolution in both the *Origin* and the *Descent*. Rarely was discomfort expressed with the idea of human descent, or the great expanse of time required by Darwin's evolution, or the pattern of struggle and survival.[48] On the other hand, this early journalism displayed not the slightest familiarity with the biological literature explicating or criticizing Darwin, nor what evolution meant to contemporary biology.[49]

For that level of sophistication, one had to wait until about 1915, when young biologists trained abroad began to return and to write for *Kexue*, which, during its first years of publication, provided Chinese readers a world of new insights into evolutionary theory unburdened by social philosophy. It introduced the currently debated "saltationist" theories of evolution—that is, evolution in sudden jumps; and it explored the ongoing discourse on natural selection and variation led by biologists like T. H. Morgan. Botanists, zoologists, and human biologists took turns explaining the development of evolutionary theory within their specialties. For the first time, Chinese readers were provided with detailed, analytic discussions of the components of Darwinian evolution. And a number of historical pieces outlined and compared major evolutionary theories from Lamarck's ideas in the eighteenth century to the present.[50]

### Zhou Jianren and Lamarckism

Chen Zhen's biology colleagues, Bing Zhi and Hu Xiensu, bore much of the burden for writing these informative pieces for subscribers to *Kexue*, but technical writing by the life sciences journalist Zhou Jianren (1890-1984) reached a far greater, more general audience over the following decades. In time, the National Central biologists and Zhou Jianren respectively came to represent an important polarity in science discourse on evolution and heredity. The former tightly adhered to a mainstream "neo-Darwinist" position (including the exclusion of all possibility of an inheritance of acquired characters, and the combination of the theory of natural selection with the discoveries of Mendelian genetics).[51] By the time Chen Zhen published his textbook in 1924, Zhou Jianren had become China's most thoughtful and articulate exponent of Lamarckian principles (emphasizing instability of genetic material from generation to generation, the central role of environment in variation, and the inheritability of acquired characteristics).

Zhou Jianren was, in effect, one of China's first and most influential professional science journalists. His two elder brothers—Lu Xun (1881-1936) and Zhou Zuoren (1885-?)—were among China's most prominent literary intellectuals. Like them, he studied in Japan as a youth; but where they turned to writing fiction and criticism, he went on for a degree from the agriculture school of Tokyo Imperial University. By 1921, Zhou Jianren was a science editor for the burgeoning Shanghai Commercial Press and for *Dongfang zhazhi*. At the press, he edited and wrote for the largest series of popular life-science books before 1949. For the journal, he wrote on every aspect of modern biology and was in a position to influence the journal's life-sciences content. In all of his publications, Zhou Jianren upheld the family literary tradition by writing very effectively in the new "vernacular" (baihua) idiom promoted and practiced by literary reformers like his brothers.[52]

Zhou's publications were able to summarize accurately the neo-Darwinian positions on inheritance and natural selection, and he knew his Mendel and his Morgan. Like many intellectuals in his generation, however, he believed that environment was the fundamental source of organic change and variation, and that these in turn were inheritable.[53] Already in the early 1920s, in articles on the tensions between biologists who emphasized a stable hereditary material and the "environmentalists," Zhou was optimistic that definitive evidence would be found to validate the latter Lamarckist position.[54] In a 1923 coauthored primer on evolution, he concluded that environment—by itself—can produce fundamental and inheritable changes, and that acquired characters can sometimes be inherited.[55] By the end of the twenties, *Dongfang zhazhi* had published fifteen of his pieces on similar topics.

Chen Zhen's textbook straightforwardly espouses the neo-Darwinian position on evolution as it had on heredity; and it explicitly rejects all elements of the neo-Lamarckist position on heredity and evolution. Though the book is not polemical, as a textbook it is in effect laying out for China's first generation(s) of biology students, the central dogmas of a distinct paradigm. In doing so, it set the standard for mainstream modern biology in China before 1949, and it represented the formal antipode to neo-Lamarckism in China—in whatever form it took.

## Evolutionary Theory in Chen's Textbook

The approach to evolutionary theory taken by Chen Zhen's textbook is quite similar to the one it takes on heredity. It moves through bodies of evidence for evolution, beginning with comparative anatomy, embryology, and paleontology. Then it illustrates the difficulty of obtaining direct evidence for a model of evolution, like Darwin's, that expected the changes leading to new species to occur gradually, in small increments, and over a relatively long span of time. Pointedly, Chen notes, however, that Hugo de Vries' research for his mutation theory, much debated between 1900-1910, not only alleged to provide direct evidence of evolution, visible in one's own laboratory, but evidence that has been thought sufficient itself to explain the evolution of new species.

Chen's discussion of the "mechanics of evolution" begins with a polite summary and rejection of Lamarck's theory, unacceptable to Chen because it depends on the inheritance of acquired characters. Then he argues the validity of Darwin's "survival of the fittest" formula (so famous now in Chinese literature). If fitness is understood as adaptation, Chen suggests, then it is demonstrable as a law: Organisms that adapt better to their environments will compete better for food, air, light, and space. These organisms will persist while others become extinct.

Chen concludes his discussion of mechanics by asking the very important question about the source of the variations that are necessary for adaptation and natural selection. Darwin had few and inadequate answers to this basic question, Chen acknowledges; however, Chen suggests that de Vries' mutation theory—granted its disputed status—has perhaps provided the basis for a scientific answer.

This treatment of mechanics by Chen inexplicably holds back from the more sophisticated discussion that would have been possible had Chen fully exploited Morgan's work as he did in the section on heredity. For example, he draws superficially upon Morgan's 1916 *A Critique of the Theory of Evolution*, which he cites in his bibliography.[56] In particular, Chen barely links his discussion of natural selection to his earlier discussion on the physical basis of heredity.

The burden of de Vries' theory, which seems to have arrested Chen, was that speciation could be accounted for by mutation, that is, discontinuous, sudden changes. Morgan himself, between 1904 to 1910, had accepted de Vries' position; but by 1916, he had moved well beyond it. Not only had many of de Vries' conclusions been invalidated, but his general mutation theory had been widely discredited.[57] Chen Zhen's narrative only briefly alludes to the contributions that Morgan's study of Drosophila mutations had made to evolutionary theory; but Chen does not suggest that this work might be creating a new understanding of mutation, one that was consistent with the basics of chromosome theory and with the theory of natural selection.[58]

Chen ends his discussion of evolution (and the main narrative of the textbook) with the observation that the source of mutation is thus far unresolved and remains a most important lacuna in our understanding of the mechanics of evolution. It is a conclusion consistent with the main themes and pedagogy of the book: science is an ongoing process, based on and driven by empirical evidence. But it is also a conclusion that wants to indicate the gradual unfolding and strengthening of the positions of neo-Darwinism.

Chen's research during the 1920s and 1930s provides further insight into his thinking on the issues of natural selection and mutation theory in a Mendelian framework.

## Chen Zhen and the Introduction of Mendelian Research

From all accounts, conditions in National Central's biology department were quite good for research. In addition to a light teaching load, biologists had access to special facilities and outside funding. In 1922, the Science Society of China established its Biological Research Laboratory in Nanjing, convenient to the university. In addition to good, new facilities, the laboratory had a respectable science library. Chen Zhen and Bing Zhi were the first and only researchers to work in the facility's zoological section during its first year. Thereafter, the laboratory became an active research arm of the university's biologists as well as a haven for visiting researchers from all over China. Chen's first research was published by the laboratory's distinguished research series.[59]

From 1926, the China Foundation heavily subsidized plant, equipment, and special research projects at the National Central's biology department and the Biological Research Laboratory. At the same time, the foundation initiated a special Science Professorship program, which, over the next decade, provided generous salaries and research budgets for renewable one-year periods to seventeen "mature scientists." Sometimes, the program brought a scientist to a campus or institute where the foundation thought his presence would be most influential. In other cases, the foundation simply took over the expenses of a scientist in his current location. Chen Zhen was the first recipient of one of these professorships, which he held at National Central from 1926 to 1928.[60]

All of Chen's research aimed at establishing Mendelian patterns of inheritance, using the common goldfish as an experimental animal. Often, he openly took pride and delight in his goldfish because he was able to use textual sources—sometimes very old ones—to document when, where, and under what conditions a particular variety had developed. The fish had begun to be bred a millennium earlier in China, and detailed, beautifully illustrated records had been kept. On the basis of these records, Chen argued that the goldfish did indeed originate in China (not Japan or elsewhere as some contended), and by implication he suggested that his use of this Chinese animal for research was a contribution to the "nationalization" of science. Further, he claimed that the combination of intense breeding and record keeping had made the goldfish even more valuable than Darwin's pigeons for the study of variation and evolution.[61]

Chen's citation and use of this earlier Chinese tradition of controlled breeding is one of only a handful of historical references to Chinese precedents that I have found in the literature consulted for my present study. The citation is remarkable for the respect it pays to the earlier material and for its refusal to indulge in self-congratulation about how far Chinese science has come since the old days. To compare this old literature favorably with Darwin's celebrated observations on variation and evolution was the highest compliment, but it was only a momentary glance at the question of China's premodern scientific tradition. I can only speculate on this lack of interest in earlier traditions of science by the Chinese science community (and by the intellectual community in general). These communities considered science to be both symptom and source of social progress and evolution. If China is considered to have lacked progress for some time, then one would have to conclude that China's earlier equivalents to the Western life sciences had long ago bogged down and failed to develop. And instead of a viable scientific tradition to build on, the moderns inherit—in Chen Zhen's terms—only "metaphysics" and "superstition." The millennial goldfish literature would then have to be considered an exceptional instance of a useful scientific legacy.

Despite Chen's nationalistic appreciation of the goldfish, he felt compelled to admit that they presented the researcher with some serious problems. Unlike Morgan's Drosophila, the fish were finicky about their diets (except when they cannibalized each other) and required a full year to breed. Nevertheless, Chen was one of a number of scientists in China and elsewhere around the world who were using the goldfish successfully for biological research. For example, in 1929 Chen published with bravado the results of his first major project, claiming that his was only the second study of the inheritance of a character in goldfish, and probably the first case of Mendelian inheritance observed in the animal. He later learned (and publicly noted) that someone in Europe had simultaneously carried out a very similar study with basically the same results. [62]

During the 1920s and 1930s Chen's research was primarily devoted to two similar projects, characteristic of neo-Mendelian genetics. According to Kohler, "the neo-Mendelian system of classifying genetic factors into organ group systems—eye color, wing shape, body color, . . . etc.—embodied the basic purpose of neo-Mendelian heredity, which was to determine how many genetic factors were involved in the formation of each morphological feature."[63] In Chen's first project, he analyzed transparent scales and mottled patterning in some goldfish; in the second, he dealt with blue and brown coloring. The standard procedure he followed involved the very careful breeding of fish exhibiting the normal character with those exhibiting the "mutant" character (transparency, blue/brown color, etc.). "These genetic crosses," Kohler explains, "were designed to synthesize all possible combinations of present or absent factors. The end products of these experiments were families of genetic formulas, one for each known mutant factor."[64] In turn, these formulas "constituted the design of an experimental instrument whose purpose was to reveal the genetic causes of morphological development."[65]

While this description certainly captures the nature of Chen's experimental work, that work had an additional feature that indicates his growing sophistication. By the time Chen completed his second project, in the mid-1930s, he was using with sensitivity the tandem concepts of "genotype" and "phenotype." The former refers to an organism's genetic inventory, inheritable by its next generation. The latter refers to its visible or measurable adult character, such as color or organ shape. Chen's research reports deal clearly with a concern that Morgan had begun to articulate in 1915, that is, that the relationship between genes (factors) and characters was complex. Morgan wrote that "a single factor may have several effects, and . . . a single character may depend on many factors."[66] Failure to recognize this could lead to a confusion of gene and character relationship.

In Chen Zhen's research, he is mindful of these caveats both for conducting his analyses and explaining his results. For example, when he discusses the difficulty of working with goldfish, he notes that "most characters in goldfish depend on multiple factor pairs; hence it is difficult to get an invariable and regular result which may be fitted to Mendelian ratios."[67] Nevertheless, by being sensitive to the complex relationship of genotype and phenotype, Chen claims to have arrived at some special findings. Thus, in the project which was duplicated abroad, only Chen's research noted and successfully explained the appearance of three different phenotypes from a set of factors that were conventionally expected to yield only two.[68]

Considering the audience that Chen had in China, his research publications had considerable pedagogic potential for deepening understanding of basic issues in genetics.

## Chen Zhen and Genetics at Qinghua University

Chen was gradually forced to abandon research due to his administrative and teaching duties and then by the war with Japan.[69] In 1928, he accepted the chair of the biology department of National Qinghua University in Beijing, making him one of a number of National Central biologists called upon to develop new departments throughout the growing national university system. Chen joined a new department at a new university, since Qinghua had only begun to make the transition from a college to a full-scale university in 1925. This was made possible by the reassignment of the Boxer Indemnity's considerable endowment—part going for the endowment of the China Foundation, part for the endowment of Qinghua University. The Rockefeller's China Medical Board was keen to help this new biology department develop, and so, Chen was in a uniquely creative position. Between 1928 and 1935, he increased the department's faculty from six to thirteen, and there were at least three American Ph.D.'s among them (from Chicago, Yale, and Cornell).[70]

At Qinghua, Chen's teaching role as well as the subject of genetics and experimental biology were prominent in the curriculum requirements for biology majors. He taught the core courses in cell biology, experimental genetics, and genetics and evolutionary theory. The latter course is one of the few in Qinghua's course catalog that lists its texts (popular titles used throughout American biology courses) and its main topics: "The variability of living things; Mendel's laws of inheritance and their adumbration; the material basis of heredity; the determination of sex; evidence for the theory of evolution—mutation theory, theory of use and disuse [Lamarckism], natural selection."[71]

Chen's excellent situation at Qinghua was made even better in 1933 with the opening of a massive new biology building, paid for by the China Medical Board and a matching grant from the central government. Chen's second and final major study of Mendelian inheritance in goldfish was completed at Qinghua in 1934. It was about this time that the Rockefeller Foundation evaluated Qinghua as part of a larger national survey of university-level biology education. W. E. Tisdale, the very demanding author of the evaluation wrote: "As far as potential possibilities exist, this department would seem to be the strongest I have seen yet in China."[72]

## The University and the Political Turmoil of 1925-1929

Midway through Chen Zhen's tenure at National Central, the university had met with the first of a number of serious disruptions that plagued the school until the Guomindang took over the central government at the end of 1928. These troubles contributed to faculty turnover. Beginning in 1925, the Guomindang Party—as a revolutionary political organization—was itself one of the major sources of that disruption. The Guomindang, under Sun Yatsen's charismatic leadership, had for years been developing a formidable power base in the south, with the ultimate goal of reunifying China after rescuing it from warlord divisiveness. In 1924, the Guomindang was reorganized and strengthened with guidance and aid from Lenin's Comintern. China's infant Communist Party formed a United Front with this new Guomindang.

In 1925, Guo Bingwen was forced to resign his presidency of National Central by pressures put on him from Guomindang sympathizers and party members at his university and in the central and provincial governments' education ministries. In the first instance, Guo was attacked because it was felt that he was overly close with the Jiangsu provincial warlord and other local notables whom the Party considered to be inimical to its revolutionary goals. More generally, Guo became a target for Guomindang ideologues and educational policy specialists because they wanted National Central to be a "party-line" school whose administration would follow Guomindang education policy and facilitate Guomindang political power. Guo Bingwen, supported by many of his faculty and students, refused to take orders from the Guomindang.[73]

The Guomindang persisted. After all, National Central was located in the wealthy province that was to be their political and economic base, and in the city that was to be their new capital. In the early 1920s, the Guomindang had already established covert "cells" on the National Central campus and begun to use the campus as a base to recruit members and distribute propaganda. And Guo's own faculty contained a powerful faction of Guomindang stalwarts, especially in the natural sciences. The Guomindang prevailed in spite of dramatic protests by students and the university board of trustees who together expressed their resentment of outside interference with the university's academic freedom. Guo himself was convinced that "Communist wing" of the Guomindang was the ultimate source of his dismissal, and he suspected that they were conspiring to take over the entire university system and the "education class."[74] Rapidly, the American old-boy network in China arranged to get Guo out of the country, to the United States, and into a comfortable administrative position funded by the China Foundation.[75]

Turmoil at National Central did not abate after the removal of Guo Bingwen. First there were demonstrations against those who fired President Guo; then against the series of people who tried to replace him. Students went on strikes and demonstrated against fees and poor accommodations; faculty went on strikes and demonstrated about unpaid back wages.[76]

National Central's internal problems were then severely exacerbated by nationwide events from which a Chinese university was no sanctuary. Within weeks of Guo Bingwen's departure from office, major cities of China exploded in a series of labor strikes against foreign-owned industry and commerce, notably in nearby Shanghai. Foreigners' violent responses provoked a general strike, and continual demonstrations against imperialism and colonialism. All of this was known as the May Thirtieth Movement. National Central's students and faculty joined in the antiforeign demonstrations, some of which were aimed at the missionary establishment both in their religious and educational roles.[77]

When the massive May Thirtieth Movement was in its eleventh month, April 1926, the Guomindang launched the "Northern Expedition," its political-military campaign to take over the central government and eliminate opposition to its reunification of the country. Since Nanjing was to be their new capital, the National Central campus was of tactical importance. The expedition sent advance intelligence and propaganda "troops" (young students from south and central China) ahead to the university where they demanded accommodations to facilitate their preparation of Nanjing for the Guomindang. By the fall of 1927, the campus was occupied by Northern Expedition soldiers and zealous camp followers who billeted themselves in school buildings which they encircled with barbed wire. One-half of the campus buildings were occupied by the three hundred uniformed young people from the "Training School for Propagandists." University students were pressured to join the Guomindang and it was rumored that some were offered money to join up.[78]

By April of 1928, the Northern Expedition had triumphed, the occupation of the campus was over and the troops were gone, having pilfered thousands of dollars worth of school equipment and caused extensive physical damage. And damage of another kind was done to the faculty and students. Part was inflicted by the tumult and demoralization caused by the occupation; part by the Guomindang's endless internal struggle to take over and control the university's administration. Many faculty sought jobs elsewhere; two-thirds of the students stayed home or transferred to other schools.[79]

## Science and Education Reforms under the Guomindang

When the Guomindang took over the central government in 1928, it moved the capital from Beijing to Nanjing. At this point, the name "National Central University" was formalized and the school was officially designated the keystone of the national university system, for which the new government had big plans.

National Central was a symbol of national independence from foreign-controlled education. Schools like it were perceived as the means to "modernize" China and strengthen it sufficiently to eliminate foreign dominance and exploitation. For the Guomindang government, the university system was a counterpart to the Academia Sinica, which it established in 1929 for the purpose of facilitating advanced research, primarily in the natural sciences. Together, they were seen as the means to centralize and control the much-desired development of science and technology, and reduce therein the role of foreigners.

Chinese schools, like National Central, prided themselves on the virtual absence of foreigners on their faculty, in contrast to missionary and other foreign schools in China. And these Chinese schools were likewise pleased that they were beginning to make it unnecessary or at least less necessary for Chinese to go abroad for a higher education. Nevertheless, dependency on a technical education abroad was still a reality; and foreign schools in China were often perceived as a source of competition for scarce resources and for the best faculty and students. Great national institutions, like the university system and the Academia Sinica, were sometimes quite dependent on funding from foreign-controlled philanthropy in the absence of sufficient public funding. Here was one among many dilemmas created by an expansion of education and research institutions beyond the capacity of the new state to pay the bills.[80]

On its accession to power, the Guomindang government launched a series of reforms that promoted science and technology education and research. Satisfied with the political credentials and aims of the Guomindang, the China Medical Board and the China Foundation cooperatively supported projects that were consonant with the reforms. Government and foundations were determined in particular to help National Central University get back to developing itself into a great educational and research center.[81]

National Central immediately profited from the new government's first year in office. The university's 1928-1929 overall budget for the sciences was double that of the previous year.[82] Some of that increase came from the foundations, now convinced that their earlier plans for National Central could be carried out successfully under the new government's aegis. The foundations' main plan, of course, was still to make National Central's medical school the public-sector equivalent of Peking Union Medical College; and to make National Central's biology department into the finest one in the university system for premedical training. That is, it was to be a feeder for the medical school the way Yanjing University serviced Peking Union. This is one reason why Roger Greene arranged to have the chair of National Central's zoology and botany departments also serve as the chair of the China Foundation.[83]

In the early 1930s, the Guomindang began to formulate an industrial development plan that called for a radical increase in college and university graduates trained in science and engineering. To that end, the Ministry of Education mandated the expansion of science and technical curricula in higher education, and a shift of student majors from the humanities and social sciences into these areas. By the time of the Japanese invasion, Government statistics record marked progress with these plans.[84] Again, this was reflected in National Central's development. The Education Ministry's 1933 overall budget for National Central's Arts and Sciences College was more than triple that of 1932. Twenty percent of the new budget was allocated just to the natural sciences, including science library acquisitions and the hiring of outstanding new faculty in each field. W. E. Tisdale, the Rockefeller Foundation science evaluator, determined that in 1933 National Central's salaries for scientists were good and that the teaching load left time for research. He found a respectable academic atmosphere and faculty morale restored.[85]

By 1933, the biology department at National Central comprised five professors, all with foreign doctorates, four part-time lecturers with varied M.S. degrees, and seven assistants with B.S. degrees from National Central. The agriculture school had its own staff of biologists as well, making this the largest concentration of biologists in the country. The biology faculty, however, was a very unstable one that started turning over continuously from about 1927, when both Bing Zhi and Hu Xiansu began to work most of the time out of research institutes in Beijing. On one hand the troubles at National Central partly accounted for this turnover in the latter 1920s. On the other, as in Chen Zhen's case, National Central also lost biology faculty because they were regularly asked to administer new departments and programs as biology research and education expanded.[86]

The quality of this biology faculty appears to have been consistently outstanding, and over time to have expanded and improved in its command of critical fields, such as physiology, endocrinology, cytology, and microbiology. Botanical taxonomy remained strong. Genetics maintained its place in the core curriculum and was occasionally represented by outstanding faculty who had received their doctorates in the United States. Nationwide, however, the field of genetics developed most fully elsewhere in China—especially at Yanjing and Nanjing universities, which are examined in the next chapter.[87]

# Conclusion

Before foreign philanthropy began to play its role, biology was established at National Central as a result of the initiatives undertaken by Chinese scientists and university administrators. Only after the pioneering curriculum was set in motion did the foundations take notice of the new national university, consider it a "good bet," and include it in their plans for the development of the life sciences. The course of biology at National Central suggests considerable agency and independence for the faculty and administration—intellectually and institutionally.

The funding for biology came from a variety of sources—public, private, and foreign: from the central and provincial governments, local industrial and agricultural business, and from the foreign foundations. It was not just that this range of funding provided necessary wherewithal for the function of the department; but also it obviated the need to shape the department and its research in response to a single patron's interests. And while the central and local governments remained the largest source of funding, there is no evidence to suggest that they ever interfered with the form or content of biology at National Central.

As the foundations began their contributions to science at the university, it is clear—at least in the case of biology—that they enabled the accomplishment of plans already laid, rather than force the school to undertake different ones. Nor did any funding agency change the course of research at the school: Bing Zhi and Chen Zhen remained steadfast in their original interests throughout their careers and succeeded in finding abundant research funding from among the variety of sources available.

Only in one important instance could it be suggested that the course of biology at National Central was redirected by the foundations. The original biology department was housed in the new agriculture school, but the foundations were not interested in supporting "practical application" of biology at National Central. As we shall see, the university's agriculture science program was weakened as it was forced apart from the biology program and was denied funding from the foundations.

Both teaching and research were expected from National Central faculty. In contrast, when the Soviet model was followed in Mao's China, scientific research was virtually eliminated from the universities and carried on exclusively in the institutes of the Chinese Academy of Science and ministries. Among the results of this division of labor (to be discussed later in detail) was a fragmented science community that had few avenues to communicate among its constituents. The character of the science community exemplified at National Central is one of openness, communication, and mobility. This was enhanced by the fact that many scientists who taught at the university were also members of the Academia Sinica.

The foundations certainly thought it important that teaching and research go together. Wherever possible, they continued to encourage the pursuit of that "Johns Hopkins" model which wanted (especially for its medical students and practitioners) education, research, and application all to take place in close proximity.

The National Central biology department did not just pioneer institutional arrangements; intellectually, it likewise displayed independence and decisiveness. Biologists like Bing Zhi and Chen Zhen performed some especially sensitive services for biology by placing and identifying it in China's "new culture." They were responsible for differentiating the scientific treatment of subjects like heredity and evolution from the social scientific and philosophic. This was done in Bing Zhi's periodical articles and by Chen Zhen especially in his general biology textbook. In their writing, as well as in their research, they resolutely designated and patiently explained the neo-Darwinian paradigm that they intended to follow. Without polemics or aspersions, they rejected the neo-Lamarckian understandings of heredity and evolution that—in spite of their efforts—were to be popular and sometimes zealously promoted in China for decades to come.

There is no doubt about the exalted place that modern science held for much of China's modern intelligentsia and political leadership. The development of National Central's biology department indicates that the exalted place of science was not an isolated place, insulated from the unpredictable events of political and cultural revolution or economic exigency.

# Notes

1. See ZXS, vol. 1:21-35; and China Foundation, *Report* 8 (December 1933): 21-33.

2. School name changes:

*1903-15*: Liang-Jiang yuji shifan xuexiao (Higher Normal School of Jiangxi and Jiangsu Provinces);

*1915-21*: Nanjing gaodeng shifan xuexiao (Nanjing Higher Normal School);

*1921-27*: Guoli dongnan daxue (National Southeastern University);

*1927-28*: During the strife of the Northern Expedition it became Disi Zhongshan daxue (Fourth Sun Yatsen University) and then Jiangsu daxue;

*1928-49*: Guoli zhongyang daxue (National Central University).

Late-imperial education reforms: See Paul J. Bailey, *Reform the People: Changing Attitudes towards Popular Education in Early Twentieth Century China* (Vancouver: University of British Columbia Press, 1990). On the development of National Central: Interviews with Professor Guo Tingyi, New York, October-November 1974. He was a student there from 1923-1926 and a faculty member from 1932-1950.

3. Guo Bingwen: see BDRC 2: 276-77; and for university expansion, see N. Gist Gee, "Southeastern University, Nanking," December 1922 (RAC: China Medical Board /ser.II.B.83).

4. BDRC 2: 276-77. And for "dominate educational policies," see John B. Grant, "Report on the National Southeastern University," November 21, 1921 (RAC: China Medical Board /ser.II.B.83).

5. For the bank loans and Guo's familiarity with the regional business community from his earlier days working with the national customs service, see: Zhu Xizu, "Guo Bingwen yu Nangao, Dongda," and Wang Chengsheng, "Zhongguo zhe ren Guo Bingwen," both in *Guobingwen xiansheng jinianji, 1880-1969* (Commemorative anthology) (Taibei: China Academy, 1971), 67-68, 92.

Private gifts: See Zhu Xizu, "Guo Bingwen yu Nangao" 66-67, and BDRC II.177 for patronage of Jiangxi military governor Qi Xieyuan. For detailed list for 1921 budget year, see John B. Grant, "Report on the National Southeastern University," 5, which corroborates Qi Xieyuan's gift to build the library. And also see Paul Monroe to Roger Greene, July 1922, "Copy: National Southeastern University, Office of the President—Donations" (China Medical Board .Ser.II.Bx.63). The latter lists the following items in addition to those listed above:

(1) From H. Y. Moh [the cotton magnate], $61,000 for scholarships for graduate students to study in Europe and America; and the construction of a farm mechanics hall;

(2) Another gift from H. Y. Moh to Cotton Investigation Fund

(3) Gifts to the women's scholarship fund

(4) Gifts to the botanical collection fund, from Commercial Press

(5) Gift from International Committee for improving sericulture

(6) Eight kinds of farm machinery, worth U.S.$8,000, from International Harvester Co., U.S.

6. Paul Monroe was director of Teachers College, Columbia University School of Education, 1915-1923, and director of its International Institute, 1923-1927. He was first

chairman of the board of the Rockefeller Foundation's Peking Union Medical College, 1920-1926; and vice chairman of the board of the China Foundation, from its inception in 1924 until 1944. For Rockefeller activities in China, see Mary Bullock, *An American Transplant*.

Target of opportunity: John B. Grant, "Report on the National Southeastern University," and N. Gist Gee, "Southeastern University, Nanking," December 1922 (China Medical Board/ser.II.B.83).

7. See Bullock, *An American Transplant* and Monroe's 1922 report on premedical science education (RAC: 601A/China Medical Board.XII.no.3); Roger S. Greene, "Proposal for Aid to the Medical School of the Central National University of China at Shanghai," March 6, 1929 (RAC: RG.1.Ser601/Bx.24).

8. For photo and description of the new science building, see "National Southeastern 'U' Metropolitan in Scope," *The China Press*, October 18, 1925 (in RAC: RG.1/ser.601). China Medical Board evaluations and funding of universities and their science programs are provided in: S. M. Gunn, "Report on a Visit to China June 9-July 30, 1931" (RAC: RG1/ser 601/ Bx.12.fldr.129); W. E. Tisdale, "Report of a Visit to Scientific Institutions in China, September-December 1933" (RAC: RG1/ser.601D/Bx.40); and S. M. Gunn, "China and the Rockefeller Foundation," January 23, 1934 (RAC: RG1/ser.601/ Bx.12/fldr.130).

9. Communist conspiracy: See Guo's Teacher's College, Columbia Ph.D. dissertation, published as *The Chinese System of Public Education* (New York: Columbia University Press, 1914). Also see his speeches in the United States from the mid-1920s in *Guo Bingwen Jinianji*.

10. See Guo's education manifesto in *The Chinese System of Public Education*.

11. Liu Boming, President Guo's dean of Arts and Sciences (see text below) was probably responsible for recruiting these Harvard graduates. One of them, Mei Guangdi (1890-1945) was his classmate at Northwestern before going on for his Ph.D. at Harvard; the other was Wu Mi (b. 1894). In 1922, they were the founders of *Xue Heng* (Critical Review), the prolific neoconservative journal based at National Central. Their efforts to preserve China's "national essence" were joined by older faculty who had no foreign experience—for example, Chen Chuping (b. 1883) a poet and veteran of the 1911 revolution. See Laurence Schneider, "National Essence and the New Intelligentsia," in *The Limits of Change*, ed. Charlotte Furth (Cambridge, Mass.: Harvard University Press, 1976), 57-89. Also see BDRC III: 24-26, 442-44.

12. Liu Boming was a native of Jiangsu Province, like Guo Bingwen. After a 1910 B.A. from the missionary Nanjing University, he got his 1912 M.A. and 1915 Ph.D. in philosophy from Northwestern University (Student Records: Northwestern University Archives). And also see: Interview with Guo Tingyi, New York, October 25, 1974; Zhang Qiyun, "Guoli zhongyang daxue," in *Zhonghua minguo daxue zhi* (Universities in Republican China), ed. Zhang Qiyun (Taibei, 1954), vol. 1: 4-5; biographical essays in *Chuang xin* (Creative weekly, Taibei), March 1, 1973; and *Huaxue yuekan* (Sinological monthly, Taibei), April 1976.

13. For Hu's biography see ZXS: 70-85. For the Science Society laboratory and the general development of botany in China, see William J. Haas, "Botany in Republican China: The Leading Role of Taxonomy," in *Science and Medicine in Twentieth-Century China*, eds. John Bowers and Nathan Sivin (Ann Arbor, Mich.: University of Michigan Center for Chinese Studies, 1988), 31-64.

14. Arnold Arboretum: For details, see Haas, "Botany."

15. See Hu's article on "Xuefa" (scholar warlords) in his university newspaper *Dongnan lunheng*, May 1, 1926.

16. See Hu's article on the anti-Christian movement in *Dongnan lunheng*, January 1, 1927.

17. Fan Institute: See Haas, "Botany," and Gunn "Report on Visit to China," 79; and Tisdale "Report," 65. The Gunn report observes the cooperation between Bing Zhi and Hu Xiansu in running the Fan Institute as well as the Science Society laboratories. All three sources agree with Gunn's observation that they were "essentially concerned with the collection and classification of botanical and zoological specimens."

Academica Sinica: See Yang Tsui-hua, "The Academia Sinica and Modern Scientific Research in China, 1927-1937," unpublished paper, New York, 1980. And also see *Guoli zhongyang yanjiuyuan zong baogao* (Acadmia Sinica annual reports), June 1935 and June 1936 for retrospectives on research activities and publications. These show that Academia Sinica biology was devoted to systematics, pest control, entomology, and forestry.

Foundation support: In addition to the support for National Central already mentioned above, the China Foundation contributed funds to the Fan Institute and Science Society laboratories; and it also gave generous stipends both to Bing and Hu through its multiyear science research professorships (see Haas, "Botany," 50-51). Also see *The Science Society of China: Its History, Organization, and Activities* (Shanghai, 1931), 19-20; ZXS: 22-25; and Laurence Schneider, "The Rockefeller Foundation, the China Foundation and the Development of Modern Science in China," *Social Science and Medicine*, 1982, no. 16: 1217-21. For the predominance of taxonomy and Hu Xiansu: Haas "Botany," 40, 45-49.

18. Such advisers were themselves working in the naturalist tradition, for example J. M. Coulter from the University of Chicago. For a description of his efforts in China, see J. W. Dyson, "The Science College," in *Soochow University*, ed. W. B. Nance (New York: United Board for Christian Colleges in China, 1956), 54-55; and see advice of J. G. Needham, Bing Zhi's mentor, in "Report to the China Foundation, June 1928," China Foundation annual *Report*, 1928, no. 3: appendix 1.

19. Gunn "Report," 14.

20. Haas, "Botany," 47-49.

21. Additionally, the China Foundation's board had Chinese members, including Bing Zhi, whose scientific interests were broader than those of the China Medical Board. There is also the influence of N. G. Gee, the science adviser to the China Medical Board and the China Foundation. His personal research interests were in taxonomy. See Haas "Botany," 48; and Haas, *China Sojourner*.

22. See J. G. Needham, "Report to the China Foundation."

23. See Yang Tsui-hua, "The Development of Geology in Republican China, 1912-1937," in *Science and Medicine in Twentieth-Century China*, 87-89; Peter Buck, *American Science and Modern China, 1876-1936* (Cambridge, Mass.: Cambridge University Press, 1980), 225-26.

24. Fu Sinian's speech, in *Fu Sinian xuanji* (Selected works), 10 vols. (Taibei, 1967), 3: 475-87. First published in introduction to *Lishi yuyan yanjiuso zhoukan* (Historical and philological research institute weekly), 1.1: October 1928.

25. Chen sometimes published under his given name Xie-san. See biographical details in Liu Xin, *Zhongguo kexue ershi nian* (Twenty-year survey of China's science) (Nanjing: Science Society of China, 1934); Chen's obituary in *Shengwu xue bao* (Biol-

ogy journal), 1958, no. 1: 64-65; and Columbia University Academic Record Cards for "Ch'en Chen."

26. T. H. Morgan, with A. H. Sturtevant, H. J. Muller, and C. B. Bridges, *The Mechanism of Mendelian Heredity* (New York: Holt, 1915); and T. H. Morgan, *The Physical Basis of Heredity* (Philadelphia, Pa.: Lippincott, 1919).

27. Garland E. Allen, *Thomas Hunt Morgan: The Man and His Science* (Princeton, N.J.: Princeton University Press, 1978), 208-9.

28. All quotes from Allen, *Morgan*, 209-12.

29. For Chen's course and the use of English, see Gee, "Southeastern University," 13-14. Also see *Guoli Dongnan daxue ijian* (Catalog of National Southeastern University) (Nanjing, 1923), 62, 66; S. D. Wilson, "Report on Science at SEU" (five-page appendix to John Grant, "Report on Welfare Activities and the Teaching of Hygiene at the National Southeastern University," November 21, 1921 (RAC: China Medical Board/ser.II.B.83). All of the scientists interviewed by this writer for this study acknowledged that science courses, at least at the university level, were commonly taught in the language of the nation that published the textbook and was apparently most prominent in the discipline. Thus, in physics and mathematics courses, German was commonly used along with Chinese; English and Chinese were used in biology and agricultural science courses.

30. *Putong shengwuxue* (Shanghai: Commercial Press, 1924).

31. The textbook's popularity: See Guo Xueting, "Meng-de-er xueshuo zai Zhongguo" (Mendelism in China) in *Meng-de-er shishi yibai zhounian jinian wenji* (Commemorative anthology for the 100th anniversary of Mendel's death), ed. Genetics Society of China (Beijing: Science Press, 1985), 12. And also interviews with geneticists Li Ruqi, Beijing 1984; C. C. Li, Pittsburgh 1985; Liu Zedong, Shanghai 1986.

32. *Putong shengwu xue*, 134. The burden of his argument about using Latin is that it should not cause China any sense of national embarrassment or dependency. The English, French, Germans, etc. all used Latin nomenclature instead of native language nomenclatures as a practical matter of efficient communication.

33. In the book's bibliography, Chen sites only two Chinese language titles, and both are biological dictionaries published between 1919 and 1921.

34. "Metaphysical" (xuan xue): Since Chen does not elaborate, I take the term as an epithet generally meaning "unscientific" or "superstitious" (as folk beliefs or religion were held to be by the New Culture intellectuals).

35. Early publications about Mendel: See Guo Xueting, "Meng-de-erh xueshuo," 9-16.

36. Introductions: For example see, Bing Zhi, "Xipao yuanli zhi yulai" (Principles of cytology), *Kexueh* 1915, no. 7: 792-98; Gaoshan, "Yichuan jinhua shuo zhe yingyong yu nongi" (The use of genetics and evolutionary theory in the agricultural arts), *Dongfang zhazhi* (August 10, 1915): 1-5; *Kexue*, 1918, no. 9: 882-90, and 1918, no. 12: 1209-14 for discussions on the Morgan school, chromosomes, and basis of heredity.

Cornell Ph.D.: See Feng Zhaochuan's research on inheritance in maize, *Kexue* 1923, no. 5: 528-37, and his Chinese lexicon of 620 genetic terms, *Kexue*, 1923, no. 7: 759-77.

37. Chen, textbook 204-5, 208.

38. Genes: textbook 209-21. Garland Allen notes that the term "gene" was purposely avoided in the 1915 *Mechanism of Mendelian Heredity*, the term "factor" being used instead. Johannsen, who coined "gene" in 1909, "originally meant [it] to be a completely abstract concept, consciously disassociated from any of the existing theories of hereditary particles." The Morgan group avoided the term initially because they wanted no misun-

derstanding of their commitment to a material basis for Mendelian factors. Allen, *Morgan*, 209. In the 1919 revision, Morgan did use the term throughout, and it was this usage that was followed by Chen Zhen.

39. Ma Zhunwu (1883-1939) translated chapter 3, "Struggle for Existence," and chapter 4, "Natural Selection or the Survival of the Fittest," in *Xinmin zongbao*, 1903, no. 3-4. In 1920, he published the first complete Chinese translation of the *Origin*.

40. Yan Fu, *Tianyan lun* (On evolution) was first published in 1895 in a periodical. Monographic editions followed in 1898 and 1901. These are the sources of numerous later printings and editions. Also see Benjamin Schwartz, *In Search of Wealth and Power: Yen Fu and the West* (Cambridge, Mass.: Harvard University Press, 1964). (Due to a change of romanization, "Yen Fu" in this title is the "Yan Fu" in the text above.)

41. Charlotte Furth, "Intellectual Change from the Reform Movement to the May Fourth Movement, 1895-1920," in *Cambridge History of China*, ed. John K. Fairbank (Cambridge: Cambridge University Press, 1983), vol. 12.1, chapter 7.

42. Peter Zarrow, *Anarchism*, and Arif Dirlik, *Anarchism and the Chinese Revolution*.

43. James Pusey, *China and Charles Darwin* (Cambridge, Mass.: Harvard University Press, 1983).

44. See Daniel Todes, *Darwin without Malthus: The Struggle for Existence in Russian Evolutionary Thought* (Oxford: Oxford University Press, 1989).

45. Kropotkin, "Modern Science and Anarchism" (1901-1923 various language editions), "Regarding Inheritance," and "Thoughts on Evolution," an essay series appearing between 1910-1915 in *The Nineteenth Century and After*. All reprinted in *Evolution and Environment*, 15-107. For Kropotkin's Lamarckist evolutionary mechanics, his rejection of Weismann's genetics, and also for his acknowledgment of the Russian biologist, Kessler, as his inspiration for the idea of mutual aid, see 111-243.

Here, "Lamarckist" refers to the evolutionary principles of Jean Baptiste Lamarck (1744-1829), the French biologist whose influential theory of evolution anticipated and rivaled Darwin's. Lamarck's theory of evolutionary change is characterized first by the principle that all life forms innately have the potential "power of acquiring progressively more complex organization," that is, to progress toward more perfect forms. And second is their "capacity to react to special conditions of the environment." In this evolutionary scheme, substantial and continuing changes in an organism's circumstances or environment bring about a real change in their "needs." In turn this brings about changes in behavior and new habits. The latter may require more or less use of certain parts of the organism, which may diminish or enlarge those parts; in some cases, altogether new parts may be produced by these new needs. (This is referred to as Lamarck's theory of "use and disuse.")

Such changes brought about in this manner, Lamarck claimed, were inheritable by the progeny of the affected individuals. This is what is meant by the inheritance of acquired characteristics (or acquired characters); it is also referred to as "soft inheritance." The implication here is that genetic material is not constant from generation to generation but may be modified by the environment, by use and disuse, etc. "Neo-Lamarckism" refers to the revival of interest in Lamarck's principles (and sometimes their misconception) after Darwin's publication of the *Origin* in 1858. See Ernst Mayr, *The Growth of Biological Thought* (Cambridge, Mass.: Harvard University Press, 1982), 353-55.

46. For the transformist tradition in Russian culture, see Douglas R. Weiner, "The Roots of 'Michurinism.'"

47. Zhang Binglun and Wang Zichun, "The Struggle between Evolutionary Theory and Creationism in China," in *Chinese Studies in the History and Philosophy of Science and Technology*, eds. Fan Dainian and Robert S. Cohen, trans. Kathleen Dugan and Jiang Mingshan (Boston, Mass.: Kluwer, 1996), 289-90.

48. There was however the minor exception of the Catholic Church in China which waged an ineffective anti-Darwinist, pro-creationist campaign for many years. See Zhang Binglun and Wang Zichun, "The Struggle."

49. According to Guo Xuecong, "Meng-de-erh xue shuo," 81-82, this was not the case with monographic literature of this period. He cites five books which he says dealt substantially with evolution and were published in China between 1907-1911: one on zoology and one on botany, two "Republic Textbooks," and one "popular lectures on evolution." I have not been able to see any of these texts.

50. In *Kexue*, see for example Bing Zhi's analyses of evolution in 1915, nos. 2, 6; Qian Zhongshu's piece on saltationists G. Bateson and Hugo de Vries, 1915, no. 7: 784-92; Hu Xiansu's translation of the Stanford biologist, Vernon Kellog, on Darwin's theories, 1915, no. 10: 1158-63, and 1916, no. 7: 770-82; Yao Jiagu on the inheritance of acquired characters, neo-Lamarckism and neo-Darwinism, 1920, no. 5: 492-95.

51. See Ernst Mayr, "Prologue: Some Thoughts on the History of the Evolutionary Synthesis," in *The Evolutionary Synthesis*, eds., Ernst Mayr and William B. Provine (Cambridge, Mass.: Harvard University Press, 1980), 1-5.

52. Lu Xun was the pen name of Zhou Shuren. Zhou Jianren sometimes published under the pen name of Ke Shi, and Zhou Qiaofeng or Zhou Songshan. Biographical details on him are sparse. See DSSI, 13; and BDCC: 1: 202-4. Brief notices appear in these biographical dictionaries: *Zhongguo jin-xiandai renming dacidian* (Beijing, 1989); *Zhongguo geming shi renwu cidian* (Beijing, 1991); *Zhongguo renming da cidian* (Shanghai, 1992).

53. The historian Frank Dikotter has asserted that between 1900 and 1911, the idea of evolution employed in social and political discourse was essentially non-Darwinian. I concur with this, but not with his later assertion that from 1915-1949 the "majority" of Chinese scientists were not Darwinians but neo-Lamarckians. Dikotter is correct to suggest the importance of neo-Lamarckism in China, but he provides virtually no evidence for how the "majority of Chinese scientists" thought. And he has nothing to say about China's biologists who were staunch Mendel-Morganists. This present study will discuss the latter issue in detail. See Dikotter, *The Discourse of Race in Modern China* (Stanford, Calif.: Stanford University Press, 1992), 107, 138-41.

54. In 1914, Zhou published one of the earliest periodical pieces on Mendel, "Yichuan shuo" (Theory of inheritance), *Zhonghua jiaoyu jia* (China education world) 1914, no. 9: 1-7. In *Dongfang zhazhi*, December 12, 1921, he wrote on the problem of the inheritance of acquired characters and the role of environment in evolutionary biology after Darwin. Zhou outlined the historical swings between the Weismann-Morgan school (emphasizing the primacy of genetic material) and the "environmentalists" such as Paul Kammerer, P. C. Vander Wolk, Hugo de Vries, Merle C. Coulter. In "Yichuan yu huanjing" (Heredity and environment), he was openly enthusiastic about foreign biologists whose findings appear to support the environmentalist position. See 1923, no. 4: 81-85.

55. See Zhou Jianren and Chen Chanheng, *Jinhualun yu shangzhongxue* (Shanghai: Commercial Press).

56. Chen's citation incorrectly dates its publication by Princeton as 1919.

57. Allen, *Morgan*, 116-24. And also see Ernst Mayr, *Growth of Biological Thought*, 546-52.

58. For example, by 1920, the ideas of gene interaction, crossing-over, and recombination, according to Allen, "showed Morgan and his students that an almost infinite number of possible combinations of characters could (and probably would) be obtained in the course of time in any population" (*Morgan*, 306). These changes did not directly result in new species, but rather they suggested themselves as evidence for the gradual, cumulative, evolutionary process of Darwin (*Morgan*, 301-17).

59. Working conditions: See N. Gist Gee, "Southeastern University," 13-14. Science Society: See *The Science Society of China: Its History, Origin and Activities* (Shanghai, 1931), 19-20. For Chen's first publication, see discussion in text below.

60. China Foundation *Report*, 1925-1926, no. 1, and also see special report on "Zoology at Tungnan University" in 1926-1927, no. 2.

61. See Chen's "Jinyu de bianyi yu tianyan" (Variation of goldfish and evolution), *Kexue*, 1925, no. 3: 304-30. And see his retrospective on his work, "A History of the Domestication and the Factors of the Varietal Formation of the Common Goldfish, Carassius auratus," in English, *Scientia Sinica* (June 1956): 287-321; original Chinese publication in *Acta Zoologica Sinica*, 1954, no. 2: 89-116.

62. Other scientists in China using goldfish: See list in *Kexue*, 1925, no. 6: 795. And in addition to those listed, goldfish were also used by two prominent Chinese embryologists: Zhu Xi and Tong Dizhou.

Scientists elsewhere using goldfish: This is evident from the research cited by Chen in his published reports. The simultaneous research was acknowledged in Chen's "The Inheritance of Blue and Brown Colours in the Goldfish, Carassius auratus," *Journal of Genetics*, 1934, no. 29: 61-74. It was in this same article, page 61, that Chen talked about the problems of using goldfish for research.

63. Robert E. Kohler, *Lords of the Fly: Drosophila Genetics and the Experimental Life* (Chicago, Ill.: University of Chicago Press, 1994), 56.

64. Kohler, *Lords*, 56.

65. Kohler, *Lords*, 57.

66. Morgan, *Mechanism of Mendelian Heredity*, 210. And again in Morgan's *A Critique of the Theory of Evolution* (Princeton, N.J.: Princeton University Press, 1916), 72; and *The Physical Basis of Evolution*, 270. Also see discussion in Allen, *Morgan*, 306-8.

67. Chen, "Inheritance of Blue and Brown Colours," 61.

68. Chen, "Inheritance of Blue and Brown Colours." and "Transparency and Mottling," 450. In the latter article, the case in point was a peculiar case of Mendelian inheritance, Chen said, "because the heterozygous condition is a new type distinct from the two homozygous types" (450). That new type was a mozaic of the two homozygous types. And this was the source of the three distinct phenotypes. This further indicated to him that he was not quite correct in his original attribution of dominance to the non-normal transparency and mottling. He found that the latter condition was "not a strictly dominant character, and that the F2 fish may be classified into three distinct phenotypes with 1:2:1 ratios" (61).

69. There are references to some further research by Chen; however, there are no publications to bear this out. In his 1956 retrospective ("History of the Domestication of the Common Goldfish"), he cites no research publication for himself past 1934.

70. See the following: *Qinghua ijian* (Catalogue) (Beijing: Qinghua University, 1927); *Beiping ge daxue de juangxiung* (An outline of universities in Beijing) (Beijing: Beijing University, 1929-1930); *Guoli Qinghua daxue ershih zhou nian jinian kan* (Twentieth anniversary volume for National Qinghua University) (Beijing: Qinghua University, 1931); *Guoli Qinghua daxue* (National Qinghua University) (Beijing: Qinghua

University, 1935); and *Qinghua daxue xiao shi gao* (Draft history of Qinghua University) (Beijing: Qinghua University, 1981), 205-9. See also the Rockefeller Foundation, *Annual Report* (New York, 1929), 226; and N. G. Gee, letter to "Tsinghua president Tsao," February 14, 1926 (attachment to Roger Greene, letter to M. K. Eggleston, March 26, 1926 in RAC: China Medical Board/ser.II.B.85).

71. "Shengwu xi" (Biology department), in *Guoli Qinghua daxue ijian* (National Tsinghua University catalog) (Beijing: Qinghua University, 1935). Among Chen's textbooks were E. W. Sinnott and L. C. Dunn's *Principles of Genetics, a Textbook with Problems* (New York: McGraw Hill, 1925, and later editions). Dunn was a student of Castle. And W. B. Scott's *The Theory of Evolution* was also used.

72. Tisdale, "Report," 27.

73. See Roger Greene, "Memo of a Conversation with Kuo Ping-wen," April 25, 1925 (China Medical Board.ser/II.B.63). And also see interview with Guo Tingyi, October-November 1974.

74. Presence of Guomindang on campus: See Lo Shishi, "Shisi nian Dongda," *Quanji wenxue* (October 1962): 27-29; Interview with Guo Tingyi, October-November 1974; and see dozens of articles by students and faculty of National Central in the school weekly magazine *Dongnan lunheng* for 1926. (First issue published March 27, 1926.) "Communist Wing": See Roger Greene, "Memo of a Conversation with Kuo Ping-wen."

75. Roger Greene, director of the China Medical Board was Guo's close adviser during these troubles. When Guo resigned, Greene offered him the presidency of the China Foundation. Guo was already serving on its first board of directors, but turned it down, he said, because that might offend his loyal supporters at the university. See memo: Roger Greene conversation with Kuo Ping-wen, May 6, 1925 (China Medical Board.Ser.II.B.63). Instead, Guo became president of the China Institute of America, funded by the China Foundation. He returned to China in 1931, and until the war was a high administrator of at least three Shanghai banks while serving in various financial posts for the Guomindang government (BDRC 2: 277).

76. See N. G. Gee to Roger Greene, March 26, 1925, and May 19, 1925 (China Medical Board/ser.II.B.63); and N. G. Gee to Roger Greene, November 22, 1927, "Brief Report on the 4th Chung Shan University, Nanking (Formerly Southeastern University)," (China Medical Board/ser.II.B.63). Guomindang occupation troops changed the school's name to that of Sun Yatsen (Chung Shan). And also see Lo Shishi, "Shisi nian Dongda."

77. The role of the university in the May thirtieth turmoil was a constant topic of discussion and debate in the school's newspaper, *Dongnan lunheng*, during 1926 and early 1927.

78. Lo Shishi, "Shisi nian Dongda"; Interview with Guo Tingyi, October-November 1974; and N. G. Gee to Roger Greene, "Brief Report on 4th Chung Shan University, Nanking," and Gee to China Medical Board, April 17, 1928, "Kiangsu University," (China Medical Board/ser.II.Bx.63).

79. Gee to China Medical Board, "Kiangsu University"; and China Medical Board 1933 Tisdale *Report*, 12-15. And also see suggestion by Zhang Qiyun that Qinghua University was a magnet for disgruntled National Central faculty (*Guo Bingwen jinianji*, 106).

80. By way of illustration: China's National Library, in Beijing, was originally built with money from the China Foundation, which also provided the National Geological Survey with annual supplements, even after it had become part of the new Academia Sinica. And after 1928, a significant amount of funding in the biomedical sciences con-

tinued to come from the China Foundation and/or the Rockefeller Foundation's China Medical Board. Such funds went to everything from bricks and mortar to specific research projects. All of this is detailed in the China Foundation annual *Reports*.

81. Education reforms are well summarized and documented in James Reardon-Anderson, *Study of Change*, chapters 8 and 9. For the foundations' determination to repair the damage at National Central and help it move on, and the foundations' desire to work with the new government toward that end, see R. S. Greene to M. K. Eggelston, June 28, 1926, and N. G. Gee to R. S. Greene, June 16, 1926 (China Medical Board.II.Bx.63); and Roger Greene, March 6, 1929, "Proposal for Aid to the Medical School of the Central National University of China, at Shanghai" (RAC: RG.1/ser.601).

82. Budget doubling: Gee to Mason, March 18, 1930 (RAC: RG1.1/ser.601).

83. Foundation plan: Best seen in Roger Greene, "Proposal for Aid to the Medical School of the Central National University."

China Foundation Chairmen: Tsai Berg [sic], a zoologist, served this dual role from 1927-1930, followed by the botanist Zang Jinshu (Tsang Kin-chou).

84. Reardon-Anderson, chapters 8 and 9.

85. Tisdale, *Report*, 12-15.

86. Biology department: summary based on surveys for the China Medical Board and Rockefeller Foundation by N. G. Gee (1931), S. K. Gunn (1931), and W. E. Tisdale (1933). These are supplemented with details from biographical collections such as the ZXN and China Foundation reports on research professorship and research project grant recipients.

87. Two examples of geneticists at National Central: Xu Xiang (Wellington Hsu) (b. 1910), received his 1929 genetics Ph.D. from Harvard, under East, working on plant chromosome structure. He was at National Central between 1930-1933 and during that time received a China Foundation research professorship. He left in 1933 for a position at the new Zhejiang University (China Foundation, *Report*, 1934, no. 6; Tisdale, *Report*, 1; Liu Xin, *Zhungguo kexue ershinian*, 210).

Feng Zhefang (b. 1899) was a cotton plant geneticist on the faculty of National Central's agriculture department after receiving his B.S. from National Central and his M.S. from Cornell in the mid-twenties. From 1931-1934 he was supported by the China Foundation for his Cornell Ph.D. in plant genetics. See ZXN: 446-56 for his work with Xi Yuanling and Yu Qipao on cotton genetics research at National Central after 1935. And see discussion in text below on agricultural genetics at Nanjing University.

# 2

# Genetics at Yanjing and Nanjing Universities

Yanjing and Nanjing universities, in stark contrast to National Central, were about as insulated as educational institutions in China could get from political and social upheaval and economic exigency. As missionary schools, they were protected by the invisible shield of international treaty, and consequently foreign philanthropies preferred them over Chinese schools. Seemingly unlimited foreign funds were lavished on both Yanjing and Nanjing science programs, and there was more than sufficient advice from foreign experts on how to use these resources. The Rockefeller Foundation China Medical Board closely guided Yanjing's science program to distinction, and Cornell University's Agriculture School did the same for Nanjing. It would not discredit the significant accomplishments of the two science programs, however, to suggest that a number of Chinese schools could have achieved the same distinction if the same resources had been provided to them. The philanthropies—not totally unaware of that possibility—nevertheless wanted as much control as possible over the creation of model science programs and the education of China's first generations of scientists. Missionary schools made that control possible and the outcome more predictable. This especially seems to be the case where the life sciences were concerned.

As a consequence of their origins, the science programs at Yanjing and Nanjing shared some broad similarities. They were initially created with mandates for practical, applied science. Yanjing's program was initially meant exclusively to prepare students admitted to Peking Union Medical College. Nanjing's was expected to produce agriculture experts dedicated to increasing the food supply and averting famine. In time, however, both programs broadened the interpretation of their respective mandates to include a range of education and research well beyond the immediate demands of medicine and agriculture.

Both programs also were similarly criticized by contemporaries for failing to adopt more holistic approaches to their mandated goals. Each was urged to teach and practice science more appropriate to China's current needs. In addition to technically advanced scientific medicine, Peking Union/Yanjing was urged to develop a broad public health program supported by relevant education and research. And Nanjing was faulted for thinking of agriculture in narrowly technical terms while turning a blind eye to social-economic reform in the countryside.[1]

## Yanjing University

If National Central epitomized the new national university, Yanjing University came to embody the highest scholarly aspirations of the missionary schools. In contrast to the urban-based, politically vulnerable National Central campus, Yanjing was tucked away in Beijing's western suburbs, seemingly insulated from the troubles that plagued the much larger university. In fact, however, Yanjing's students periodically engaged in strikes, protests, and demonstrations for nationalistic causes that aroused students in the big Chinese universities.[2] Nevertheless, where National Central appeared to thrive on creative chaos, Yanjing typically was an orderly, hothouse scholarly community, nurtured by abundant Rockefeller and China Foundation funds, in addition to the ample operating budget provided by its well-heeled American board of directors, and its exorbitant tuition. The handsome campus was a classic piece of chinoiserie, evocative of the bucolic images on trade porcelain: willows and flowering plums, a lake and pagoda, arched stone footbridges. (After 1950, Yanjing was dissolved and its campus was appropriated for the prestigious National Beijing University.) In spite of appearances, however, Yanjing was one of the most dynamic intellectual communities in the country. Its Chinese studies program, with ties to Harvard, was among the most distinguished. Its natural science program, especially biology, was in many respects peerless.

Yanjing University came into being in 1918, the product of a merger of two small mission colleges in northeast China. After locating permanently in the Beijing suburbs, it began with a student population of 160 and grew to 800 by 1937. The new school did not develop its natural sciences until after it agreed, in 1923, to house the premedical science program of the Peking Union Medical College. The plan was implemented in 1925, and from that date until 1937, Yanjing's science faculty grew in quantity and quality. In particular, its biology department developed from an original staff of just one or two foreign teachers to a superbly trained, versatile, and active research team of ten—all Chinese but one. At its peak, this group contained four biology Ph.D.'s, all trained in the United States.[3]

Through a combination of factors to be discussed below, Yanjing's biology department was home to an extraordinary trio of geneticists—Li Ruqi, Chen Ziying, and Tan Jiazhen. Between 1927 and 1937, they received their respective doctorates under Morgan's group at Columbia or at the California Institute of Technology (Caltech), and each member of the trio did postdoctoral collaborative research, some of it quite extensive, with key members of the Morgan group—C. B. Bridges, A. H. Sturtevant, and Th. Dobzhansky. Not only was the high quality of the trio's research repeatedly acknowledged throughout the field, but they are further recognized for their participation in some of the major developments of genetics during this period. Their research epitomizes the shift of experimental genetics from studying the way that chromosomes operate in general (mechanics) to their specific operation in the development of organisms, and their role in the evolution of new species. As such, these geneticists were uniquely qualified to guide the growth of genetics in China.

## Yanjing Biology and the Rockefeller Connection

This trio of geneticists and other distinguished scientists would not have become associated with Yanjing had it not been for the university's agreement with the Rockefeller China Medical Board to take over its premedical science program. Since the details of this transaction are very well chronicled, suffice it to say here that the China Medical Board, which administrated the medical college, had earlier grown frustrated with the inability to find Chinese students well prepared enough for the demands of Peking Union Medical College. For a stopgap solution, in 1918 Peking Union itself set up a preparatory science program for students admitted to medical school.

Then the China Medical Board resolved the problem in two ways. For the long range, it developed a policy of improving premedical science education in major universities, especially in Chinese universities like National Central. More immediately, it gave its premedical science program to Yanjing, lock, stock, and faculty. And the China Medical Board gave Yanjing substantial endowments and grants to maintain and develop its science curriculum, all with the understanding that Yanjing would become a major source of students for Peking Union Medical College.[4]

Because of Yanjing's relationship with Peking Union Medical College, its biology department additionally developed a close link with Suzhou (Soochow) University's biology department. Suzhou, also a missionary school, had a fine department, established and administered by N. Gist Gee, who was also the China Medical Board's influential adviser for science education in China. Throughout the twenties, Gee surveyed university science departments and advised both the China Medical Board and the China Foundation how and where to spend their funds. In addition to his influence on funding, Gee was able to shape biology at Yanjing by placing former Suzhou students in Yanjing's graduate biology program, and by occasionally placing them on Yanjing's biology faculty. Further, Gee, an accomplished freshwater taxonomist, maintained close personal relations with Yanjing's biology faculty through collaborative research work.[5]

The flow of students between Suzhou and Yanjing is marked. From 1923 to 1933, Yanjing's biology department awarded twenty-two masters' degrees (half during 1923-1930, and the other half during 1931-1933), with nine of them going to Suzhou graduates. (Another seven successful masters' students came from Yanjing's undergraduate program, and the rest from various missionary schools; none was from a Chinese school.) Of the eleven Yanjing students awarded masters' degrees in biology during 1923-1930, at least three went on for doctorates; and of the eleven during 1931-1933, at least five went on for doctorates. Four of these eight Ph.D. holders had done their undergraduate work at Suzhou.[6]

This productive record was fostered by funding from the two foundations: During the 1923-1933 period, the Rockefeller Foundation supported thirteen of Yanjing's twenty-two biology graduate students with one- or two-year fellowships, while the China Foundation gave advanced-study or research fellowships to another seven.[7] J. W. Dyson, Gee's Suzhou colleague, thought it equally significant that, over the 1920s and 1930s, eight of the Suzhou biology faculty had received Rockefeller scholarships, mainly for doctoral training at American institutions. He was quite explicit about the intended developmental role of these scholarships: They were "the foundation upon which was to be built the structure of the future staff of Doctors of Philosophy to man institutions and to head up their science departments."[8]

Gee and Dyson were proud of the role their biology graduates played at Yanjing, as well as other departments around the country. By the late 1920s, fifteen of them were either the heads or senior faculty in their departments. With the exception of two, however, all taught at missionary schools. (Of the fifteen, at least seven obtained American doctorates, three of those from Cornell.)[9]

Clearly, the major concern of biology at Suzhou was taxonomic, reflecting Gee's personal research interests.[10] Dyson reports that a formal policy to develop in that direction, rather than toward experimentalism, was established when Gee began implementing the China Medical Board's nationwide science education plan. Around 1923, Dyson notes that John M. Coulter, former head of the University of Chicago's botany department, was engaged as a consultant to Suzhou's science program. Coulter advised "more serious attention to taxonomy and studies of the local flora, as such data were basic to the botanical sciences in any country."[11] Suzhou responded successfully with specialized training of its faculty and students. Coulter's advice probably also influenced Gee's science education policy throughout China. Nevertheless, Suzhou and Yanjing's biology departments respected and fostered experimental biology and genetics, due initially to the considerable presence of Alice M. Boring (1883-1955).

# Yanjing's Biology Department and Genetics

Boring's rich experience in taxonomic work and experimental biology made it possible for Yanjing and Suzhou to teach and research well and cooperatively in both areas. Between 1900 and 1910, she received her B.A., M.A., and Ph.D. in biology from Bryn Mawr, working under T. H. Morgan and Nettie M. Stevens. She did extensive collaborative work with her mentors in experimental cytology and embryology, coauthoring numerous papers with Morgan. Her dissertation was on chromosomal genetics and designed to provide further evidence to the Stevens/Wilson hypothesis that sex determination was linked to specific chromosomes.

In 1908-1909, Boring had the distinction (rare for any biologist, and especially for a woman) of a year's research fellowship in Wurtzberg and at the renowned Naples Zoological Station, working with the histologist Theodor Boveri. Her teaching career began with a one-year appointment at Vassar in 1907-1908, which was followed by ten years on the faculty at the University of Maine (1908-1918), where she collaborated with Raymond Pearl at the Maine Experimental Station on some of his work in biometry and population genetics. When Boring accepted a teaching position with Peking Union Medical College's premedical science program, in 1918, she had already published fifteen papers under her own name, including her dissertation. She taught there for two years, then returned home to teach at Wellesley for another three. She went back to China in 1923 as a full professor and head of Yanjing's biology department.[12] She was an active member of the department until December 1941, when the Japanese closed Yanjing University.

## Chen Ziying and Developmental Genetics

From 1921-1925, Boring's most important colleague in the biology department was Chen Ziying (b. 1897). Chen received his B.S. from Suzhou University in 1921, and doubtless due to the influence of N. G. Gee, he was subsequently brought into the rigorous teaching program at Peking Union/Yanjing. Chen taught there for four years while simultaneously working under Boring to earn his master's degree, the first awarded by Yanjing in biology. His thesis project was a chromosomal study of two kinds of Drosophila mutations.[13] Then, supported by a Rockefeller Foundation fellowship, Chen was sent to Columbia University where he received an M.A. in biology in 1926. In 1929, under the supervision of T. H. Morgan and A. H. Sturtevant, he completed a doctoral dissertation that, fifteen years later, was still being referred to as a classic by scientists such as Joseph Needham, the renowned embryologist.[14]

Chen Ziying's doctoral research was one of a number of similar projects Sturtevant assigned at this time to doctoral students or persuaded younger members of the Morgan group to undertake. For example, according to Kohler, Sturtevant enlisted Th. Dobzhansky "who was a skilled morphologist, to dissect developing [mutations of a specific kind] and find out when [the mutation's] effects appeared." And he set Li Ruqi (Chen Ziying's future colleague) "the task of finding out at what stage of development lethal genes had their effect." In doing this, Kohler concludes, Sturtevant was "making use of visitors' morphological skills—essential skills, which the . . . [Morgan] group did not have—to build a major program in developmental genetics."[15] This emphasis on developmental process contrasts with the focus on structure and location that informed the group's characteristic mode of analysis up to this date, that is, gene mapping. This technique was based on a fundamental tenet of classical genetics, namely, that genes are arranged linearly on a chromosome. Here the main concern was to determine not only that a particular gene was located on a particular chromosome, but in addition, to determine the relative positions of genes sitting on the same chromosome. These were determinations intimately associated with the sources of variation and gene expression (as in the concepts of "crossing over" and "linkage"), but they were depicted as static, ordered structures—blueprints.

Chen's research was an exemplary study of how genes affect the development of characters at the earliest stages of embryonic growth. Sturtevant did not want his associates to be satisfied with the generalization that a chromosome, gene, or set of genes transmits a visible character. Now he wanted them to be cognizant of when and where in the developmental process the effects of those shaping forces were first manifest. Chen studied the expression of certain mutant genes and found that the effects on their related character expressions "could be observed in the very earliest condition of the imaginal buds." An imaginal bud or disc is an inchoate organ, visible on a developing embryo or pupa. Chen argued that before this kind of research, when an organ of the Drosophila was, for example, found to be reduced or absent, "it was not known whether its imaginal disc showed a corresponding reduction or whether it was full size in the larvae and pupae, and failed to carry through to the later stages."[16] As we shall see below, in the 1930s, Chen was able to work once more with the Morgan group and to carry this kind of developmental research one significant step forward.

In 1928, Chen Ziying returned from Columbia to Yanjing as an associate professor in its growing biology department. The department's new chair was yet another Suzhou alumnus, Wu Jinfu (b. 1896), an entomologist with a recent Cornell doctorate. Chen had a comfortable situation at Yanjing and was able to continue his research under a two-year China Foundation fellowship. The school could not hold on to him, however, when he was offered the chair of a virtually new biology department at Amoy University, one of the recent additions to the national system. Chen was charged with the task of developing it into a department of Yanjing's quality but with an emphasis on marine biology.[17]

### Li Ruqi Joins Yanjing Faculty

When Chen Ziying returned to Yanjing in 1928, his Columbia classmate, Li Ruqi (the second of our "trio") had already joined the school's biology faculty and completed his first year of teaching. Alice Boring was able to recruit Li to Yanjing thanks to Li's wife, who had studied biology with Boring at Wellesley. Up to that time, Li Ruqi had intended to use his genetics training in the practice of animal husbandry "somewhere out in Western China," as he had planned when he left for the United States a decade earlier. Boring asked him to take the responsibility for genetics classes and the supervision of senior and masters' theses on genetics topics.[18]

From 1926 onward, Boring herself turned to taxonomy, as evidenced in a series of collaborative publications on Chinese fauna with N. G. Gee, Wu Jinfu, and others. Boring's reorientation may account to some degree for the paucity of geneticists coming out of Yanjing's program. In spite of Yanjing's strength in the field, between 1923 and 1933 only four of the twenty-two graduates produced masters' theses in genetics. Of the eight who went on to earn doctorates, two specialized in genetics.[19]

The presence of Li Ruqi (b. 1895) on the biology faculty marks the beginning of the department's most mature and productive period. He brought remarkable energy to his work and, like his colleagues, a superb education and pedagogy. He arrived at his profession in science and genetics, however, through happenstance. Li was one of four sons of a "small merchant" family in Tianjin, the northeastern port city. The family provided each of the four with two years of Chinese-style higher education, but only Li Ruqi persevered and continued his education. During 1911-1918, he won a series of fellowships to Qinghua Middle School and College, where his distinguished record earned him full support for a Boxer college fellowship to the United States.[20]

Li's educational motivation at the time was nationalistic, and he initially thought to serve China through literature or historical scholarship. The tumult of the 1919 May Fourth Movement, however, convinced him that a productive career in science would be a safer bet. His Qinghua mentors preferred to see him become a banker or industrialist, but he decided to follow some of his friends to Purdue University and a future in animal husbandry, a subject about which he knew very little. He had been convinced by nationalist literature in the popular periodical press that in such a field of endeavor he could make an important contribution to China's development. Nor did he have a strong background in biology before he began to study at Purdue's Agriculture School in the autumn of 1919.[21]

By the time Li approached his senior year at Purdue, he had taken and done well in numerous biology courses beyond the requirements of the animal husbandry program. He was advised, therefore, to do a master's degree research project in addition to the required senior thesis. He took up the challenge, first by doing a senior project on "The Pathology of Hog Cholera," and then, on the suggestion of a classmate, a master's project on genetics. He undertook the latter, he said, because it looked both interesting and "easy"—all that was necessary was to pick a Drosophila mutation and then "fix the character to a chromosome."[22] And he did just that, emulating the technique of another student's research, and revisiting a problem raised by Morgan's research a decade earlier. Using statistical analysis techniques he had just picked up from reading Raymond Pearl's work, Li's project involved a massive sample, which he analyzed with great clarity, and not a little brashness, especially where he took issue with some of Morgan's observations and provided answers to questions left unanswered by him.[23]

Stimulated by his ad hoc first excursion in genetics, Li spent the summer of 1923 at the University of Michigan, Ann Arbor, in an intensive study of Drosophila genetics under the guidance of Frank Shaw. Shaw was impressed enough with Li's performance that he offered to recommend Li to Morgan's doctoral program at Columbia, if that was what Li wanted. Li accepted Shaw's offer, though it had not occurred to him until this point to specialize in genetics, let alone do it at Columbia.[24]

*Li Ruqi with Morgan's Group at Columbia*

Li, lacking the decorum he should have known from China, just showed up without warning at Morgan's office at the end of the summer and insisted on handing him Shaw's letter of reference and discussing the doctoral program. After reading Li's master's thesis, Morgan was apparently impressed enough to accept him into the M.A. program. Li had two year's remaining of his Boxer fellowship to cover the expenses of the one-year M.A. and the first year of a doctoral program. Morgan promised to cover the remaining expenses if Li were advanced to the Ph.D. program.[25]

He was indeed advanced to the doctoral program, where Morgan assigned C. B. Bridges to supervise his dissertation, which was completed in 1926.[26] As indicated above, Li's research was part of a design by Sturtevant to encourage a program of developmental genetics. Thus, Li's doctoral research was closely akin to that of his Columbia fellow-student, Chen Ziying, who completed his dissertation shortly before Li. Again, the general research goal was to relate genotype and phenotype through the mediation of embryonic development. The specific goals of Li's work grew out of the work of C. B. Bridges, one of Morgan's key students and coworkers at Columbia.[27]

Bridges had Li explore the effects of various kinds of gene deficiencies, especially on live birth and longevity. Like Chen Ziying, Li paid attention to the chronology of embryonic development vis-à-vis the development of imaginal discs. In addition, Li varied environmental factors (e.g., food, temperature) and genetic deficiencies (by eliminating sections of a chromosome or a whole chromosome). The interaction of these factors revealed—in greater and more accurate detail than previous research—the degrees of independence characteristic of various developmental processes. It indicated how these processes depend on different genetic and environmental factors. Most important for Sturtevant's agenda, Li's project introduced a concern that would become central to developmental genetics in the late 1930s, namely, the effect of genes on the process through which an embryo is formed and developed (embryogenesis).

After returning to China, Li Ruqi continued to do a series of research projects aimed at further amplifying the developmental problems he addressed in his dissertation. One series, on gene deficiencies, was a long-distance collaborative effort with C. B. Bridges, carried out between 1931-1935.[28] Another was devoted to a more embryological study, looking at the effect of temperature on the development of a particular Drosophila character. It was carried out with graduate students at Yanjing, under the sponsorship of the China Foundation.[29] And a third series, devoted to the embryological development of the Chinese "Rain Frog," was accomplished programmatically between 1934 and 1941.[30]

Li's teaching responsibilities now regularly included courses on experimental biology, general embryology, and genetics.[31] By 1933, he was supervising eight senior theses and five masters' theses, a number of which were exercises in experimental embryology or in experimental genetics, using Drosophila.[32]

## Yanjing Geneticists at Caltech

In 1935-1936, both Li Ruqi and Chen Ziying received Rockefeller research fellowships that gave them furloughs from their respective universities for the purpose of working with Morgan's group at Caltech. Here, both of them participated in the next stage of research leading to what Kohler designates as "the real beginning of embryological genetics in Drosophila."[33] Chen collaborated with Th. Dobzhansky, and Li with Bridges in projects that contributed to working out the basic embryology of Drosophila, among other things.[34] According to Kohler, this in turn led, at the end of the 1930s, to research that "dealt with embryological processes and with the genes that determined" the development of form and structure (morphogenesis). "These new experimental methods," he concludes, "were true hybrids of genetics and embryology, and unlike earlier improvisations they actually did produce knowledge of how embryogenesis worked."[35] Unfortunately, further such collaborative opportunities for Chen and Li were obviated by the onset of the war and the death of Bridges in 1938.

### Tan Jiazhen

During the 1935-1936 academic year, the third and youngest member of the Yanjing trio was also working at Caltech with the Morgan group. Tan Jiazhen (b. 1909), Li Ruqi's most accomplished student, was there completing his doctoral dissertation under the direction of Th. Dobzhansky. These were the years that Dobzhansky was writing his seminal 1937 book, *Genetics and the Origin of Species*, in which he forged an "evolutionary synthesis," as it came to be called, bringing together the research traditions of the naturalist and experimentalist to bear with great consequence on the fields of genetics and evolutionary theory.[36] Tan Jiazhen was right in the middle of it all, making very significant contributions while building up a great, invaluable fund of research experience, and an international reputation.[37]

Tan was born in the coastal port city of Ningpo, a hundred miles south of Shanghai, in Zhejiang Province. He was third eldest in a family of seven children and the only one in the family with an education beyond middle school. His father, a post office master in the British-run postal system, was often shifted to different provincial assignments because he lacked the knowledge of English required to work in Ningpo's main office. Tan Jiazhen's early education was therefore rather itinerant and did not settle down until he began his foreign-style education in a Ningpo Methodist missionary junior middle-school. His father had first thought to have Tan trained to work in treaty-port commerce and thereby bring some wealth to the family. A close relative, however, recognized Tan's gifted intelligence and made a convincing case to send the boy on through the general curriculum of foreign schools and thereby prepare him for a more lucrative "high official position."[38]

Thus, Tan was permitted to complete his secondary education at another missionary school, in Huzhou, further south along the coast. Here he began his English language training and was required to spend considerable time in Bible study. He was also regularly dunned to convert to Christianity, but he "obstinately refused" to do so. Nevertheless, Bible study piqued his curiosity, though perhaps not in the way that his teacher anticipated: He was seriously impressed by "Genesis" and it left him puzzling for some time to come over the validity of special creation. This, as much as his meager science training, he recalls, was an early source of his interest in biology.[39] Notwithstanding Tan's resistance to conversion, his mission school recommended him strongly and successfully for a full fellowship at Suzhou University, which he entered in the autumn of 1926, without need of the entrance examination.

Suzhou's little biology department continued to maintain its high standards, in spite of the departure of N. G. Gee to the Rockefeller Foundation and Wu Jinfu to Yanjing. It continued to use up-to-date English language textbooks in the original, and to provide ample laboratory work. Tan recalls a curriculum of courses emphasizing morphology—in cytology, histology, and comparative anatomy. His instructors were educated at Hopkins and Cornell and, in retrospect, he remembers them teaching up to contemporary Western standards. In his junior year, Tan was greatly impressed by the lecture course on genetics and evolution. The texts used were H. H. Newman's 1921 edited book, *Readings in Evolution, Genetics and Eugenics* and T. H. Morgan's 1927 *Theory of the Gene.*[40] Not only did the experience of this course lead Tan "to reject the Christian doctrine of special creation," it also decided him on graduate study in genetics.[41]

Tan completed all of his Suzhou B.S. requirements a semester early, but he stayed on for the spring 1930 session as a teaching assistant in general biology laboratory sections. Then he applied to Yanjing for graduate work. Wu Jinfu, the Suzhou and Cornell-trained entomologist, was chair of Yanjing's biology department. He recommended Tan to N. G. Gee for a Rockefeller fellowship. Gee was then in charge of the fellowships; and in consequence, Tan was given two years of full support by the foundation.

### Tan at Yanjing

Though he took Alice Boring's course on invertebrates and another in cytology, Tan's M.A. program mainly comprised independent reading and a major research project. Wu Jinfu made the fateful suggestion that Tan work with Li

Ruqi on some breeding and inheritance-pattern experiments using the abundant local lady-bird beetle (*Harmonia axyridis*) as his experimental organism. The beetles were particularly useful because they displayed quite a range of shell colors and an even greater range of polka dot patterns on the shells. Using a massive sample, Tan and Li first studied the color variations quantitatively; then looked for irregularities; and finally tried to distinguish environmental from hereditary sources of variation. (For example, they found that the number of spots was a function of temperature during pupation.) From the first phase of this project there followed two papers in which Dobzhansky's unique work was cited and discussed with discernment.[42] In none of Tan's studies at this time had geography yet been used as a variable. Within a few years, Dobzhansky taught him its critical importance.

The third paper described with some finesse how inheritance of beetle shell colors and patterns comported with Mendelian laws. So well did the latter turn out, that Li Ruqi sent a draft of the paper to T. H. Morgan (now at Caltech) for possible publication in the United States. Morgan in turn showed it to Dobzhansky (an expert on the biology of lady-bird beetles), who got Bridges to help him edit the paper for the *American Naturalist*, where it appeared in 1934.[43]

As these papers were appearing, between 1932-1934, Tan graduated Yanjing and received a Rockefeller fellowship to continue his research and teach part-time at Suzhou University. Tan's work received special notice in the 1933 Tisdale report for the Rockefeller Foundation on the state of science in China. In addition to Tan's published pieces, Tisdale notes that Tan was becoming nationally famous because of a handsome teaching exhibit he had created for the Suzhou-based Biological Supply Service. In little boxes, Tan arranged a dozen or more lady-bird beetles to illustrate the inheritance factors of color and polka-dot patterns.[44] Within a few years, Tan used this technique, in stunning color photographs, to illuminate some of his most important publications.[45]

As Tan Jiazhen worked at Suzhou, he determined that he wanted to go on for a doctorate in genetics at Caltech. Fortunately, Dobzhansky was impressed enough with Tan's work that he got Morgan to invite Tan to take his Ph.D. with them. Tan accepted, but then learned that a fellowship would not be available to him for his first year; there were just too many more senior students in line for support. Not wanting to postpone for a year, Tan patched expenses together from a gift from his father and loans from his brother and friends, and sailed for southern California in the summer of 1934.[46]

*Tan Jiazhen at Caltech*

Tan moved through a demanding doctoral program in just two years, the second of which was supported by a Rockefeller fellowship. There was course work in addition to the dissertation: organic and biochemistry; Morgan's evolution and genetics course; Sturtevant's advanced genetics; Ernest Anderson's plant genetics; and because Tan's minor was embryology, Albert Tylor's experimental embryology.[47] Dobzhansky directed his dissertation.

Dobzhansky pointed Tan to a research program that was redolent of the former's earlier interest in evolutionary theory, and his rich experience as a naturalist and taxonomist and, more recently, as an experimental geneticist. Steven J. Gould notes that "[Dobzhansky] began his scientific career [in Russia] as a classical systematist, a specialist on . . . ladybird beetles. Most of his early papers are descriptive and taxonomic. Later his interests enlarged to include genetics and evolution."[48] By the time that Tan came to Caltech to work with him, Dobzhansky (along with Sturtevant) was engaged in a project that required a fusion of all of these scientific traditions, that is, the traditions of the field naturalist and of experimental research. Dobzhansky's project provided Tan "a rare opportunity to study experimentally the process of variation and evolution in nature."[49] Using Drosophila, this became possible with the "thorough analyses of the genetic composition and changes in composition of natural populations over time. These analyses required "a combination of field sampling studies and subsequent laboratory experiments" including analysis of chromosomal structure.[50]

Initially, it was as an unexpected result of field sampling that Dobzhansky was led in this direction. Samples of a naturally prevalent type of Drosophila (the "*pseudoobscura*") were gathered from far ranging and diverse biogeographical areas, from California to Texas. It was then discovered that these phenotypically identical Drosophila populations from different regions "had distinctive and characteristic combinations of chromosomal inversions," that is, genes were arranged in a different order on some chromosomes. And further, "these regional populations were not separated by geographical barriers, yet seemed to be in the process of evolving into distinct species."[51] Assaying the chromosomes of these large field samples was only possible due to the recent discovery of Drosophila's giant salivary gland chromosomes. These made it possible actually to see chromosomal structure, and so to identify more efficiently the inversions that were to become markers for distinct geographical populations.

In his chronicle of Dobzhansky's momentous research, William B. Provine suggests that the arrival and work of Tan Jiazhen at Caltech galvanized the collaboration of Dobzhansky and Sturtevant and helped to bring it to "a new and deeper phase." They assigned Tan a dissertation project with two related goals: First, to make genetic and cytological maps of *D. pseudoobscura* chromosomes; and then to examine the chromosomal inversions of two types of *pseudoobscura* (known as races A and B) to see if there were any systematic differences between them. It was already demonstrated that a hybrid of A and B produced only sterile offspring, that they were in effect incipient species. And Provine notes that Sturtevant had already determined "that inversions could be found in the various laboratory stocks of races A and B; the question for Tan was whether races A and B differed consistently by the same inversions."[52]

Tan's dissertation research found that A and B did consistently differ by six inversions; and it further determined that other inversions occurred *within* A and B, but that none of these were identical with the six that, Tan concluded, were racial markers. Further, he concluded "racial differentiation partook in this case of both sources of evolutionary variability—gene variability and variability of the gross structure of the chromosomes."[53] This latter formulation in turn indicates that Tan was already thinking like Dobzhansky about evolution. In particular, Tan's conclusion implies that there are two sources of evolutionary variability and both are genetic. Dobzhansky put it this way in the introduction to his 1937 *Genetics and the Origin of Species*: "Observable genetic phenomena are the source of all evolution."[54] Or, as Gould summarized it: "the principles of genetics, accessible through experimental work in the laboratory or short-term study of natural populations, suffice to explain evolution at all levels [e.g., individual organism or species], despite the distinctness of problems at different levels."[55]

## Tan's Postdoctoral Research

Tan completed his doctoral dissertation in the spring of 1936 and then, sponsored by a Rockefeller postdoctoral fellowship, he feverishly continued to do research at Caltech until the summer of 1937. For the most part, his projects continued to attack the problem of the genetic basis of variation and species differentiation. From this period, he published a dozen papers, either under his own name, or respectively with Dobzhansky, Sturtevant, or other postdoctoral students.[56]

Collaborating first with Dobzhansky then with Sturtevant, Tan worked on a series of studies devoted to comparing the chromosome structures and/or gene arrangements of different Drosophila species. The results were perhaps not quite as stunning as his doctoral research, but Dobzhansky repeatedly cited them in his *Genetics and the Origin of Species*, and found them to be profound confirmations of his fundamental argument about the relationship between gene arrangements and species, between genetics and evolution.[57] When Dobzhansky commented privately on the outcome of his 1936 study with Tan, he exalted, "Here at last you have real differences in chromosomes of different species."[58]

Tan himself tried to accumulate as much experience as possible in research design and technique. And when Dobzhansky suggested that he stay more narrowly focused, Tan explained that on his return to China he was going to be expected "to lead the way" in experimental genetics and therefore it was desirable to have hands-on experience with the latest techniques.[59] This is why he collaborated with other postdoctoral students on technically demanding transplantation experiments.[60]

In May of 1937, Sturtevant and the Rockefeller field representative agreed that efforts should be made on Tan's behalf, before he returned to China, to secure him a position that would permit the maximum time and facilities for continuing his research. Tan had earlier been informed that there would be a position for him in the Biology Institute of the Academia Sinica; but, for budgetary reasons, he would have to wait a year before joining the institute. It was quickly arranged by the foundation to provide the institute with Tan's first year of wages and research expenses, so that he would not have to be in limbo for a year. The institute's director, however, would not accept Tan as a regular member of the staff, but only as a visitor, and only as long as the foundation paid for him. The institute, said the director, was having a severe budget problem and had a three-year freeze on new membership. And further, the institute was not equipped to accommodate a geneticist since its specialties were currently limited to research on "fisheries, oceanography, and plant pathology."[61]

Frustrated by the bureaucratic runaround, Tan then considered offers from three universities. He accepted an associate professorship from Zhejiang University in Hangzhou, a recently established national school with impressive plans to specialize in science and technology. The Rockefeller Foundation had already given the highest marks to the university's science faculty and administration. Tan was promised a light teaching load, no administrative duties, and ample research time. And so the foundation was pleased to send Tan back to China with their blessings and a sizable research grant for establishing the research facilities he needed to continue his work without interruption.[62] Unfortunately, Tan began his work in China at the outset of the Japanese invasion. His wartime experience is discussed in the next chapter, where it will be seen that Tan persevered in his brilliant work on evolutionary genetics throughout the Japanese occupation. Ironically, it would be in peacetime that his research was halted—by Soviet Lysenkoism and Chinese Communist Party science policy.

## Plant Genetics and the Nanjing-Cornell Program

From 1925 to 1950, Nanjing University, a missionary school, had the biggest and best agriculture program in China; in turn, the program was home to the country's largest concentration of geneticists and genetics research. In contrast to Tan Jiazhen and his Yanjing colleagues, the Nanjing group specialized in plant genetics, with a decided orientation toward agricultural application. Some of Nanjing's profuse research was outstanding, though perhaps not as momentous as that of the Yanjing trio. The biology education at Nanjing was likewise outstanding, and apparently provided an excellent foundation for dozens of students who went to the United States—mainly to Cornell University—for graduate education. Where the Yanjing geneticists had developed a bond with Columbia and Caltech, Nanjing, in a much more formal way, created a program in "plant improvement and genetics" with Cornell.

The Nanjing-Cornell program and the considerable funding that made it possible were fuel for the heated competition that existed between Nanjing University and the neighboring Agriculture School of National Central University. Their rivalry came from more than sharing the same city. It was an expression of foreign-Chinese, private-public tensions, and ultimately, it was about attracting the best faculty, the best students, and winning more scarce resources for program development. The rivalry was not about educational philosophies: Both schools sought to provide the full range of training required for making a difference in China's agricultural productivity—from theoretical genetics, to controlled selection techniques, to agricultural extension. Nanjing and National Central differed in their approaches to agriculture where the latter expressed concern with rural social and economic issues beyond the technical problems of agricultural science. The dean of National Central's Agriculture College advocated, fruitlessly, a larger role for agriculture colleges that included the development of rural education and social reforms.

Randall Stross, in his study of the often quixotic American "agricultural missionaries" in China, acknowledges—without providing any details—the "extraordinary" accomplishments of Nanjing's program. He then goes on to criticize the program for limiting itself to a technical approach to agriculture and failing to become involved with rural social-economic reform. His observation indicates an instance of a greater and persistent issue in China's discourse on development: Could scientific and technical means alone solve China's problems, and was an appeal to them exclusively just a device for obviating social revolution? Stross' narrative leads to the impression that the Nanjing agriculture program (like other American agricultural missionary efforts in China) was a failure. The following discussion suggests a different assessment, though it acknowledges the purely technical nature of the program.[63]

Since foreign-controlled philanthropy was behind the development of Nanjing, to the detriment of National Central, the story of the rivalry provides particular insights into problems of developing native Chinese educational institutions. Whatever problems the growth of Nanjing's Agriculture School may have caused, however, there is no denying the extraordinary technical success of its plant improvement and genetics program. Not only did it quickly produce a much needed core group of successful educators, but this same group was responsible for substantial research contributions that eventuated in new varieties of higher yielding or more disease resistant food and industrial plants. In part through the efforts of some Nanjing agriculture faculty, the Nationalist government was made to recognize the importance of agricultural research and to establish the first government institutions devoted to that end. Nanjing plant geneticists were closely associated with those institutions from the early 1930s, through the war, right up to 1950.

## Origins of the Plant Genetics Program

In 1923-1924, Nanjing University's plant genetics program had its origins in a U.S.$675,000 windfall grant from the American Committee for China Famine Relief. The committee, largely made up of American missionary organizations, had originally raised relief funds for the disastrous Huai River floods. When its work was somehow completed earlier than expected, it was left with a hefty surplus which it applied to the prevention of future famine in China. The dean of Nanjing University's Agriculture School, John Reisner, took advantage of his connections with the missionary community, and successfully lobbied for the funds.[64]

Nanjing University's College of Agriculture and Forestry, as it was formally known, was founded in 1914, in response to the Huai Valley flood and famine of that time. The college was one of a series of efforts by American "agricultural missionaries" to alter the nature of farming in China (and thereby end famine) through the introduction of American science and technology knowhow. Sometimes these efforts were motivated by quixotic zeal, ending in failure and embarrassment. John Reisner's zealous direction of the college, from 1917 to 1931, involved no windmill tilting and put to good use a torrent of funding that began with the 1923 Famine Relief windfall and continued with sizable contributions from the Rockefeller and China Foundations, and the British Boxer Indemnity Fund.[65]

## Rivalry between Nanjing University and National Central

The funding of Reisner's college became a zero-sum game between itself and the Agriculture College of National Central University, culminating in dynamic expansion for the former and arrested growth for the latter.[66] The dean of National Central's Agriculture College was Zou Bingwen, who, like Reisner, held a Ph.D. from Cornell's agriculture science program. Zou was a tireless advocate of agricultural science education in China, and for decades he lobbied local and central government agencies to give priority to the creation of strong agriculture schools within the new national university system. He especially wanted to see Boxer Indemnity educational funds carefully awarded according to national priorities—such as the modernization of agriculture.[67]

By 1923, Zou Bingwen had become so frustrated with China's inattention to agricultural education and research that he approached Reisner with a merger proposition. At the time, Zou's college was bigger and better funded than Reisner's. Rather than continue to use scarce funds inefficiently, however, Zou wanted to create one large model college by combining his and Reisner's institutions as well as a number of smaller agriculture schools in the Nanjing area. Shortly after Zou's proposition, Nanjing received its immense gift from the Famine Relief Fund, thereby becoming fiscally quite independent and with nothing to gain from a merger.[68]

His overtures to Nanjing rejected, Zou Bingwen then turned to the China Foundation for funding. In the early 1920s, he had prevailed upon his university president, Guo Bingwen, to use his connections with the China and Rockefeller Foundations for funds to develop the agriculture college. When Guo did so, he was told by the foundations that "rural education" should be left to other and lesser schools. Roger Greene and other foundation policymakers had plans for National Central that did not include such practical and utilitarian pursuits. The development of basic science, especially of a biomedical nature, was the goal.[69]

Zou Bingwen did not discourage easily. On behalf of his college, between 1923 and 1926 he filed proposals with the Rockefeller Foundation's International Education Board and then with the China Foundation for relatively modest support of various departments and research projects. During this period, the Education Board determined to concentrate all of its support for agricultural education in Nanjing University; and so Zou's proposals were rejected. The China Foundation responses to Zou Bingwen were shaped by China Medical Board policy. (Remember that the Rockefeller and China Foundations had interlocking directorates). Those departments of Zou's college that were responsible for formal science disciplines received financial support. (Bing Zhi's zoology department and Hu Xiansu's botany department were housed in Zou's college and were cases in point.) The departments or research projects that dealt with agricultural practice received little or nothing.

There were secondary factors that affected foundation decisions to fund Nanjing and ignore National Central. First, Nanjing's fiscal soundness (thanks to the Famine Relief money) made it a sure bet compared to any government agriculture school whose resources were always in doubt. Secondly, National Central suffered for years from the disruptions of campus and national politics while Nanjing University enjoyed a stable administration and virtual freedom from political disruption. Thus, foundation favoritism combined with Nanjing's status as a treaty-protected foreign entity to insulate it from its rival's problems.[70]

In the early 1930s, Rockefeller Foundation China policy did eventually reorient itself toward agricultural development and rural reform (the so-called "China Program"). Nevertheless, Nanjing University's agricultural program continued to receive the lion's share of its funding, to the exclusion of National Central's program, among others.

## The Cornell-Nanjing Program and Plant Genetics

Flush with new resources, Nanjing University and its Agriculture College grew in size and developed in quality. Just before the 1923 Famine Relief grant, the university had around one hundred students altogether. By 1929, the number was over five hundred, one-third of whom were in the Agriculture College. Those numbers doubled by the beginning of the war and remained at that level until 1949. The faculty size and composition likewise changed: In 1925 there were forty-one foreign faculty and nine Chinese; six years later, there were sixty-four Chinese faculty and only four foreign. A 1931 tally showed that in the previous six years the college had graduated over two hundred students, forty of whom went abroad for advanced training.[71]

The developmental core of the college was its new plant genetics and improvement program. Its structure was straightforward: For five years, beginning in academic year 1925-1926, three senior faculty from Cornell's plant breeding and improvement program would take turns working at the college—advising, organizing, teaching, researching. These advisers arranged for the best of the students to be admitted and funded for graduate work, primarily at Cornell. The link with Cornell grew stronger with time, and by 1946 the college had sent a total of sixty-two students to study there in its Department of Plant Breeding.[72]

The curriculum was rigorous, and became even more so when the college and its Cornell advisers chose a new Ph.D. from Cornell, Shen Zonghan, to shape and eventually lead the program from 1926.[73] Shen was the first Chinese to acquire a doctorate in the field of plant breeding and his influence on its development was as significant as the Cornell advisers. This was evident early on in his efforts to compensate for the dearth of Chinese language textbooks and the failure of English language texts (reflecting education in the United States) to address the problems of cultivating food plants favored in China (like rice, sorghum, millet, and sugar beets). Shen insisted that "text material be integrated with field experience,"[74] and that the basic genetics course require a research project devoted to the genetics of one food or industrial plant. The Cornell advisers found no fault with Shen's efforts to acclimatize the program.[75]

Genetics was taught at Nanjing with at least the same attention to fundamentals that characterized the curricula at Yanjing and National Central—notwithstanding Nanjing's ubiquitous concern with "plant improvement." From the testimony of alumni, catalog descriptions, and the assigned textbooks, Nanjing's core genetics course appears to be much like the best of those taught elsewhere in China. For example, the first year of the program, the course catalog described the "Genetics and Eugenics" course to be about "phenomena of development, variation, heredity, Mendelian and Neo-Mendelian mechanisms of inheritance, together with their relation to plant and animal improvement. . . . There are laboratory studies in variation and the Mendelian laws as seen in corn, cotton, etc. [sic]." This course initially used popular American college textbooks, in their original English.[76]

## New Textbooks from the Program

By the late 1930s, the plant improvement and genetics program was creating its own very comprehensive and up-to-date Chinese language textbooks, based on faculty lectures. This in part reflected wartime unavailability of foreign books; but it also indicates a desire to decrease dependency on the English-language originals, as Chen Zhen had done in the early 1920s with his pioneering general biology text. It also indicates the availability of standardized translations for basic terminology, the result of a decade's efforts started in the early 1920s by a group of Chinese students in the United States enrolled in the Cornell plant improvement program. They had been successful in launching a project to create a standardized Chinese glossary of genetics terminology that continued to be developed well into the future.[77]

The geneticist, Fan Jianzhong, published his 1942 *Yichuanxue* (Genetics) for the Nanjing program to illuminate the internal history and development of genetic science itself, rather than the practical potential of genetics. If the book has a central argument, it is that the future of genetics lies in the field of cytology and the study of the gene. It begins with a long, detailed, well-illustrated exposition of Mendel's laws, followed by an equally extensive introduction to the descriptive statistics necessary to evaluate quantitative features. A short but dense section covers the interplay of evolution and genetics, with emphasis on the Morgan school since the late 1920s. Notable is a survey of the role of radiation as a research instrument. Throughout, the author carefully introduces the extensive technical vocabulary with the standard English term in parentheses next to the Chinese. At nearly four hundred pages in length, this was by far the most comprehensive and sophisticated text devoted to genetics, to date. Because of the war and then Lysenkoism, it would take twenty years for a better introductory survey of the field to be written in Chinese.[78]

Before the war, perhaps the most prestigious Chinese language textbook to emerge from the Nanjing faculty was written by Wang Shou.[79] Derived from years of classroom lectures and personal research work, Wang's 1936 *Zhonguo zuowu yuzhong xue* (Plant Improvement in China) distilled the development of genetics from Mendel to Dobzhansky and, through indigenous case studies, the techniques of applying genetics to the improvement of food and industrial plants important to China. Wang enjoyed some fame at the time for his development of a variety of barley that bore his name and had spectacular success in China and the United States. The book is notable for citing the work of Chinese plant geneticists as well as standard American authorities. It is the first major Chinese-language textbook that gives practical instruction for the use of statistical techniques. The book is written in a very straightforward, lucid, modern vernacular idiom that Wang says he purposely used to communicate with the common farmer. At the conclusion to the book, he unconventionally expresses a deep concern to convey plant breeding science to the masses and to put that science into practice.[80]

Wang's textbook was based on lectures he delivered in a special two-year certificate program that Nanjing offered for farmers, those planning to farm, or government officials wanting to learn about farming. The course was taught completely in Chinese and was one of a number of educational and research efforts that were carried on outside of the conventional college classroom.[81] Between 1926 and 1931, in addition to the two-year program, summer institutes were sponsored by the Nanjing-Cornell program. These were intensive three-week programs aimed at the staff of agriculture extension stations and government officials. Over the five-year period, about two hundred students were exposed to subjects ranging from genetics, to statistical techniques, to the study of the special problems of open-fertilized crops.[82]

Modern genetics/plant improvement was provided in two other venues for communication outside of Nanjing University classrooms. The Agriculture College had the wherewithal to acquire rights to establish experimental farms in a half-dozen different climate and soil zones throughout China. Local governments owned and controlled many of them, while others belonged to Christian missions. Students and faculty had access to them for the extended length of time required to carry out experiments and tests—often four to six years. And once the work on the experimental farm proved satisfactory, its results were shared with agriculture extension stations, whose job it was to broadcast the new technique or new seeds for use in the farming community itself. Some new seeds were produced in significant quantities and distributed gratis; however, it was not easy convincing the farmers to use them.[83]

## Love in Bloom: Nanjing Agricultural Science and the State

As the Nanjing-Cornell program approached its fifth and final year, Shen Zonghan had become convinced that Nanjing University's efforts at plant improvement, seed distribution, etc., could become more effective only if these efforts were part of a national program sponsored by the central government. In 1928, Shen proposed to the new Guomindang government that they establish a "national agricultural research bureau." For the next seven years, he and various government agencies negotiated endlessly, eventually creating new research institutes, and seriously affecting the development of agricultural research a few years before the war's outbreak.[84] When the Japanese invasion did come, the government and Nanjing University, like many other institutions, packed up and fled west, to safety. It was in its western sanctuary, Sichuan Province, that the Nanjing plant geneticists, the National Agriculture Research Bureau, and scientists from other universities and plant improvement programs finally did coalesce and carry out a very robust research program that lasted until 1949, producing notable contributions both to the benefit of genetics and agriculture.

Shen Zonghan initially turned to his Cornell mentor, Harry H. Love (1880-1966), to design an agriculture improvement program for the Nationalist government. Love had been a leading figure in the Nanjing-Cornell program, and was bringing the program to a close in 1931, when he accepted the government's offer to act as adviser on agriculture policy for three years. Technically, Love worked in a cooperative program with both the national government and the Jiangsu and Zhejiang provincial agriculture bureaus. His main task was to draw up a national plan that would result in, among other things, the reduction of China's heavy reliance on imported grain.[85] In January of 1933, he published a grandiose "National Program for the Development of the Agriculture of China."[86]

Love's program featured a highly centralized agency that had the political and economic power to dictate all aspects of agricultural development over the seven or more years required to arrive at improved varieties of plants. The scientific core of the program was to be a greatly enlarged plant genetics and improvement effort that in turn would be carefully articulated with a variety of other institutional developments. The latter ranged from the training of many more technical personnel for experimental and extension work, to the creation of local cooperative farm credit programs, and local stations for the free distribution and/or sale of select seeds.

Clearly, one motive for government action at this time was the success of the Communist Party in its rural soviet in southeast China. Love addressed that issue as well in his program proposal. He argued that agricultural improvement would be cheaper and more effective than military efforts at winning the peasantry away from communism; and he wondered aloud how much China's agriculture could be improved with just one month's military budget of the national and provincial governments.[87] In 1934, the year that the Guomindang destroyed the Communist Jiangxi Soviet, Love returned permanently to Cornell, his proposal circulating through the government and distributed in English and Chinese through Nanjing University's prominent agriculture science journal. Piecemeal, various parts of the proposal were realized (though not necessarily because Love had suggested them). For instance, a farm credit program was instituted and remained one of the most significant social-economic rural programs the Guomindang was willing to sponsor.[88] Institutionally, the central government's new Ministry of Agriculture and Forestry was given the authority of centralized planning and coordination for agriculture science research and education, with the goal of linking these to production.

Love's efforts may well have strengthened Shen Zonghan's hand in the development of government sponsored agriculture research. In 1934, Shen left Nanjing University and became chief technician of the National Agriculture Research Bureau, whose resources and personnel finally began to grow. Shen also saw the creation of a National Rice and Wheat Improvement Institute, ostensibly a freestanding government agency, but in fact one whose administration, personnel, and resources overlapped with Shen's Research Bureau.[89] The research of Nanjing University plant geneticists and other faculty of the Agriculture College was regularly supported by these government agencies, thus providing further resources and research outlets for them as well as for faculty at other agriculture colleges, including those of National Central and Hangzhou Universities. All of these developments appealed greatly to the Rockefeller Foundation, whose field agents had, since the early 1930s begun to lobby the foundation for rural and agricultural programs. And so, from 1935, the foundation's "China Program" began to add still more resources to Shen's agricultural research enterprises.[90]

In the summer of 1937, the intellectual momentum that Shen's efforts had created, the extensive field tests and laboratory research projects that his one-hundred-plus staff had begun—all abruptly stopped, as the Japanese invasion of China moved devastatingly toward the Shanghai/Nanjing area.

# Notes

A preliminary discussion of some of the material in chapters one and two was made in my essay "Genetics in Republican China, 1920-1949," in *Science and Medicine in Twentieth Century China*, edited by John Bowers and Nathan Sivin (Ann Arbor, Mich.: University of Michigan Center for Chinese Studies, 1988), 3-30.

1. For criticism of Peking Union Medical College/Yanjing, see Mary Bullock, *An American Transplant*. And see text below for criticism of Nanjing.

2. Students and campus politics: Philip West, *Yenching University and Sino-Western Relations, 1916-1952* (Cambridge, Mass.: Harvard University Press, 1976). "Yenching" is the former romanization.

3. Dwight Edwards, *Yenching University* (New York: United Board for Christian Colleges in China, 1959); *A Brief History of the Department of Biology of Yenching University* (Beijing: Yenching University, December 1931) (RAC: RG.1/ser.601/B.41).

4. Peking Union Medical College-Yanjing: *China Medical Board Historical Record*, vol. 1A (RAC: RG.1.1, ser.601a.Bx.1a); *A Brief History of the Department of Biology of Yenching University*; Mary Bullock, *An American Transplant*; M. Ogilvie, "Alice Middleton Boring (1883-1955) an American Scientist in China," *MSS* (Shawnee, Okla.: 1984), 15-20.

5. Haas, *China Voyager*.

6. J. M. Dyson, "The Science College," in *Soochow University*, ed. W. B. Nance (New York: UnitedBoard for Christian Colleges in China, 1956), 50-56; and *A Brief History of the Department of Biology of Yenching University*.

7. *A Brief History of the Department of Biology of Yenching University*.

8. Dyson, "The Science College," 52-53.

9. Dyson, "The Science College," 52-53.

10. Haas, *China Voyager*, passim.

11. Dyson, "The Science College," 54-55.

12. Marilyn Baily Ogilvie, "The 'New Look' Women and the Expansion of American Zoology: Nettie Maria Stevens (1861-1912) and Alice Middleton Boring (1883-1955)," in *The Expansion of American Biology*, eds. Keith R. Benson et al. (New Brunswick, N.J.: Rutgers University Press, 1991), 52-79; and M. Ogilvie, "Alice Middleton Boring." Also see Alice M. Boring, "Publications of Alice M. Boring" (one-page typescript, Bryn Mawr Archives); Dwight Edwards, *Yenching University*, 160; *A Brief History of the Department of Biology of Yenching University*; James McKeen Cattell, ed. *American Men of Science*, 5th ed. (New York: Science Press, 1933), 14; and N. G. Gee, letter to R. M. Pearce, February 5, 1929 (RAC: RG.1/ser.601/B.40).

13. *A Brief History of the Department of Biology of Yenching University*; and "Genetics of Two Mutations in the Fruit Fly, Drosophila m." (in English), *China Journal* (Beijing), 1923, no. 6: 593-603; 1924, no. 2: 143-57.

14. Columbia University, Graduate Student Record Cards of Chen Tse-yin. (This romanization of Chen's name is used throughout his records and publications in America.) Chen's dissertation was published as "On the Development of Imaginal Buds in Normal and Mutant Drosophila Melanogaster," *Journal of Morphology and Physiology* (May/June 1929): 135-200. And also see Joseph Needham and Dorothy Needham, eds., *Science Outpost: Papers of the Sino-British Science Cooperation Office 1942-1946*, (London: Pilot Press, 1948), 226.

15. Robert E. Kohler, *Lords of the Fly: Drosophila Genetics and the Experimental Life* (Chicago, Ill.: University of Chicago Press, 1994), 188. According to Li Ruqi, himself, C. B. Bridges was his dissertation adviser and suggested his dissertation topic (interview, Li Ruqi, Beijing, August 1984). Bridges of course may well have been working within Sturtevant's larger plans.

16. Chen, dissertation, 136.

17. Details of Chen's career: N. G. Gee to R. M. Pearce February 5, 1929 (RAC: RG.1/ser.601.Bx.40: "Biology Department," 2). Also see China Foundation *Report*, 1929-1930 and 1930-1931. The *Report* indicates that Chen's fellowships were for "A Physiological Study of Chinese Crabs."

18. Interview with Li Ruqi, Beijing, August 1984.

19. *A Brief History of the Department of Biology of Yenching University*, 10-19; and Yuan Tung-li, *A Guide to Doctoral Dissertations by Chinese Students in America 1905-1960* (Washington, D.C.: Sino-American Cultural Society, 1961).

20. Li biography: Interview with Li Ruqi.

21. Li education: Interview with Li Ruqi.

22. Li's senior thesis for Purdue University Agriculture School: "The Pathology of Hog Cholera," 1923. Research for master's thesis: Interview with Li Ruqi.

23. Master's thesis: See Li Ruqi, "A Study of the Vermilion Stock of Drosophila Melanogaster with Special Reference to the Recurrence of Balloon Wings and the Appearance of the 'Matted' Bristles," Purdue University Agriculture School, June 1924.

24. Interview with Li Ruqi.

25. Interview with Li Ruqi.

26. Morgan regularly turned dissertation supervision over to Sturtevant and Bridges. The latter, however, was not on the Columbia faculty, but on the payroll of the Carnegie Foundation and so Morgan was pro forma Li's dissertation supervisor (interview with Li Ruqi; and Kohler, *Lords*). Li's dissertation: "The Effect of Chromosome Aberration on the Development in *Drosophila m.*," Ph.D. diss., Columbia University, 1927, and published in *Genetics* (May 1927): 1-59.

27. Source of research problem: In the dissertation, Li cites a series of seven experiments conducted by Bridges between 1913-1923.

28. Li Ruqi and C. B. Bridges, "Genetical and Cytological Studies of a Deficiency (Notopleural) in the Second Chromosome of *Drosophila m.*," *Genetics* (November 1936): 788-95. And also see summary in Yanjing University, *Science Notes* (April 1934): 3.

29. Li Ju-ch'i and Tsui Yu-lin, "The Development of Vestigial Wings under High Temperature in *Drosophila m.*," *Genetics* (May 1936): 248-63. Reprinted in Li Ruqi, *Shixian shengwuxue lunwen xianji* (Anthology of experimental biology research) (Beijing: Science Press, 1985), 107-19. This collection of Li's research contains many English-language pieces not published outside of China.

30. See "Studies of the 'Rain Frog' Kaloula Borealis" (in English), parts 1-4, *Peking Natural History Bulletin*, 1934-1935, no. 1 and 1940-1941, no. 4. Reprinted in Li, *Shixian shengwuxue*, 226-68.

31. Courses: Yanjing University *Bulletin, College of Arts and Sciences*, no. 15.20 (June 1932): 8.

32. Students: See *A Brief History of the Department of Biology of Yenching University*; and Tisdale, *Report*, 58.

33. Kohler, *Lords*, 244.

34. Kohler, *Lords*, 244; C. B. Bridges, *Genetics* (March 1938): 110.

35. Kohler, Lords, 244-45.

36. Theodosius Dobzhansky, *Genetics and the Origin of Species* (New York: Columbia University Press, 1937; 1980 reprint with introduction by Stephen Jay Gould). Subsequent editions in 1941, 1951, and 1970 as *Genetics and the Evolutionary Process*.

37. Tan Jiazhen: Alternative romanization: T'an Chia-chen. This form is used on his English-language publications before 1950. He is also known as C. C. Tan.

38. Tan biography: See ZXS: 373-74. And also Interview with Tan Jiazhen, Shanghai, August 16, 1984.

39. Interview with Tan Jiazhen.

40. Interview with Tan Jiazhen; also see ZXS: 375-76; and William B. Provine, "Origin of *The Genetics of Natural Populations* Series," in *Dobzhansky's Genetics of Natural Populations 1-43*, eds. R. C. Lewontin, et al. (New York: Columbia University Press, 1981), 37-38.

41. Interview with Tan Jiazhen; and Provine, "Origin," 37-38. (Provine incorrectly cites Newman's title as *Evolution, Genetics, and Heredity*. And in our 1984 interview, Tan cited the 1921 first edition title as the one he read.)

42. Two papers: Tan Chia-chen and Li Ju-ch'i, "Variations in Color Patterns in the Lady-bird Beetles, Ptychanatis axyridis Pall.," *Peking Natural History Bulletin*, 1932-1933, no. 7: 175-93. Reprinted in Li Ruqi, *Shixian shengwuxue*, 64-83. And Tan Chia-chen, "Notes on the Biology of the Lady-bird Beetles, Ptychanatis axyridis,," *Peking Natural History Bulletin*, 1933-1934, no. 8: 9-18. And also Interview with Tan Jiazhen; and Provine, "Origin," 38.

43. Tan Chia-chen and Li Ruqi, "Inheritance of the Elytral Color Patterns of the Lady-Bird Beetle, *Harmonia axyridis pallas*," *American Naturalist*, 1934, no. 68: 252-65. Also see Interview with Tan Jiazhen; and Provine, "Origin," 38.

44. Tisdale, "Report," 56.

45. For example, see Tan's "Mosaic Dominance in the Inheritance of Color Patterns in the Lady-Bird Beetle, Harmonia Axyridis," *Genetics* (March 1946): 195-210.

46. Interview with Tan Jiazhen.

47. Interview with Tan Jiazhen. Ernest Anderson (b. 1891) was a 1920 Cornell Ph.D. under Emerson. Tylor was Morgan's student from Columbia and became a close protégé at Caltech (see Allen, *Morgan*, passim).

48. Steven Jay Gould, 1982 introduction to Dobzhansky, *Genetics and the Origin of Species*, xxii. And also see Garland Allen's characterization of Dobzhanky's work in *Life Science in the Twentieth Century* (New York: Wiley, 1975), 140-41.

49. Kohler, *Lords*, 267.

50. Allen, *Life Science*, 140-41.

51. Kohler, *Lords*, 264

52. Provine, "Origins," 37-38.

53. Tan, "Salivary Gland Chromosomes in the Two Races of Drosophila Pseudoobscura," Ph.D diss., California Institute of Technology, 1935. Published in *Genetics* (July 1935): 392-402.

54. Dobzhansky, *Genetics and the Origin of Species*, xxx.

55. Gould, "Introduction," xxvii.

56. Tan's 1936-1937 post doctoral research fellowship: "Summary of Grant-in-Aid Action (N.Y.) July 21,1937," attached to correspondence C. C. Tan to S. M. Gunn, Oct. 5, 1943 (RAC: "Chekiang University Genetics 1937-1945" R61/ser.601D/Bx.40. fldr.328).
Tan publications: "Genetic Maps of the Autosomes in D. pseudoobscura," *Genetics* (November 1936): 796-807; "Translocations in D. pseudoobscura," *Drosophila Information Service*, 1937, no. 7: 67-68; "Linkage Maps of D. pseudoobscura," *Drosophila Information Service*, 1937, no. 7: 68-70.

57. (1) Dobzhansky and Tan: "A Comparative Study of the Chromosome Structure in Two Related Species, D. pseudoobscura and D. miranda," *American Naturalist*, 1936, no. 70: 47-48; and "Studies on Hybrid Sterility. III. A Comparison of the Gene Arrangement in Two Species, D. pseudoobscura and D. miranda," *Z. Induct. Abstamm.*, 1936, no. 72: 88-114. (2) Sturtevant and Tan: "The Comparative Genetics of D. pseudoobscura and D. melanogaster," *Journal of Genetics*, 1937, no. 34: 415-32.

58. Cited in Kohler, *Lords*, 268.

59. Interview with Tan Jiazhen.

60. (1) Tan and D. F. Poulson: "The Behavior of Vermilion and Orange Eye Colors in Transplantation in D. pseudoobscura," *Jor. Genetics*, 1937, no. 34: 415-32. (2) Gott-

schewski and Tan, "The Homology of the Eye Color Genes in D. melanogaster and D. pseudoobscura as Determined by Transplantation, II." *Genetics* (March 1938): 221-38.

61. S. M. Gunn to Warren Weaver, May 21, 1937, and Chia Chi Wang (Biology Institute) to S. M. Gunn, June 23, 1937 (both in RAC: "University of Chekiang Genetics").

62. Acceptance of Zhejiang offer and RF grant: See correspondence of S. M. Gunn to Warren Weaver (July 16, 1937); and RF Grant-in-Aid RA NS 158 (July 21, 1937) for U.S.$1,500 (both in RAC: "University of Chekiang Genetics," SS). For evaluation of Chekiang University, see Tisdale, "Report" 1-6.

63. See Randall Stross, *The Stubborn Earth: American Agriculturalists on Chinese Soil* (Berkeley, Calif.: University of California Press, 1986).

64. These funds were administered in China by a "China Famine Fund Committee," appointed by the United States ambassador to China. See Harry H. Love and John H. Reisner, *The Cornell-Nanking Story* (Ithaca, N.Y.: Cornell University Press, 1963), 5. Also see "Report of the President and Treasurer," University of Nanking *Bulletin*, 1923-24, no. 18: 30. For a detailed narrative and analysis, see Stross, *Stubborn Earth*, chapter 4.

65. Stross, *Stubborn Earth*, passim. And Love and Reisner, *Cornell-Nanking*. The latter source says that the Agriculture College received U.S.$750,000 from the Famine Relief Committee, as opposed to U.S.$675,000, the figure reported in an official university document, cited in footnote no. 1, above. Also see Chunjen C. Chen, "The History and Present Status of Agricultural Education in China," Tsinghua College, October 20, 1925 (in RAC: China Medical Board.II.85). This source (p. 3) says the amount given by the Famine Relief Committee was U.S.$700,000. Annual reports of the British Boxer Indemnity Fund (*British Boxer Report*) from 1935 to 1941 show C$10,000 average annual grants to Nanjing's Agriculture College. The British Boxer fund was established in 1925 and began to distribute funds in 1931. Between 1931 and 1945, it had 11.1 million British pounds to distribute. Its plan was to distribute 30 percent for agriculture education and improvement (see Wang Chen-chun, "The Work of the British Indemnity Advisory Committee," *Chinese Social and Political Science Review* (Shanghai), 1927, no. 3: 361-72.

66. Stross, *Stubborn Earth*, 147-50.

67. Zou Bingwen, from Jiangsu Province, received his doctorate in agricultural science from Cornell in 1916. The next year he became the first director of agriculture studies at National Central when it was still a provincial teachers college. Zou's advocacy of agricultural education closely linked with practical application can be seen in these examples of his advocacy: *Kexue*, 1916, no. 7: 615-22 argues for the unity of agricultural education and practice; *Nongxue zazhi* (Agricultural studies) published by the Agriculture College of Dongnan daxue, October 15, 1924: 1-4 and 5-10, argue for the use of Boxer Indemnity Funds for agricultural science research, and the necessity of promoting agricultural education. Also see strong criticism of China's lagging agricultural education and its failure to affect practical farming in Chunjen C. Chen, "History and Present Status."

68. Stross, *Stubborn Earth*, 139.

69. For Roger Greene and China Medical Board policy on "rural education": Roger Greene, "Memo of Conversation with Kuo Ping-wen, May 6, 1925" (RAC: China Medical Board/ser.2/Bx.63). For Greene's reluctance to entertain new programs devoted to vocational and rural education (especially at the cost of basic sciences development), see memo from Greene to Houghton, "American Indemnity Fund," October 22, 1924 (RAC: China Medical Board/ser.2/Bx.53).

Also see grant proposal "Letter to the Committee of China Foundation" from P. W. Tsou, Dean, College of Agriculture National Central University, December 27, 1926 (RAC: China Medical Board/ser.2/Bx.38).

70. Stross, *Stubborn Earth*, chapter 6. It is noteworthy that major Rockefeller Foundation field reports and policy suggestions for China do not begin to address agricultural science education until 1934. In their respective 1931 and 1933 reports, neither Gunn nor Tisdale address the subject (Gunn, "Report on Visit to China"; Tisdale "Report"). In Gunn's 1934 "China and the Rockefeller Foundation," he argues the need to shift funding from missionary to Chinese schools, and to target agricultural science education. He nevertheless notes that Nanking University has by far the best Agriculture College in China and that the foundation should continue to support it, at even higher levels than the International Education Board did. Among the Chinese schools, he says, National Central's college is the best of a very poor lot. He also says that the latter had had a good reputation at one time, "but through political complications it sank to a low level" (26). Gunn, however, does not speculate whether Rockefeller Foundation funding policies also may have contributed to that decline.

71. Statistics: W. P. Fenn, *Christian Higher Education in China, 1880-1950* (Grand Rapids, Mich.: Eidermans, 1966): 92, 239; Love and Reisner, *Cornell-Nanking*, 6.

72. The program's development and accomplishments are detailed in Love and Reisner. The flow of Chinese students to Cornell is discussed on page 67.

73. Shen Zonghan (1895-1980), from Zhejiang Province, received his 1918 B.A. from the National Agricultural College in Beijing, M.A. in agricultural science from the University of Georgia in 1924. He worked on his doctorate at Cornell from 1924-1927, and interrupted his studies there for seven months in 1926 to return to China and help develop the Nanjing-Cornell plant breeding program. The Rockefeller International Education Board defrayed costs of this 1926 China trip, and then contributed to the program directly. Shen left China for Taiwan in 1949.

74. Stross, *Stubborn Earth*, 195, and chapter 8, for detailed narrative of Shen's life and work, which is well documented in Shen's autobiography *Shen Zonghan zishu*, 3 vols. (Taibei, 1975); and in the festschrift, *Shen Zonghan xiansheng jinianji* (Taibei, 1981).

75. Interview, Li Jingzhun (C. C. Li), Pittsburgh, January 10, 1985. Li (b. Tianjin, 1912) received his B.S. under Shen's guidance in 1936, and a Cornell plant breeding Ph.D. in 1940. Li was one of Shen's most successful students and ardent disciples. He was professor of genetics and biometry at Nanjing's Agriculture College, 1943-1946; professor and chair, Beijing University agronomy department, 1946-1950. From 1951, an expatriate, he was on the faculty of the University of Pittsburgh Department of Biostatistics of which he later became chair. More will be said about Li in the text below.

76. Interview with Li Jingzhun. Catalog description: University of Nanking, *Bulletin*, 1924-25, no. 1: 77. The use of the term "eugenics" at this time was sometimes synonymous with "improvement" through controlled breeding of plants and animals and did not necessarily indicate any interest in the controlled "improvement" of humans. In some cases, however, it did indicate just that: For example, see translation of human eugenics' discussions from Herbert E. Walter's textbook, cited next, in *Kexue*, 1924, no. 8: 920-33.

Textbooks: Herbert E. Walter (b. 1867), *Genetics: An Introduction to the Study of Heredity* (New York: Macmillan, 1922 and many later editions). Walter was an early collaborator with W. E. Castle in mammalian genetics, and obviously shared Castle's interest in human eugenics. See Daniel J. Kevles, *In the Name of Eugenics: Genetics and the Uses of Human Heredity* (Berkeley, Calif.: University of California Press, 1985).

Also see Edmund W. Sinnott and L. C. Dunn's *Principles of Genetics*. This is the same text used by Chen Zhen at Qinghua University.

77. Glossary: See Feng Zhaochuan's 650-term glossary and discussion of the project in "Yichuanxue mingzi" (Genetics terminology), *Kexue*, 1923, no. 7: 759-75. Feng was a 1922 Cornell Ph.D. in agriculture science.

78. Fan Jianzhong, *Yichuanxue* (Chengdu: Ministry of Culture, Education, and Industry), 2 vols. Fan's identity is unclear. No one with his personal name can be identified; however, a Fan Te-shen seems to be the same person. He received his B.S. from Nanjing University in the late 1920s and his Ph.D. (studying silkworm genetics) from Berkeley in the early 1930s, a student of Eakin and Schectman. During 1931-1934, the China Foundation supported his research on two consecutive projects at Berkeley: a comparative study of germ cells in Chinese orthoptern insects, and a study of parasitic protozoa of goldfishes. In 1940-1941, Fan was a professor at Nanjing University, in Chengdu, and the foundation funded his research on a genetical study of poultry in Chengdu. See China Foundation *Report*, 1931-1932, no. 6; 1933-1934, no. 8; and 1940-1941, no. 15.

79. Wang Shou (b. Shanxi, 1897). Shanxi University B.S. 1919; Nanjing Agriculture College M.S., 1919-1923; on faculty 1924-1932. He was trained at Cornell via the Nanjing program, 1932-1934, and was on the Nanjing faculty 1935-1950. He was director of Shanxi Academy of Agriculture Science, 1950-; president, Shanxi Agriculture College and director CAS Institute of Crop Breeding and Cultivation, 1958- (ZXN: 90-103).

80. For the "Wang Barley" story, see Love and Reisner, *Cornell-Nanking*, 79-81; and Wang Shou "Inheritance in Barley," *Zhonghua nongxue huibao* (Chinese agriculture), no. 148 (May 1936): 1-16.

81. Interview with Li Jingzhun.

82. Love and Reisner, *Cornell-Nanking*, 63-65.

83. Interview with Li Jingzhun; Love and Reisner, *Cornell-Nanking*, 7, 32-33. For success in wheat breeding, see Ramon Meyers, "Wheat in China—Past, Present, and Future," *China Quarterly* (July 1978): 310-15.

84. Stross, *Stubborn Earth*, 196-99.

85. Love and Reisner, *Cornell-Nanking*, 68.

86. Original English language publication of the proposal in *Zhonghua nongyeh xuebao* (Chinese agriculture studies), no. 108 (January 1933): 1-46.

87. "Proposal," 3, 18, 21.

88. Rural cooperative credit program: See Chen Yixin, "The Guomindang's Approach to Rural Socioeconomic Problems: China's Rural Cooperative Movement, 1918-1949," Ph.D. diss., History Department, Washington University, 1995.

89. Stross, *Stubborn Earth*, 197-99.

90. Rockefeller support: Gunn, "China and the Rockefeller Foundation," section 15: "Conclusions," 57-58. And see brief references to the new policy in Rockefeller Foundation *Annual Report* (1935): 321, 329. Also see James C. Thomson, Jr., *While China Faced West: American Reformers in Nationalist China, 1928-1937* (Cambridge, Mass.: Harvard University Press, 1969), 125-29.

# 3

# War, Revolution, and Science

The war with Japan affected all of China, fragmenting the country into the considerable areas occupied and controlled by the Japanese, the northwest areas under Communist control, and the southwestern areas under the Guomindang. This survey of wartime work in biology deals first with Guomindang, then the Communist areas, emphasizing science education, research, and policy that would play prominent roles in the Peoples Republic after 1949. In doing so, I am arguing that, perhaps for the first time in China, there emerges a discourse on the fundamental relation of science and social revolution, and the kind of science appropriate for a socialist society. This discourse, however, does not take place solely between the Guomindang and the Communist Party. It also occurs among Communist Party cadre themselves, and in some of its most contentious arguments it anticipates the Party's ambivalence toward science in Mao's China.

During the summer and autumn of 1937, universities, research institutions, and the central government itself all fled the advancing Japanese armies. Over the course of many months they were forced to retreat to China's far west and southwest. The government relocated in the city of Chongqing, Sichuan Province, nearly a thousand miles up the Yangzi River from Shanghai and Nanjing. Academia Sinica, National Central and Nanjing Universities, among others, resettled there too, as did Shen Zonghan's National Agrricultural Research Bureau and affiliated organizations. In Sichuan Province, exigencies of war were used by the Guomindang government to advance the centralizing policies it had begun to implement in the early 1930s. Gradually, substantial efforts were made to plan and direct the entire wartime economy under a National Resources Commission, an agency at the highest political level, made up of talented technocrats with a penchant for nationalizing basic industry and coordinating all scientific research efforts under a national emergency plan.[1]

By the time the Guomindang moved west, the Communist Party had established itself in northwest China, where—just a few years earlier—it took refuge from the Guomindang's campaign to destroy it. The capitol of the Communist area was Yan'an, which, like Chongqing, was more than a military command center. Yan'an had become a living symbol of the Party's will to survive and to continue the socialist revolution in the unlikely midst of one of China's poorest and most inhospitable rural areas.

While the Communist Party committed itself fully to expanding its rural base and to resisting the Japanese, it continuously debated and experimented with policies and institutions that would shape the socialist society that it was determined China would become. This was especially true of education. Yan'an was, after a fashion, a college town, with a variety of schools devoted not only to training cadre for Party service, but also devoted to technical education and research. These institutions were regularly contested because of their dual purpose: They were needed to serve immediate revolutionary and wartime needs; and they were also expected by many to act as models or prototypes for the future socialist society.

From the late 1930s until the end of the Japanese occupation, there was an especially revealing controversy over the science education and research most appropriate for present and future Communist China. The antagonists in the ensuing debates and reforms anticipated some basic controversies about science after 1949. In particular, the central figure in the Yan'an science debates became the leading Chinese promoter of Soviet Lysenkoism.

The wartime efforts devoted to science in Communist and Guomindang areas were quite independent of each other. In 1949, however, the two quickly came together. In the case of biology, it is more accurate to say that the two collided. The biologists described in this chapter were all victims of that collision.


## Science in Guomindang-Controlled Areas

The Guomindang's policy toward the science community was conspicuously protective during the war. Basically, it sought "to keep the flame of science burning," to keep it intellectually alive by fostering research and facilitating communication among its scattered members; and to keep it developing by providing the wherewithal for science education at all levels. China's survival during the war as well as after the war was considered to be dependent on science. During the war, the Guomindang hoped that science could help support the military effort through clever use of scarce resources and that science could find new means to feed the displaced population. By all accounts, no pressure was needed to rally the science community to these projects. Nevertheless, basic science education and research continued and were encouraged. Scientists who were students or already practicing their craft during this period agree that, to maintain standards, science curricula in the universities remained essentially unaltered, and that it was not necessary to justify their research activities with applied goals. Nor was it necessary to justify research projects in terms of their great moment; it was quite enough just to keep one's hand in the game, no matter how modest the work.[2]

## "China Scientization" Movement

This passionate commitment to science by the Guomindang was already evident by the early 1930s in the establishment of the Academia Sinica and in the educational reforms that demanded a greater role for science and technology. It was never more clearly articulated than in the "China Scientization Movement," promoted by Guomindang Party intellectuals in the years just before the war. Alternately called the "science salvation movement," the immediate goal of this often melodramatic, sometimes apocalyptic public relations campaign was to justify the Guomindang's science-related reforms. There was, however, an ideological subtext that spoke to the Communists' understanding of what was broken in China and how to fix it. In effect, by arguing the centrality of science for China's salvation, Guomindang intellectuals denied the priority of social-economic reform, let alone a socialist revolution.[3]

Between 1933 and 1937, Chen Lifu (b. 1900), Jiang Kaishek's house intellectual, published the journal *Kexue di Zhongguo* (Scientific China) to promote the fundamental notion that science and its offspring, technology, were at the base of a society's progressive evolution.[4] Taking issue with Marxist materialist theory, the journal argued that science and technology were the true base of political economy.[5] Communist intellectuals soon responded volubly with caveats about the class nature of science and scientists, and the need for China to differentiate between the bourgeois science of the West that was inappropriate for China, and the proletarian science of the future. The latter were issues that would have grave consequences for the science community after 1949; but in the short run, especially during the war they largely remained philosophic abstractions.[6] Chen Lifu's Guomindang colleagues meanwhile translated his ponderous message into practical issues: The promotion of science must be accommodated with institutional support for extensive translation projects; and science must be broadcast to the masses through "vernacularization." Additionally, the growth of science must not be hobbled with a false dichotomy between pure and utilitarian science.[7]

These attitudes expressed before the war clearly anticipated the Guomindang's aforementioned policies toward science and scientists during the war. They likewise account in part for special wartime efforts on behalf of science. Among these was the state-sponsored national translation project that began in Nanjing in 1933 with a staff of eighty and continued in Chongqing reduced by twenty workers, but nonetheless devoted to diminishing the dependency on foreign languages by Chinese science and other fields of learning. Among its ambitious projects was a series of lexicons, on every subfield of science, to provide nationwide standards of usage. A 1940 publication by the project lists the scores of books translated and lexicons compiled to date, and also indicates broad cooperation from the science community.[8]

The central government also accommodated the science community, and helped it to remain a community, by subsidizing the publication of science journals that were distributed all over unoccupied China, wherever there was a school or research institute. And it contributed further to this enterprise by encouraging or permitting foreign agencies to publish periodicals devoted to the natural sciences. One such publication, jointly sponsored by the British Council and the United States Department of State, was *Acta Brevia Sinensia*, a forum for reporting on scientific research conducted in China. Another, also sponsored by the United States Department of State, was *China Science Service*, whose purpose was "to present to the scientists in China important papers on science and technology from Western technical periodicals."[9]

The Chongqing government's concern with the wartime state of science is further evidenced in a survey of life-science education from middle schools through college-level schools in Sichuan Province. This 1941 project details every aspect of the biology curriculum of each school, including the numbers of teachers and their names, class hours per subject, lecture and laboratory hours, and textbooks used. It also provides a model curriculum of rather demanding standards, which it expected the schools to meet.[10]

Guomindang concerns about science are also illustrated in the conferences it sponsored and organizations it encouraged during the war. One of the earliest conferences, held in early 1941, was devoted to the problem of nutrition. The subject was examined from multiple points of view ranging from human physiology and growth studies to plant physiology and genetics. It was reported to have had very positive responses throughout the science community.[11] Another conference, in 1943, brought more than three hundred scientists together from all over unoccupied China to discuss "how to advance science and technology in China," and "how to promote international science cooperation."[12]

Throughout the war, the Guomindang encouraged the operation of scientific professional societies, and then, in typical corporatist fashion, it brought representatives of those societies together in Science and Technology Councils. Consistent with his earlier science salvation campaign, Chen Lifu brought one council under his Ministry of Education "to promote science and technology research, spread scientific knowledge, method, and spirit among the people; and to introduce scientific techniques to industry and scientific management to manufacturing."[13] Whatever the Guomindang's political motives may have been for such efforts, the science community appears to have welcomed the opportunities to congregate and to get some relief from their harsh sanctuaries.

## Wartime Experimental Biology

It was in the context of this Guomindang nurturance of science that three exemplary centers of experimental biology flourished. They not only "kept the flame of science burning," but made substantial advances in a number of fields which directly or indirectly enhanced genetics research in China, and sometimes had resonance in the international science community. The first of these three was guided by Li Xianwen and devoted to plant genetics. It was based in Chengdu, the capital of Sichuan Province, at the relocated National Rice and Wheat Improvement Institute. The second center of research was located to the south, in Yunnan Province, where Beijing, Nankai, and Qinghua Universities established a consortium they called Xinan Lianda (Southwest Associated University).[14] Here, a center for plant physiology research developed around the formidable Tang Peisong. And to the east a third center was formed at Zhejiang University, now relocated in the remote and rugged isolation of Guizhou Province. Tan Jia-zhen gathered a talented group of young geneticists there, engaging in research that built on his previous work in population and evolutionary genetics.

In Sichuan Province, the National Rice and Wheat Improvement Institute was part of a sizable accidental community of agricultural scientists whom the war had displaced from many universities as well as government and private research organizations. The institute was still nominally a subsidiary of the National Agriculture Research Bureau, now in Chongqing and reporting to a new and stronger Ministry of Agriculture and Forestry. In fact, the institute was subsumed and funded by the Sichuan provincial government. These agencies all found themselves able to lobby successfully for agricultural research funding throughout the war.[15]

The National Agriculture Research Bureau—in addition to its own staff—was also able to draw on the energies of specialists in plant improvement and genetics who had come west with their colleges, universities, or institutes. These agricultural scientists numbered at least two dozen Western-trained Ph.D.'s in experimental biology, and on the order of one hundred technicians and graduate student assistants. These numbers include the workers at the Rice and Wheat Improvement Institute.[16] The critical importance of agriculture research was readily acknowledged by all political entities, and by the early 1940s the prolific work of the research community was continuously surveyed in two journals published out of Chongqing, and in prominent Western journals when communications permitted.[17]

### Li Xianwen and Plant Genetics

Li Xianwen (b. 1902) was director of the National Rice and Wheat Improvement Institute and he drew upon these aggregated resources to develop one of the most active and creative wartime plant genetics research groups. Li's overseas training had been at Purdue University and then at Cornell, where he received his doctorate in 1930. His doctoral research was on the inheritance and linkage relations of two dwarf types of maize. The China Foundation supported his research at Cornell, and also, just before the war, his work on the induction of mutations through centrifugal force. Before moving to Sichuan, Li Xianwen had been on the biology faculty at National Wuhan University.[18]

With an eye for exceptional talent, he appointed as his first assistant Li Jingxiong (b. 1913), a former student of his at Wuhan University. Li Jingxiong was originally trained in plant genetics in the early 1930s at Zhejiang Agriculture College. His mentor had been one of the first Chinese to receive a Ph.D. from Cornell in agricultural genetics.[19] The second appointment was Bao Wenkui (b. 1916), a precocious graduate of National Central's biology department.[20] Both assistants were working on the equivalent of a master's degree in anticipation of doctoral work in the United States and both went on to distinguished careers in China after the war.

From the late 1930s, Li Xianwen primarily focused his team's research on a long series of studies on polyploidy, using millet and wheat.[21] Broadly, polyploid organisms (usually plants) have supernumerary chromosomes and thereby have potential for producing significant variation and even new species in a single generation. In the mid-1920s, emerging international data revealed an unsuspected plethora of plants that were polyploids and suggested that controlled polyploidy might be used to alter and improve cultivated plants such as wheat, cotton, potato, oat, or sugarcane. One type of polyploid—the allopolyploid—has particularly great potential for change because it possesses sets of chromosomes that are present separately in two or more species that went into its makeup. Such individuals could occur through hybridization or, by the 1930s, polyploidy could be induced with the chemical colchicine. The latter technique was commonly used by Li and others at the institute. In his 1937 *Genetics and the Origin of Species*, Dobzhansky effused that "it is no exaggeration to say that the production of allopolyploids is the most powerful tool yet available to a geneticist for molding living matter into new shapes."[22]

While Li Xianwen and his associates may well have shared Dobzhansky's transformist vision, their publications exhibited no enthusiasms save those for meticulous, patient technique, and a critical attitude toward the research of predecessors. Notably, there was never a hint that they were under pressure from their sponsors to justify their budgets with results that could be measured directly in production. Clearly, however, very practical issues were involved in their genetic studies of debilitating abnormalities such as "dwarfism" in local wheat crops and "polyhusk" in rice, or in their studies of haploid breeding. Haploids have only one set of unpaired chromosomes in each nucleus. Their potential advantage, in contrast to polyploids, lies in the creation of immediate genetic uniformity, uniform lines, and easily identified genotypes.[23]

Li's papers are also notable for expressing an awareness of the leading scientific literature in the field and a willingness to engage it. For example, in their studies of polyploidy, they regularly cite and speak to the work of prominent Japanese and Russian geneticists.[24] During the difficult war years, access to current scientific literature and to journal back issues would have been impossible without the respective efforts of the British and American agencies devoted to keeping the Chinese scholars in unoccupied China in touch with technical literature from the West. Chinese scientists were supplied with requested books and microfilms of journals as well as with scientific instruments and chemical reagents. Joseph Needham, the renowned embryologist, was in charge of the British effort. He visited virtually every school and laboratory that was not in occupied territory, and recorded his observations in a series of articles for *Nature*. He was consistently impressed with high morale of the scientific community, and its ability to carry out useful research by making do with available resources. Needham's narratives and photographs reveal the utterly primitive conditions in which these scientists worked and the daily hardship of their personal lives.[25]

## Tang Peisong and Plant Physiology

At this time, Joseph Needham's survey of China's wartime science expressed no greater appreciation for any scientist in China than for Tang Peisong (b. 1903). While Tang was a plant physiologist, not a geneticist, he and his work are very important to the story of genetics in China. The quality and prolific output of his research raised experimental biology in China to higher levels of sophistication. With his lively intellect he maintained an open dialogue with geneticists that was quite valuable for the development of both fields. More generally, Tang was keenly sensitive to the increasing dependence of experimental biology on chemistry and physics and the need to avoid artificially segregating new biological subfields.

Tang's undergraduate work was completed in 1925 at Qinghua University. Then, there was graduate work in botany at the University of Minnesota, followed by a 1930 doctorate in plant physiology from Johns Hopkins, under the direction of Burton Livingston. For the next two years, Tang was a research fellow and instructor in physiology at Harvard. His summers were spent at Woodshole, assisting R. S. Lillie and collaborating with R. W. Gerard. Tang returned to China in 1933 to join the biology faculty at National Wuhan University, where he was a colleague of Li Xianwen.

In 1938, Tang moved to Kunming and set up the Physiological Laboratory of Qinghua University, one of the three divisions of the university's Institute of Agriculture Research. His senior colleagues there worked respectively in biochemistry and biophysics, primarily dealing with problems of human nutrition and the growth of food and industrial plants. Tang also continued to work on cell respiration, the subject area of his doctoral and postdoctoral research.[26]

Throughout the 1940s, Tang himself also led a series of studies on polyploidy that complemented the work of Li Xianwen's group in Chengdu. With practical agricultural issues in mind, Tang used colchicine or indol acetic acid to induce polyploidy. He demonstrated some of the possibilities and limitations of polyploidal food plants. Tang was aware that polyploid plants are usually larger and sturdier than normals, and their rate of growth is slower. Polyploids obtained through inducement, however, are usually unstable, so that in a generation or two they revert to normal sets of chromosomes. He argued that "most plants of economic value are already polyploid and their numbers of chromosomes are already optimum."[27]

During this same period, when the government in Chongqing promoted research on human nutrition, Tang's group took this charge quite seriously and used their unique skills to follow up a variety of leads promising low-cost means to increase nutritious food plant productivity. Among these was the technique called "vernalization," popularized and promoted by the Soviet agriculturalist T. D. Lysenko. Tang's laboratory found some of Lysenkos vernalization technique to be useful, but also found Lysenko's theoretical explanations of the process to be thoroughly without foundation. Within just a few years, Lysenko would become the bane of Soviet and Chinese biology, and Tang Peisong would be among the first victims of the oppressive policies that bore Lysenko's name.[28]

## Tan Jiazhen: Evolutionary and Population Genetics

The last of the three experimental biology research centers under consideration was developed by Tan Jiazhen. In July 1937, Tan returned to China from Caltech, expecting to begin his career at Zhejiang University. The Japanese invasion, however, had already driven the university out of its home in Hangzhou, a hundred miles southwest of Shanghai, and into the interior of Zhejiang Province. As the Japanese advanced, the university and its six hundred students moved five times between 1937 and January 1941, when they finally settled down in Zunyi, Guizhou Province.[29] Altogether, the university community had trekked close to a thousand miles to reach the relative safety of this far southwestern town. The last of their moves seems to have been the most difficult. Tan wrote that they had no transportation and so "our retreat was slow and mixed with horror. Both the university itself and the faculty members suffered a great loss."[30] Tan had with him his wife and three children. He had initially left them behind with his family in Ningpo, and they had only been able to join him a year later.[31]

During the three-year trek, Tan determined to make the most of his situation and began a project that took advantage of his movement through a wide variety of biogeographical areas. With the aid of his talented graduate students, he began a geographical distribution survey of both Drosophila and a species of lady-bird beetle called *Harmonia axyridis*. Little if any such study had been made of Drosophila in China, according to Tan. Doing so would provide information valuable for taxonomy as well as for experimental biologists using it in their research.

Within the species of lady-bird beetle, which he had studied at Yanjing, there were a great many color pattern variations; but most interesting to Tan was the fact that "different geographical regions are inhabited by different representative types." By the time Tan was two years into the trek west, he had collected five thousand samples and determined that they exhibited an entirely different series of variations from those that had been studied earlier by others in Siberia, north and east-central China, and Japan.[32]

After finally being able to settle down in 1941, Tan used his Rockefeller money and a small government grant to assemble a group of ten young aspiring geneticists to work with him on a nonstop series of studies, clearly inspired by his Caltech work, in the genetics of natural populations (using the Drosophila), and the evolution of natural populations (using the lady-bird beetles).[33]

Zhejiang's biology department was able to provide a solid foundation to Tan's genetics' students thanks to other experimental biologists on the faculty, especially Bei Shizhang in embryology and Lo Zonglo in plant physiology.[34] By all measures Tan was quite successful in his work from 1941 to 1945. He had begun to place his five senior students in graduate programs in the United States. By the end of 1945, he had placed all of them, with fellowships: two at Columbia, in the genetics of microorganisms and plant genetics, respectively; one at the University of Texas, Austin, in Drosophila population genetics; one at the University of California, Berkeley in the population genetics of grasses; and one in Drosophila genetics at Rochester. They all completed their doctorates by 1953.[35]

With these students and undergraduate assistants, Tan published at least ten papers during the war, in university and government journals, and all the while kept the Rockefeller Foundation apprised of his activities. He made no claims to exceptional contributions, but rather took encouragement from this work because of the great promise it held in store for the future development of genetics in China. His detailed reports never complained about the primitive conditions and lack of proper equipment. On the contrary, they were always ebullient, radiating his energy, self-confidence, and optimism.[36]

For the 1945-1946 academic year, the Rockefeller Foundation provided Tan generously for a research sabbatical at Columbia University, where Dobzhansky was now working. The two had corresponded throughout the war and Dobzhansky was quite willing to find local funds as well, to help with sabbatical costs. Zhejiang University continued Tan's salary for the support of his family, who stayed in China. Tan left for the United States before the war with Japan ended, and because of all the hazards and travel problems, it took him two months of determined effort before reaching his destination in September 1945.[37]

Tans work at Columbia refined his wartime studies, and it resulted in two papers, published in *Genetics* in 1946. The first paper reported his study in the very new area of the genetics of sexual isolation. Two species of Drosophila were used. In particular, the study addressed recent work by Dobzhansky and Ernst Mayr and took issue with current explanations of sexual preference patterns.[38]

If this first paper was a careful step into a new frontier, the second paper was a tour de force, bringing together Tan's studies of color pattern variations in lady-bird beetles. His massive sampling and precise analysis led him to demonstrate soundly that all the racial variations expressed in color patterns were a function of how often the alternate forms (the alleles) of just one gene combined with each other. In addition, he demonstrated that there was a peculiar Mendelian inheritance rule, "mosaic dominance," that governed this process. Dobzhansky incorporated Tans study into later editions of *Genetics and the Origin of Species*. He considered the phenomenon that Tan analyzed to be paradigmatic of race variation in general.[39]

During the sabbatical year, Tan took every opportunity available to travel to other centers of genetic research. On one trip, he visited Caltech and University of Texas at Austin, to renew old acquaintances, discuss research trends, and to open doors for his students. Toward the end of his sabbatical, the Rockefeller Foundation made it possible for him to travel to Europe to attend a conference on nucleic acids at Cambridge University. With characteristic energy and enthusiasm, he stretched that trip to include visits to the biology departments of Oxford and Edinburgh, and then to the Pasteur Institute in Paris, and the University of Geneva.[40]

This European side trip completed, Tan returned to New York to finish up his research work and to lobby the foundation for a grant-in-aid to refurbish the genetics laboratory of Zhejiang University, recently returned to its war-devastated campus in Hangzhou. A truck for making Drosophila and lady-bird beetle field surveys, a greenhouse, and incubators were the major items requested; but Tan departed for China before the foundation was able to evaluate his request.[41]

A year later, the Rockefeller Foundation was still dickering with Zhejiang University over the equipment grant. The university president assured the foundation that Tan had been given prime space for his laboratory and the aid of three of the graduate assistants who had worked with him in Meitan. In addition, Tans teaching load was minimal and he would have no administrative duties. And further to demonstrate that Zhejiang's biology department was a good investment, it was noted that two new biologists were joining the staff: Li Huilin, a 1942 Harvard Ph.D, specializing in plant geography; and Gao Shanyin, a former Rockefeller Fellow, just returning from the Rockefeller Institute at Princeton, where he had worked on viruses.[42]

The Rockefeller Foundation was holding back because of the civil war in China, not because there were any doubts about Zhejiang University. If the Communists won, then foundation investments in China would be considered a loss. A decision was delayed until January of 1948, and then Tan was notified that no support would be forthcoming "until there is some improvement in both financial and political stability" in China. Earlier and to no avail, a foundation field agent had advised that the support for Zhejiang University was a safe bet: Hangzhou was remote from the areas of "unrest" as were the locations where the fieldwork was to be done; and "the work was not of a controversial nature." "If work of this sort cannot be done," the agent concluded, "one might as well decide to abandon all constructive scientific work in China."[43]

Tan had one more round with the foundation. He was invited to attend the International Genetics Congress in Stockholm, July 7–14, 1948, and needed supplementary travel support. After much effort, he got enough funding from the foundation and other agencies to attend the congress as well as two other European conferences, and to spend seven weeks at the Naples Zoological Station.

He received further support to return to China via the United States where he hoped to raise money for laboratory equipment.[44]

It was in Stockholm that Tan first heard about "Lysenkoism" and the politicization of biology in the Soviet Union. By mid-August 1948, news was available about the July 31-August 7 meeting of the Lenin Academy of Agricultural Sciences, wherein T. D. Lysenko had initiated a comprehensive attack on genetics.[45] This together with news of the Communist success in China undermined Tan's optimism about his future in China. In October, he talked in New York with Warren Weaver, director of Natural Sciences for the Rockefeller Foundation. Weaver noted Tan's acknowledgment that "the situation is deteriorating in China, and it may be necessary for him to leave his post [at Zhejiang University]." Weaver then records the following:

> [Tan] points out, with good humor, that it is somewhat bizaar [sic] that he, as a geneticist, has more to fear from the Communists than his scientific colleagues in chemistry, or physics, or mathematics. Indeed these fields are quite likely to be appreciated by the Communists, and given good support. [Tan] is, however, well known as a Morgan geneticist, and he would certainly fare badly under the Communists. Thus he will very likely have to get out somewhat in advance of their possible entry. . . . [Tan] also asks whether we would conceivably help him to get out of China, if that necessity should arise. He has had some tentative proposals from the University of Rochester that he become a successor to Curt Stern.[46]

Weaver then, in his words, "tells the old, old story": that the foundation trains people for service in their own country and that "we wish to avoid whenever possible helping a person to move away from his own country; but that that necessity sometimes is overpowering, as was the case in our European refugee scholar program." The talk ended with Weaver's advice that Tan do his utmost "to remain in the Orient"—perhaps in Formosa or Hong Kong where there would presumably be protections for him.[47]

One need not accuse the foundation officials of racism or even disingenuousness to wonder aloud about their sense of contemporary East Asian geography, culture, or politics. It is not at all clear why the officials did not consider a move by Tan to a British colony or a former Japanese colony any less a failure of their policy to train scientists for "service in their own country," or any less a move away from his own country than a move to the United States. And their myopia did not end there.

In mid-November, Tan was on his way back to China, responding to an urgent plea from his family, as conditions deteriorated in the Shanghai-Hangzhou area. Weaver wrote him a formal letter, reiterating his position in their conversation, but adding that if it turned out to be the case that it is "physically and intellectually impossible" for Tan to carry out any effective work in China, the foundation will reconsider its position. Weaver asked a colleague to comment on the letter to Tan. The colleague thought it was very good, and concluded: "My guess is that Chinese Communists may not take too seriously Marxian theory as applied to genetics. . . . We should not anticipate by helping Chinese to decide to become refugees."[48]

He of course was dead wrong; and Tan was unfortunately quite prescient. In less than a year, Soviet Lysenkoism had established itself in the People's Republic and immediately began a campaign to eliminate Western genetics and evolutionary theory from all education, research, and publication. As one of China's most eminent geneticists, Tan soon became a specific target, as we shall see in the next chapter. Tan's nemesis was Luo Tianyu (1900-1984), the heavy-handed and zealous leader of the initial campaign to promote Lysenkoism in China.

The following discussion on science in the wartime Communist-controlled area is devoted to what is known about Luo Tianyu. His career is especially relevant to the Yan'an science debates and to the post-1949 fate of science in general and biology in particular.

## Science in Communist-Controlled Areas

In Yan'an, the exigencies of warfare and day-to-day administration naturally dictated constraints on time and energy available to read, debate, and implement ideas about science. Nevertheless, the CCP leadership encouraged a small group of Party veterans with science training to build science research and educational institutions to address current needs and to act as prototypes for the future Communist society. In the early 1940s, the veteran cadre and agricultural scientist Luo Tianyu was at the center of one of the most divisive controversies about the future course of science development. What made this controversy especially notable is that it became bound up with Mao's 1942-1944 Party "Rectification" (zheng feng) which used political purges and reeducation campaigns to establish orthodox policy, largely based on his own thought, to govern the subsequent activities of the Party. The nativist and populist quality of Mao's Rectification teachings and policies is shared by Luo Tianyu's approach to science. Luo uncompromisingly rejected science policies of other veteran cadres who advocated a formal, specialized approach to science, quite like the American institutions emulated by the Guomindang.

### Luo Tianyu and Yan'an Science

Luo Tianyu was born in Dingyuan xian Hunan Province. He entered Changsha No. 1 Middle School in 1916, and four years later he began his agricultural science training, with an emphasis on forestry, at the old Agriculture College of Beijing University (Beiping nongyeh daxue). At the same time he joined the school's Socialist Research Cell. In 1922, he joined the China Socialist Youth League (Zhongguo shehuizhuyi qingnian duan), and then the CCP, in 1924. He appears to have spent only two years at the college. His first CCP assignment was working with the peasant movement in the western suburbs of Beijing, and after study in a CCP school, he became a ward secretary for the western section of Beijing. In 1927, Luo returned to Hunan to join CCP efforts at organizing the peasant movement. He was associated with the CCP's military in Changsha and an editor of *Hongzhunribao* (Red Army daily) in 1930. For the next six years, he had teaching positions at agriculture and peasant schools in Honan and Hunan Provinces. From 1936 to 1938, he made his way to the northwest and finally to Yan'an where he remained throughout the war.[49]

In 1939, a year after Luo Tianyu arrived in Yan'an, the Yan'an Natural Science Institute (Zeran kexue yanjiuyuan) was established and he was appointed director of its biology department.[50] Like the other three departments—physics, chemistry, and mining—Luo's was initially "designed to produce immediate, practical results," and "to attract scientific and technical personnel from other parts of China, to use them to train more technicians, and to channel this resource into production."[51]

Luo was apparently the right man for such a job. From 1940 through 1946, his accomplishments are cited in *Liberation Daily* (Jiefang ribao), the CCP's central newspaper, and recalled with satisfaction in memoirs of students and colleagues. He was a very pragmatic, hands-on agriculturalist, closely attuned to the changing needs of the CCP's northwest border region (the Shen-Gan-Ning). Broadly speaking, his goal was to manage a balance between the ecology of this area and the heavy demands of a rapidly expanding population.

His work began with a coup that even his detractors agree earned him the status of local hero and minor legend. Luo was responsible for establishing a beet-sugar manufacturing facility in Yan'an, based completely on locally grown beets and a handmade refinery.[52] Most of his efforts, however, were not quite so immediate in their results. For example, he believed in teaching through problem solving and was consequently a regular participant in Yan'ans Guanghua Experimental (work-study) Farm, whose students warmly appreciated his teaching.[53] Additionally, he was a regular contributor to the "science column" in the *Liberation Daily*. One series he wrote for the column was devoted to introducing local flora and suggesting medicinal or industrial uses for them.[54]

Introducing local flora to immigrants from east and south China was part of a larger campaign for Luo Tianyu. He feared that the immigrant's ignorance could have dire consequences for their ability to feed themselves and for the precarious ecology of the northwest. He patiently dealt with amateur or newly arrived farmers through a set of step-by-step guidelines that began by admonishing them to match their crops to the location, climate, water, and soil. Southerners needed to acquaint themselves with different temperature variations and hours of sunlight in the northwest; and they could not take for granted that crops familiar to their previous experience would continue to be suitable.[55]

Luo was especially concerned that newly established collective farms and military farms be aware of the regions serious problems with soil erosion, and he became the recognized leader of a strong campaign for forestation and soil erosion control.[56] He preached a solid message on the relationship between deforestation, floods, droughts, and pests; and he made a sound case as well for the possible long-term ill effects of desiccation on total climate change.[57] To deal with these problems, he was personally involved in tree planting programs in Yan'an's environs. And he waged a special campaign against the common practice of clearing waste- or wilderness-land by burning.[58]

To this point, Luo Tianyu's activities in Yan'an do not display any predisposition for Lysenkoist biology or for any particular Lysenkoist theory or attitude. Far from Lysenkoisms aggressive and domineering orientation toward nature, Luo advocated an attitude of respect and harmony, though he by no means suggested a passive relationship. He further differed from Lysenkoist biology by eschewing theory and generalities, preferring to deal with the immediacies of agricultural production rather than with agricultural "science." It was here that Luo began to have problems with colleagues at the National Science Research Institute, and here that he began to articulate a doctrine that gave some coherence to what were otherwise desultory activities.

Within a year or so of its founding, the National Science Research Institute had a new director who began to reshape its original utilitarian mission. By 1942, a considerable number of the institute's faculty seemed intent on moving the curriculum and educational philosophy to a position very close to that of American research universities and to that of the prewar Guomindang university system. This position asserted the fundamental need for specialized knowledge in China's future society, and in turn a need for an educational system to train the specialists (a term that connotes "professionals" as well).[59]

Institute policies wanted a broad-based science curriculum that covered the full range of standard natural science disciplines from their theoretical foundations, through laboratory research, and then—but only then—to practical application. This policy criticized those in Yan'an who saw distinct contradictions between theoretical and applied science, foreign and Chinese science, or "today's" science and the scientific needs of tomorrow. And it argued that if Yan'an only promoted a utilitarian science focused on what Yan'an thought to be China's current needs, then China's scientific community would not be capable of dealing with future and different problems.

## The 1942-1944 Rectification Campaign

Discussions of institute educational policy began intelligently and amicably enough; soon, however, they were engulfed in the strident and doctrinaire forces of what became the CCP's Rectification campaign. The 1942-1944 Rectification was the result of a number of converging issues. First, Mao and the CCP concluded that the recent great influx of new Party members was a problem for Party unity and discipline. Secondly, recent severe setbacks militarily and economically required a tightening of discipline and careful management of resources. And finally, the Rectification campaign was a medium through which Mao Zedong eliminated remaining rivals and challengers to his supreme leadership. In the process, Mao claimed to be domesticating Marxism, and to be bringing the Party into its proper dialectical relationship with the Chinese peasant masses.

In 1942, the CCP Central Committee issued a Rectification resolution on education in Yan'an.[60] It motivated Luo Tianyu to challenge his institute colleagues and attempt to impose his will on the institutions form and content. He based his position on that section of the resolution which sought to end "the discrepancy between what is studied in theory and applied in reality," and he clearly was inspired by other Rectification texts, some of them written by Mao Zedong, which urged the unification of thought and action "by relying less on abstract knowledge of the type found in books—particularly foreign books—and more on the practical experience of the Chinese people—particularly the peasantry."[61]

Luo Tianyu wanted to do more than just return the institute to its utilitarian beginnings. He made a strong case for prioritizing different kinds of scientific work during wartime, but he also argued that scientists' thinking should be raised to the highest political level, and that the strength of their Party spirit (dang xing) and class spirit should be tested. He wanted the science education curriculum guided solely by the realities of the war and Party policies. All else he trivialized: Why, he asked, should time be wasted studying such irrelevancies as the theory of the electron or the molecule?

Luo cited his own pedagogy as the appropriate one for the institute. He had done away with textbooks and with classrooms; and his students no longer studied conventional "botany," because local Chinese flora were never in the books. Instead, they learned their botany in the field the way they learned their agriculture at the Guanghua Experimental Farm.[62] This exemplified Luo's understanding of the Rectification campaign's goals to unify theory and practice, education and research, research and production.

He wanted this reflected in the structure of the institute, which he proposed to consolidate into two units: his biology-agriculture unit (which would subsume the arboretum, the experimental farm, and experimental biology) and a physical science unit (subsuming physics, chemistry, and engineering). These units would be organized not around specialized disciplinary criteria, but rather around practical and pressing issues. And, Luo further argued, since his biology department had already distinguished itself in its practical accomplishments, it would receive a healthy increase in funding and act as a model for the other unit's reform.[63]

Institute faculty and students, even some from Luo's own department, strongly rejected his ideas and his reorganization scheme. Luo was accused of having a narrow sense of reality like that of a skilled technician; and he was criticized for encouraging the inflexible mind-set of those farmers or engineers who are bound to a "that's the way I learned it" tradition. All this is the result, it was argued, of rejecting the overview and long-range thinking made possible by the use of theory.[64]

The victory of Luo's colleagues was short lived. Within a year, influential forces came to his support, giving Luo's work public praise, and echoing his arguments, which by then had become indistinguishable from the position of the Rectification campaign. In April 1943, the institute was absorbed into Yan'an University, where Luo Tianyu became head of the Department of Agriculture. The new department was devoted very much to the kind of curriculum and activities earlier practiced and advocated by Luo, including a full survey of the region's flora, reforestation, growing sugar beets, and cultivating medicinal plants. The experimental farm remained active as well.[65] From the time that Luo entered the faculty of Yan'an University until the end of 1948, reliable foreign observers noted that in form and content all education in the Communist "base areas" closely resembled what Luo advocated and practiced.[66]

## Evaluating Luo Tianyu's Career

In summary, the basic elements of Luo's Yan'an career are clear, but the sparseness of data make it impossible to reach a definitive judgment about his motivation and his relationship to the "leftward" movement of CCP science policy. The historian and archivist Gong Yuzhih suggests a number of feasible interpretations. Luo may have been using the rising tide of "left Maoism" (that is, the extremes of the Rectification campaign) to stage his minicoup at the institute; or perhaps he was just more sensitively attuned to the course of Mao's thought and the Rectification campaign. Then again, Gong suggests that it is entirely possible that the Rectification campaign caught up with a position that Luo had held for some time.[67]

Whatever the case might be, Luo's record in Yan'an is sufficient explanation for his important postwar administrative appointments: On leaving Yan'an, he was made dean of the North China Agriculture University (Huabei nongyeh daxue), near Beijing. In 1949 he was appointed dean of the new National Beijing Agriculture University (Guoli Beijing nongyeh daxue), keystone of the new agriculture education system. The more interesting issue is Luo Tianyu's embrace of Lysenkoism. Using Gong Yuzhi's apt questions, it can be reasonably asked if Luo was merely riding the pro-Soviet tide so as to enhance his own power beyond his new academic administrative job. Or was it rather that Luo was significantly ahead of the pack in anticipating the force and direction of the pro-Soviet policy and the inordinate role that Lysenkoism would play in it? Or again, did the Party's policy on biological science in China "catch up" to Luo's position when it adopted Lysenkoist biology? The following chapters explore these issues in detail.

# Notes

1. Wartime administration of science: See Needham, *Science Outpost*, 21 for contin-

uation of Shen as head of National Agricultural Research Bureau; also see William Kirby, *Germany and Republican China* (Stanford, Calif.: Stanford University Press, 1984) for National Resources Commission and centralized industrial plan.

2. Interview with C. N. Yang, co-conducted with John Israel, New York 1980; and interview with Li Jingzhun. During the war, Yang, a future Nobel Laureate, was an undergraduate physics student at Lianda, in Kunming. Li taught agricultural science subjects in Chongqing. Needham's reports in *Science Outpost* provide an extensive picture of this mix of basic and applied science throughout unoccupied China.

3. For an overview of this campaign see Liu Xinming, "Guanyu 'Zhongguo kexuehua yundong'" (The China scientization movement), *Zhongguo kezhi shiliao* (China historical materials of science and technology), 1987, no. 2: 21-29.

4. *Kexue di zhongguo* was published in Nanjing: vol. 1.1 (January 1933) to 10.3 (August 1937).

5. For instance, see Chen Lifu's editorial introduction to the journal's first issue, "Zhongguo kexuehua yundong," January 1, 1933; and his "Qian-hou zuo-you tan" (Discussing past-future, right-left), May 15, 1933: 6-9. Also see Zhang Hozhi, "Jiqi jiuguo" (Machines to rescue the nation), June 15, 1933: 3-12.

6. Communist responses to Chen Lifu, et al., were led by Ai Siqi in such pieces as "Kexue di guofang dongyuan" (Sciences national defense mobilization) in Ai Siqi, *Shijian yu lilun* (Time and history) (Chonqing: n.p., 1939), 245-50; and the 1939 "Shehui zhui geming yu zhishifenzi" (Socialist revolution and intellectuals) in Ai Siqi, *Lun Zhungguo di shuxing* (On China's exceptionalism and other essays) (n.p., 1946), 46-59. Also Ai Siqi and Wu Liping, *Kexue lishiguan* (Shanghai: n.p., 1939), 210-11.

7. Zhang Qiyun, "Kexue yu kexuehua" (Science and scientization), *Kexueh de Zhongguo*, January 1, 1933: 4-11. The author was a historical geographer on the faculty of National Central University.

8. Translation project: See *Guoli bianyiguan gongzuo gaikuang* (Overview of the work of the National Translation Bureau) (Chongqing, 1940). And Needham, *Chinese Science* (London: Pilot Press, 1945), 26-27. This is a collection of photographs documenting Needham's tour of China's science community in 1944.

9. *Acta Brevia Sinensia* was published out of Chongqing in eight mimeographed issues, January 1943 to December 1944. Issues ranged between fifteen and thirty-seven pages. It was edited by the Natural Science Society of China, a group of scientists and engineers themselves mainly concerned with industrial applications of science. After the war, this society published the bimonthly journal *Science and Technology in China*, based in National Central University in Nanjing.

*China Science Service* (originally titled *Science Service*) was published monthly from March 1944 to August 1945 in Chongqing. This journal was a spin-off of the International Cultural Service of China, the United States government agency devoted to providing Chinese scholars—mainly scientists—with microfilms of technical journals. (For this microfilm project and other related American wartime efforts, see Wilma Fairbank, *Americas Cultural Experiment in China*, publication no. 8839 (Washington, D.C.: Department of State, 1976).

10. See *Shengwu xueke jiaoxue shicha baogao* (Investigative report on biology education curricula in the schools), Yu Han, comp. (Chengdu, June 1941), 78 pages. No follow-up to this study could be found. There may have been similar studies for other fields of science.

11. Nutrition conference: See Tang Peisong, "Biology in War-Time China," *Nature*, 3897 (July 8, 1944): 43-46. Tang goes on to note a proliferation of organizations and projects devoted to nutrition studies as a result of the conference.

12. Nineteen forty-three Conference: reported in *Acta Brevia Sinensia* (April 1944): 9. In all of the wartime documents I consulted, the term "unoccupied China" appears to refer only to Guomindang areas, exclusive of Communist areas.

13. Councils: Another council fell under the direction of Hu Xuhua, director of the Sanmin zhuyi Youth Corps and former president of Hunan University. See *Acta Brevia Sinensia* (April 1944): 9.

14. For a comprehensive history of this consortium see John Israel, *Lianda: A Chinese University in War and Revolution* (Stanford, Calif.: Stanford University Press, 1998).

15. Needham, *Science Outpost*, passim. Interview, Li Jingzhun.

16. Needham, *Science Outpost*, Articles 3 and 5. Interview, Li Jingzhun.

17. Journals: (1) *Nongbao* began publication in 1935, out of Nanjing University. In 1938, after the university's arrival in Sichuan, it continued publication under government auspices. (2) *Kexue nongyeh* (Scientific Agriculture) was a new quarterly published out of the Ministry of Agriculture and Forestry, Sichuan Province, from September 1943. Its editor was Ling Li (b. 1910), a 1937 Ph.D. from University of Minnesota. Among its distinguished editorial board members were geneticists Zhao Lianfang (b. 1894), a rice specialist who received his Ph.D from Wisconsin in 1927, and Feng Zhefang (b. 1899), a cotton specialist who received his Cornell Ph.D. in 1935.

Western journals: Thanks to the wartime work of Joseph Needham and the British Council and to similar American efforts, research articles by Chinese scientists were transported overseas and published in mainstream journals. Needham lists 139 such papers in *Science Outpost*, 287-94.

18. China Foundation, *Report*, 1927-1928, 1937-1938. And Needham, *Science Outpost*, 115; Interview, Li Jingzhun. In 1949, Li Xianwen moved to Taiwan where he had a long, successful research career devoted to the development of Taiwan's agriculture.

19. Li's mentor was Feng Zhaochuan (Ph.D., Cornell 1922). He was responsible for the most sophisticated publications on genetics in the early 1920s, including the first extensive glossary of genetics terminology.

During 1944-1948, Li Jingxiong became a recognized expert in maize and radiation genetics during his graduate work in the United States. Along the way to a doctorate at Cornell under L. F. Randolph, he studied with L. J. Stadler at University of Missouri, and C. R. Burnham at Minnesota. During this period, Li was also invited to Caltech twice to work with E. G. Anderson to assist with the Bikini atomic bomb radiation study. Most of his post-1949 career in China was based at the CAS Institute of Agriculture and Forestry, and National Beijing Agricultural University—agricultural science's most important institutions. He quickly became one of China's most prominent and enduring experts on maize. A picture of him, standing in a field of corn, appeared on the cover of the June 24, 1985, *Beijing Review*. It indicates that he had recently won the first class prize for hybrid corn development. ZXN, 464-71; Harlan, "Plant Breeding and Genetics," in, *Sciences in Modern China*, ed. Leo Orleans (Stanford, Calif.: Stanford University Press, 1980), 306.

20. Bao Wenkui received a Lend-Lease Fellowship to pursue his doctorate in biochemistry and genetics at Caltech, 1947-1950. In the late 1950s, he, like Li Jingxiong, was appointed to the CAS Institute of Agriculture and Forestry and to the faculty of Beijing Agricultural University, where he spent the rest of his career. ZXN, 505-18. See biography by Yu Youlin in ZBT 3 (1979): 85-93. And see next chapter for Bao's post-1949 career.

21. For example: Li, Bao, and Li, "Interspecific Crosses in Setaria: 2. Cytological Studies of Interspecific Hybrids," *Journal of Heredity* (October 1942): 351-55; Li and Li, "On the Inheritance of Pentaploid Wheat Hybrids, a Critique," *Kexue nongyeh*, (Septem-

ber 1943); Li and Li, "Supernumerary Chromosomes in Pearl Millet," *Kexue nongyeh*, (December 1943): 139-42; Li, Li, and Bao, "Cytological and Genetical Studies of the Interspecific Cross, Setaria Italica S. Viridis," *Kexue nongyeh*, (June 1944): 229-48.

22. *Genetics and the Origin of Species*, 207, 210. Also see Dobzhansky's continued enthusiasm for the subject in *Genetics and the Evolutionary Process*, 382-83. For earlier awareness of the subject, see for example H. H. Newman's *Evolution, Genetics, and Eugenics*, 515-16.

23. For example, Bao, Li, and Li, "Studies on the Inheritance of Dwarfness in Common Wheat," *Kexue nongyeh*, (September 1943): 1-12; Li and Li, "Cytological Studies of a Haploid Wheat Plant," *Kexue nongyeh*, (March 1944): 183-89.

24. For example, the work of H. Kihara and S. Matsumura is closely evaluated in "On the Inheritance of Pentaploid Wheat Hybrids," *Kexue nongyeh*, 1943, no. 3.

25. See Needham's *Science Outpost*, and his *Chinese Science*.

26. For overview of natural science at the three-university consortium, see Israel, *Lianda*, 203-24.

27. Tang, *Green Thraldom: Essays of a Chinese Biologist*, with an introduction by Joseph Needham (London: George Unwin and Allen, 1949), 73-74.

28. See next chapter for "Lysenkoism" in China.

29. Because Zunyi lacked sufficient housing for the science departments, they had to establish themselves in Meitan, a village seventy-five kilometers to the east. See description of Meitan in Needham, *Science Outpost*, 206-07.

30. Tan to M. C. Balfour, March 1, 1940 (RAC: "University of Chekiang Genetics").

31. Interview, Tan.

32. Tan to Rockfeller Foundation, July 10, 1939 (RAC: "University of Chekiang Genetics").

33. Tan to Warren Weaver, May 1, 1946 (RAC: "University of Chekiang Genetics," 2).

34. Bei Shizhang (b. 1903): 1928 Ph.D. from University of Tubingen; chair of Zhejiangs biology department, 1933-1949; director CAS Institute of Experimental Biology, 1950-1954; founding director of CAS Institute of Biophysics, 1959-. Lo Zonglo (b. 1898): 1931 Ph.D. Hokkaido Imperial University; director of CAS Institute of Entymology, 1951; director CAS/Shanghai Institute of Plant Physiology, 1957.

35. Tan to Warren Weaver, Director, Division of Natural Sciences, Rockefeller Foundation, May 1, 1946 (RAC: "University of Chekiang Genetics"). Five students: Sheng Tsu-chia (micro-biogenetics), Ph.D. Columbia, 1951; Sze Li-chi (plant cytogenetics) Ph.D. Columbia, 1951; Hsu Tao-chueh (Drosophila population genetics), Ph.D. Texas, 1951; Liu Tse-tung, who began at Berkeley, got his Ph.D. in human genetics at Michigan, Ann Arbor, 1953; Ms. Fung Sui-tong Chan (Drosophila genetics), who began at Rochester, got her Ph.D. at Iowa State, 1953. (See CASS: 333; Tung-li Yuan, *Guide to Doctoral Dissertations*. And Interview, Tan.) Name transcriptions are those used on dissertations.

36. Research reports: See 1941-1943 abstracts of Tan's genetic studies in RAC: "University of Chekiang Genetics"; and Tan, "A Report of the Laboratory of Genetics and Cytology, Biological Institute, National University of Chekiang, Meitan Kweichow, 1942-1944," *Acta Brevia Sinensia* (December 1944): 5-22.

37. "Note: FBH," April 14, 1944; Correspondence: Tan to L. C. Dunn (Zoology Department, Columbia University) June 1, 1944; "Grant in Aid to Columbia University" (Oct. 13, 1944); Tan to F. B. Hanson (Rockefeller Foundation), April 14, 1945. All in RAC: "University of Chekiang Genetics."

38. Tan Chia-chen, "Genetics of Sexual Isolation between Drosophila Pseudoobscura and Drosophila Persimilis," *Genetics* (November 1946): 558-73.

39. Tan Chia-chen, "Inheritance of the Elytral Color Patterns of Harmonia axyridis and a New Phenomenon of Mendelian Dominance," *Chinese Journal of Experimental Biology* (Chungking), 1942, no. 2; "Mosaic Dominance in the Inheritance of Color Patterns in the Lady-Bird Beetle, Harmonia Axyridis," *Genetics* (March 1946): 195-210. And for Dobzhansky's citation of this work, see his *Genetics of the Evolutionary Process*, (New York: Columbia University Press, 1970), 270. A mosaic occurs when only one strand of a DNA molecule is mutated. Upon replication, two DNA molecules will be produced—one mutant and one nonmutant; and thus the mosaic organism has tissues that are of two or more genetically different kinds.

40. Memo: "Dr. C. C. Tan, Columbia University," July 18, 1946 (RAC: "University of Chekiang Genetics").

41. Tan to Warren Weaver, May 1, 1946 (RAC: "University of Chekiang Genetics").

42. Chin Wang, Acting President Chekiang University, to Warren Weaver, April 20, 1947 (RAC: "University Chekiang Genetics"). Li Huilin (b. 1911): Arnold Arboretum Harvard Ph.D., 1942. Gao Shangyin (Harry Zanyin Gaw) (b. 1908): Yale Ph.D. in physiology 1935.

43. Roger Ennus to Marshal Chadwell, Associate Director, Rockefeller Foundation, May 23, 1947, and May 24, 1947; and Warren Weaver to C. C. Tan, January 8, 1948. (All in RAC: "University of Chekiang Genetics").

44. Tan to Warren Weaver, May 24, 1948; L. C. Dunn to Warren Weaver June 3, 1948; Notes of Tan Interview with Warren Weaver, New York, October 22, 1948. (All in RAC: "University of Chekiang Genetics").

45. (Author's) Interview, Tan Jiazhen.

46. Interview: Warren Weaver with Dr. C. C. Tan, October 22, 1948 (RAC: "University of Chekiang Genetics").

47. Interview: Warren Weaver with Dr. C. C. Tan, October 22, 1948.

48. Weaver to Tan, November 15, 1948; and comment on letter by CBF, November 23, 1948. (Both in RAC: "University of Chekiang Genetics.")

49. Biographical sources for Luo Tianyu are almost nonexistent. The above outline is for the most part drawn from an obituary for him, circulated anonymously in Beijing on the customary single sheet of paper and dated August 15, 1984. He died on July 15. Some of the information in the obituary was supplemented or corroborated in interviews with Li Jingzhun, Pittsburgh 1985; Tang Peisong, Beijing 1986; and Wu Zhongxian, Beijing 1986.

50. This follows James Reardon-Anderson's translations of organization names as well as his fine exposition of the development of Yan'an science and the debates on science policy in *The Study of Change*, chapter 14.

51. *Study of Change*, 342.

52. Interviews: Tang Peisong; Wu Zhongxian. Later, Luo is reported to have helped another group of peasants build a granular white sugar processing facility, using sugar cane (JFRB, March 27, 1946).

53. KRZZ, 1: 148; and Peng Erming, "Yan'an zeran kexue yuan—wodi mujiao" (The National Science Research Institute, my alma mater), in KRZZ, 3: 332-43.

54. See the column "Kexue yuandi" (Science garden), JFRB, 1942, no. 8-12, and in KRZZ, 5: 61-77.

55. Luo Tianyu, "How to select land locations for public farms and military farms (tuntian)," JFRB, January 3, 1943.

56. See letter from Li Fuzhun (founder of the National Science Research Institute) regarding Luo's forestry work, dated September 22, 1940 (KRZZ, 2: 85). And see also recollection of his forestry work in Peng Erming, "Yan'an zeran kexue yuan"; and KRZZ, 5: 194.

57. Luo Tianyu, "Bianqu di nongyeh qihou yu hulin zaolin yundong" (The weather for agriculture on the border, and the "Protect forests-create forests" movement), JFRB, November 13, 1945.

58. Luo Tianyu, "Fang huo-shao shan dahai" (Against the dangerous burning-to-clear-land), JFRB, May 26, 1946.

59. Unsigned articles, "Huanying kexue ishu renzai" (Welcome talent in science) JFRB, June 10, 1941, and KRZZ, 1: 58-61; and "Tichang zeran kexue" (Promote natural science), JFRB, June 12, 1941, and KRZZ, 1: 62-65.

60. The resolution was aimed specifically at the Yan'an Cadre School, not at science education, per se. See "Guan yu Yan'an ganbu xuexiao di queding" (Resolution of the Central Committee of the CCP on the Yan'an Cadre School), in Boyd Compton, trans., *Mao's China: Party Reform Documents 1942-44* (Seattle, Wash.: University of Washington Press, 1952), 74-79.

61. *Study of Change*, 353.

62. Luo, "Du 'Guan yu Yan'an ganbu xuexiao di queding'" (On reading the resolution of the Central Committee), JFRB, "Xuexi" column no. 15, July 23, 1942. More fully summarized in *Study of Change*, 353-54.

63. Luo, "Du 'Guan yu Yan'an ganbu.'"

64. Kang Di, "Dui Luo Tianyu tongzhi 'Du Guan yu Yan'an ganbu xuexiao di queding'" (Regarding comrade Luo's article 'On reading the resolution of the central committee'), "Xuexi" column no. 21, JFRB, September 25, 1942. And *Study of Change*, 354-55. A 1990 study of Lysenkoism in China claims to see a precursor to the genetics controversy in some of these 1942-1944 debates between Luo Tianyu and his colleagues. I see no evidence that genetics was at any time the substance of these debates. (See Jiang Shihe, "Michulin xueshuo zai zhongguo 1949-1956: Sulian di yingxiang" [Michurin theory in China: Soviet influence], ZBT, 1990, no. 1: 18-24).

65. *Study of Change*, 358-59; and Israel Epstein, "Scientific Research and Education in the Border Region," dated January 1945 and reprinted in Needham, *Science Outpost*, 201. There is some confusion with the latter. While it lists a curriculum that sounds exactly like Luo Tianyu's work, this article does not cite him by name. It refers only to the agriculture department's head as "a former professor of Chekiang University." I have seen no suggestion that Luo was ever associated with that school.

66. See Michael Lindsay, *Notes on Educational Problems in Communist China, 1941-1947* (New York: Institute of Pacific Relations, 1950). The recurrent theme in this collection of on-site observations is the respective relationships between "theory and practice" and the short- and long-term goals of education.

67. Gong Yuzhi, "Zhongguo Gongzhandang di kexue zhengce di lishi fazhan" (Historical development of CCP's science policy), ZBT, 1980, no. 6: 6-11. This article is a thoughtful reconstruction and analysis of the Yan'an science policy debates and Luo Tianyu's role in them.

# Part II

## Mao's China, 1949-1976

**4**

# Learning from the Soviet Union: Lysenkoism and the Suppression of Genetics, 1949-1951

That Rockefeller Foundation official could not have been less prescient in his November 1948 evaluation of Tan Jiazhen's prospects under a Communist regime. In fact, the Chinese Communists took very seriously what the official referred to as "Marxian theory as applied to genetics"—that is, Lysenkoism. Within a year after Tan returned to China the Communists had triumphed and Lysenkoist biology had begun to be proselytized under the aegis of the CCP's comprehensive pro-Soviet policy. Passionate Chinese advocates of Lysenkoist biology, aided by its Soviet evangelists, rapidly established it as the exclusive approach to biology in China, and banished classical genetics from teaching, research, and publication. In 1952, the CCP formally espoused Lysenkoist biology and banned Western biology related to genetics. The ban was quite effective for the next five years, until it was revoked as a result of the CCP's liberalizing "Hundred Flowers, Hundred Schools" policy. From that point, until the Cultural Revolution in the late 1960s, genetics and experimental biology in general enjoyed a renaissance, led largely by scientists who were introduced in earlier chapters of this study.

From the time of its arrival in China, Lysenkoism was dictatorial and dogmatic. It was both a reflection and source of divisive tensions between the CCP and the Chinese intellectual community; it politicized the science community; and it was an early source of a growing displeasure with the role of the Soviet Union in China. The following discussion of Lysenkoism in China looks at its institutional and organizational bases, and examines its effects on scientific education, research, and publication. Throughout, it notes the respective roles of the CCP, Soviet experts, and Chinese scientists.

## What Is Lysenkoism?[1]

In the *Communist Manifesto*, Marx encouraged his readers to abandon the philosophy that saw itself merely as a means of understanding the world for that philosophy which saw itself as a means of changing the world. Echoing Marx's comments on philosophy, Lysenkoist biology relentlessly reminded its new audience in China that the goal of biology should be to change and control nature, not merely to understand it. Lysenkoist biology was premised on a trans-formist belief that all of organic nature was infinitely malleable and subject to human manipulation. At the heart of this belief were two principles that clashed most violently with classical genetics.[2]

The first principle was that the effective source of all organic change and development—including evolution and new species—was the environment, that is, forces outside of organisms, not within them (like chromosomes or genes). Second was the corollary that humans could fundamentally reshape organisms (especially food plants) to their own benefit solely through the manipulation of the organism's environment. These Lysenkoist principles were necessarily extended to the realm of heredity, which was defined as "the effect of the concentration of the action of environmental conditions assimilated by the organism in a series of receding generations."[3] Heredity, in this construction, is therefore quite simply "internalized environmental conditions."[4] Lysenkoism asserts that acquired characters can be inherited, and so, cumulative changes in an organism caused by the environment are the substance of its "hereditary nature" and the legacy its progeny will inherit.

Lysenkoism was built on claims that, following these principles, food plants could be improved so that they would grow in previously hostile environments, or grow more abundantly in native settings. Lysenkoism's self-styled "agrobiology" consisted of a set of transformist techniques to exploit what it called the phasic or stagelike development of plants. By manipulating the temperature or available light at the appropriate time of a plant's development, it could be changed in a desired direction, and even changed in its "heredity nature." Winter-ripening wheat could be changed into spring-ripening, and vice versa. Particular characters of a food grain could be changed so that those originally requiring a moderate climate could be made to grow in harsh northern climates.

T. D. Lysenko (1898-1976), who gave these ideas his name, was trained in the provinces as an agronomist during the 1920s. Throughout the 1930s, he was a most ambitious careerist, climbing with ruthless energy and Machiavellian acumen to the pinnacle of the Soviet science bureaucracy. In 1938, with the support of Stalin, Lysenko became the president of the Lenin Academy of Agricultural Science. A decade later, at the infamous July-August 1948 meeting of that academy, Lysenko and his followers completed the hegemony of their approach to biology, banning classical genetics altogether.

By this time, Lysenko had expanded his basic set of biological ideas in two directions. He took guidance respectively from I. I. Prezent (1902-1970?), an influential ideologue with pretensions of expertise in dialectical materialism, and I. V. Michurin (1855-1935), a homegrown horticulturist iconized by Lenin. Their respective roles warrant a brief summary here because the "Lysenkoism" that was brought to China in 1949 was an amalgam of their ideas and those of T. D. Lysenko.

Michurin was valuable to Lysenko for his nativist and populist image as well as for his alleged role as an earlier practitioner of Lysenko's biological principles. Working with the fruits and vegetables of his locale, Michurin had reputedly bred scores of new varieties that were exceptionally hardy and prolific. His much-quoted motto was co-opted by Lysenko to set the tone for the new biology: "Don't wait for favors from nature; we must triumph over nature." Lysenko went so far as to designate Michurin as the true successor to Darwin, based on Michurin's understanding of the environmental sources of evolution and on his aggressive intervention in the evolutionary process by way of creating (supposed) new species with his breeding techniques. Never at a loss for a catchy label, Lysenko designated Michurin's (and his own) school of evolutionary thought "Creative Darwinism."

Michurin used intuitive, trial-and-error techniques that he said were successful because of his intimate familiarity with his local environment and his personal, hands-on involvement in the selection and breeding process. Moreover, Michurin pooh-poohed formal agricultural science taught in big cities and in universities where no one got their hands dirty, no one was involved directly in production, and everyone was caught up in fancy theories, mathematical techniques, and statistics that bore no relationship to reality. Lysenko approvingly echoed Michurin's anti-intellectualism and mockery of establishment biology, and he used Michurin to legitimize his own famous lack of experimental rigor. By the time Lysenkoism was introduced to China, Michurin had become its patron saint and chief exemplar. Lysenko and his apostles introduced their work to China under the name of "Michurinism" (Mi-chu-lin-zhuyi), a name that stuck throughout the entire period of Lysenkoist dominance there.

I. I. Prezent was a graduate of Leningrad State University's Faculty of Social Sciences in 1926, and soon specialized in the philosophy of the life sciences. He seems to have anticipated many of Lysenko's fundamental attitudes toward science in general and biology in particular through his devastating attacks on the Russian ecology movement in the early 1930s. Though he began his

career as a partisan of Mendelian and Morganist theories of heredity and speciation, he had reversed that position by the time he established a power base in a
variety of academic positions. He aided Lysenko's rise to dominance in Soviet
biology and provided him with the trappings of ideological legitimacy—a dialectical materialist and properly socialist rationale for attacking and rejecting
classical genetics. Lysenko took from Prezent the argument that science inevitably has a class nature: If Marxism is correct, then how could science developed
within bourgeois-capitalist society be appropriate for a revolutionary socialist
one? Science produced by and practiced in a bourgeois-capitalist society could
only be idealist, formalistic, metaphysical. Only science that evolved out of a
socialist society, like Lysenko's biology, was valid.[5]

It will be seen in detail below that this class approach to science judged
classical genetics with particular harshness. Geneticists were decried for the
isolation of their laboratories from nature and production. Their concept of the
gene was condemned as a mere figment of bourgeois idealism. Lysenkoism saw
hostility to dialectics and to transformist principles where classical geneticists
claimed the basic material of heredity to have a fixed structure, function, and
location. Lysenko summarized all of these ideas in his July 31, 1948, opening
speech to a session of the Lenin Academy of Agricultural Sciences. This speech
became the locus classicus for the teachings of Chinese Lysenkoites and their
Russian mentors. It was translated many times into Chinese and was used endlessly as a template for explanations of Michurin biology and for arguments
against classical genetics and against Darwin's approach to evolution.

It should not go without saying here that Lysenko's speech (along with others delivered at the academy session) was in a very real way also an indictment
of biologists who practiced the "old biology" and/or criticized Lysenko's new
biology. Some of those thus indicted had already suffered ill consequences; others would meet severe punishments ranging from the loss of positions at universities and the academy, to jail and death. Stalin had not only approved of
Lysenko's hegemony and edited his speech, he obviously also served Lysenko
as a political rolemodel.[6]

Lysenko's speech was more than an indictment of Soviet transgressors; it
was a grandiose declaration of war against all practitioners of classical genetics
from Weismann, to Mendel, to Morgan. And just as Lenin's Comintern had
brought its design for revolution to China in the 1920s, Lysenko determined to
bring his biological revolution there a generation later.

## Lysenkoism in China

Beginning in the early 1930s, Chinese biologists were familiar with Lysenko's
work before it was incorporated into a full-blown ideology and became canonical. In 1934, for example, Nanjing University's prominent agricultural science
journal published a long, careful review of "recent Russian developments of
'vernalization.'" The latter term was used by Lysenko in particular for a tech-

nique to alter the time required for a seed to sprout by subjecting it to controlled periods of light and darkness or cold and heat. German, American, and Soviet research is cited in this article, along with Lysenko's work at the Odessa Plant Breeding Institute, his base at the time. The review summarizes without comment Lysenko's theoretical explanations for his research. There is not a hint of controversy suggested.[7]

By far the most sophisticated pre-1949 Chinese encounter with Lysenko's work took place during the war, in the plant physiology laboratory of Tang Pei-song. Tang and other well-trained physiologists successfully re-created some of Lysenko's vernalization experiments whose goal, in Tang's words, was "to hasten the maturation of certain plants." During unoccupied China's wartime quest for quick, cheap, and easy methods to increase food plant yields, vernalization seemed promising. Going well beyond Lysenko's limited use of cold treatment of seeds, Tang's laboratory successfully shortened time to maturation by subjecting seeds to high temperature ultrahigh frequency electromagnetic waves as well as certain chemicals. It also developed an analysis much more complex than Lysenko's of phasic development and of the most effective combinations of treatments for the experimental plants. Tang, however, expressed no familiarity with Lysenko's broader theoretical principles.[8]

In June of 1948, Lysenko's vernalization work was cited prominently in a bibliographic review published in the research journal of the Guomindang Ministry of Agriculture. Without editorial comment, Lysenko's work and general approach are discussed along with those of biologists from all over the world. The motivation for the essay clearly is to chart the progress toward increasing wheat production by enabling earlier planting or hardiness in colder climates. The essay concludes with a caveat: Until a firm grasp of the actual genotypes of major wheat varieties is available, experimental data from around the world will remain confused by names (like "winter wheat" or "spring wheat") that scientists like Lysenko had incautiously given to their experimental plants.[9]

These samples illustrate the limited, technical interest shown Lysenko's work in China before 1949. For lack of information or interest, neither Lysenko's career moves nor his wider biological ideas were reported until well after the August 1948 academy meeting. Thus, Chinese biologists and the Chinese science community in general were virtually unprepared to deal with the militant promotion of Michurin biology by Chinese and then Russian Lysenkoites, within months of the Chinese Communist formal accession to power on October 1, 1949.

## Lysenkoism's Institutional Base in China

Chinese Lysenkoites used the new agricultural schools of northeast China as their earliest base of operations. From January 1949, the CCP controlled the Beijing area, where the largest agriculture schools were located. The Party lost no time implementing an ambitious reorganization plan for a higher educational system which comprised an incoherent and often redundant multiplicity of schools and colleges created respectively by the early Republican-era state, by missionaries, and by the Guomindang government. Major agricultural schools and research institutes were very high on the CCP agenda, and by autumn of 1951 a network of these institutions had been established in urban centers from Beijing to Canton. Associated with the large schools and institutes was another network of small, local agriculture colleges, extension programs, and research stations. Formerly prominent agriculture schools, like the one at Nanjing University, were absorbed into this new system and lost their identities. This entire agricultural education and research structure was controlled until the mid-1950s by a new Ministry of Agriculture, quite sympathetic to Lysenkoism and active in its promotion.[10]

Early in 1949, the Ministry of Agriculture consolidated three schools in the Beijing area to create what was meant to be the country's largest and leading agricultural science institution, Beijing Agricultural University. The small and ailing agriculture colleges of Qinghua, Beijing, and North China Universities melded their faculty into a substantial, thoughtfully organized institution.[11]

Directly under the new school's dean were three program directors. These positions were initially filled with brilliant young biologists who had patiently waited throughout war years for such an opportunity. Li Jingzhun, who had been chair of Beijing University's Department of Agronomy, became codirector of the new school's agronomy program. Li, a 1940 Cornell Ph.D. in plant genetics, was a disciple of Shen Zonghang. He was also an outstanding biostatistician, and unusually competent with the statistics required to use Seawall Wright's population genetics. Tang Peisong became Li's codirector. He had been head of Qinghua's College of Agriculture, and his exceptional work in plant physiology has been noted above. To direct the new Department of Animal Science was Wu Zhongxian, who had held that position at Beijing University. Wu was a 1937 Edinburgh Ph.D. in animal genetics and an adept in biostatistics.[12] All three directors were deeply trained in classical genetics; all three were in China by choice; none was a CCP member nor ambitious to become one.

Over the course of 1949, the CCP's presence grew at the new agricultural university. Under the guidance of dozens of party cadre assigned to the school, small study groups ("cells") on Marxism and for the dissemination of CCP policies were required of all faculty. Even when faculty failed to participate, however, they were treated with respect and made to feel that the Party valued their work. Li Jingzhun recalls how this attitude was already in evidence in December of 1948, when Communist troops advancing on Beijing came to his college to help it pack up and move out of harm's way, until the Guomindang was driven out and the city secured.[13]

The Lysenkoites seemed to become a presence at the new Beijing Agricultural University "instantaneously" with the October 1, 1949, promulgation of the new People's Republic. Their arrival marked the end of CCP solicitude for the faculty and the beginning of a massive pro-Soviet campaign to "learn from the advanced experience of the Soviet Union." First, the dean of the school was replaced unceremoniously by a man in a Yan'an-style uniform (a "Mao suit") un-known by any of the faculty. This was Luo Tianyu, the CCP veteran whom we met earlier as the agriculture department chair at Yan'an University, where he advocated the application of "Rectification" reforms to science. Since the war, he had been the dean of the North China Agriculture College, one of the three consolidated schools. Luo would coordinate the development of Lysenkoism in China during the following two years. Next, the faculty were told that in addi-tion to their study cells, they must attend separate study groups devoted to Michurin biology.[14]

Luo Tianyu quickly made it clear in aloof public speeches and through his minions (but never through personal contact) that the biology faculty were ex-pected to denounce classical genetics and to embrace Michurin biology. Never-theless, Li Jingzhun refused to attend the Michurin study sessions, and openly tried to engage in debate with Michurin exponents. His students soon apprised him that his conversion was becoming a major issue in the study sessions, and that, in effect, his recalcitrance was denying Luo Tianyu a trophy convert and an exemplar for the rest of the school. Unlike typical CCP thought remolding, Luo's campaign did not require Li to denounce his family's relatively wealthy, bourgeois status nor his foreign education. His "Mendel-Morganism" was the only issue.[15]

By February of 1950, Li was forced to resign his codirectorship and was not permitted to teach his genetics or biometry classes. A number of colleagues and students warned him that if he did not relent and convert to Michurinism, Luo Tianyu appeared determined to punish him. Li and his close colleagues by now had heard how Lysenko dealt with his enemies; and they were also aware how the CCP was capable of dealing with dissident intellectuals. It was therefore arranged by friends to get Li, his wife, and child swiftly out of Beijing, then out of the country. For weeks they followed a complicated escape route, and on March 12, 1950, they safely crossed the border to Hong Kong. By the end of the year, Li's story had become a Cold War cause célèbre within the Western biol-ogy community, thanks to a letter of his published in the *Journal of Genetics*, entitled "Genetics Dies in China."[16] Geneticist and Nobel Laureate H. J. Muller personally took up Li's cause, found him an important position in the United States, and arranged for his immigration.[17]

Wu Zhongxian found it necessary to protect himself through more passive techniques. He did not express his dissent publicly, nor did he protest too loudly when he, along with his entire department, were signed up as members in the local branch of the newly established national "Michurin Society." It was fast becoming evident that it would be impolitic to remain outside of such organiza-tions. Consequently, permitting one's name to be listed and "wearing the cloak of Michurinism" became common subterfuges. Still, Wu's courses in bio-statistics were abolished, and he defensively shifted his work out of genetics and into animal nutrition.[18]

Tang Peisong, the third of the program directors, maintained stubborn resis-tance to Luo Tianyu's campaign until the winter of 1950, when he was subjected to a public thought remolding session. In the auditorium of the university he was accused before two thousand faculty and students of criticizing and resisting the Party's policy of learning from the Soviets. Many hundreds of faculty from other Beijing area schools sat outside the building where they had been required to listen to the proceedings on a loudspeaker system. When all was said and done, Tang suffered neither physically nor financially, his formal punishment being ostracism for the rest of the academic year. His classes were canceled and he was sent home. By this time, however, political turmoil on the campus was so rife that few classes were able to meet regularly.[19]

Tang's ostracism lasted only a few months, thanks to the intervention of the CCP. Higher up its hierarchy, a former student of Tang Peisong's famous father intervened on his behalf and determined that the university's Revolutionary Committee had gone too far. Shortly thereafter, Tang was invited to join the Chinese Academy of Sciences Institute of Biology. In his eyes, this was a CCP public relations tactic to mollify and win over prominent scientists. Nonetheless, he accepted the offer, if only to get away from the university, which (like many other schools) had become a political football and constant target for CCP campaigns.[20]

In just a short time then, Luo Tianyu had brought turmoil to his university, alienated some of its star faculty, and, only by force, converted the curriculum to Michurin biology. Elsewhere he displayed rather more entrepreneurial skill and savoir faire. For example, in late 1948, at the inception of four other major agricultural research organizations in the Beijing area, he had managed to pack them with zealous Michurinists. It was characteristic of these four that few of their faculty or researchers had graduate degrees, and that all had been trained in China or Japan (not Europe or North America). Their senior workers were largely CCP members who had been active since at least the beginning of the anti-Japanese war; and many junior workers had recently joined the Party. One of these organizations reported to the Chinese Academy of Sciences; the others to the Ministry of Agriculture.[21] The work of these four organizations along with Beijing Agricultural University was constantly cited in national publications to illustrate the accomplishments of Michurin biology.[22] The research of virtually all other plant biologists was ignored.

Another organization important to Luo Tianyu's campaign was the China Michurin Study Society, which was devoted to spreading Michurin biology to the masses. With Luo as its national president, the organization had a modest beginning in the suburbs of Beijing in the spring of 1949. It had its official inauguration at a big public ceremony in February of 1950. On the campus of the new Beijing Agricultural University, a large platform was erected and decorated with banners depicting Mao, Lysenko, and Michurin. Speakers included the society's coordinators, some scientists who were Party members, and hero workers and peasants. This was meant to illustrate how the Michurin Study Society was a new kind of organization for revolutionary China, wherein all social elements worked together and followed Mao's "mass line." The celebratory speeches had a lot to say about the amazing fruits and vegetables Chinese had already grown, using Michurin techniques.[23]

The Michurin Study Society went on to advertise itself and spread the good news about Michurin biology through its official journal.[24] On its masthead, it carried the following extract from the society's bylaws:

> This society is established to study the construction of agro-biological science and the correct way to approach its methods; and to direct workers, peasants, teachers, students, agricultural workers, etc. collectively to study and research the most advanced, most popular agro-biological science, and to popularize it amongst the masses along with the natural laws of agricultural production. Thereby, the masses [will be enabled] to realize sure control of increased production and to transcend the old confines of "depending on heaven for food."

The cover of the journal's September 1951 issue bears the standard iconic depiction of Michurin—gray-bearded, wearing a battered fedora, his skin weathered, his fingers gnarled from hard work. Below his portrait, one of his mottos: "Don't wait for favors from nature; we must triumph over nature." Inside, there are lists indicating the Michurin Society's expansion during its first year: fifty branch societies, all over China, including the far northwestern and southwestern provinces.[25] The journal's hundred pages of text are largely filled with translations from Russian reports on the results of using Michurin's and Lysenko's techniques with wheat, cotton, and livestock. There are also little introductions to various Russian agronomists, letters from new and happy members of the society, and reports on Michurinist curricula in secondary education. The tone is positive but there are also occasional denunciations of one or another aspect of the old biology.

Within the larger community of biological scientists, Luo Tianyu used every available venue to promote Michurin biology and to assert its hegemony. Sometimes he chose a standard, professional one, like the Beijing Biological Studies Society, which had a week-long conference for four hundred members in July 1950.[26] Subdisciplines of biology offered state-of-the-field panels with representatives of a dozen different institutions, including CAS institutes. Biochemistry received the most attention, and Tang Peisong (this was before his ordeal) chaired one of its sessions. Back to back with biochemistry, Luo Tianyu and his colleagues were given an entire day to introduce Michurin biology. His panel topics included "The Inheritance of Acquired Characters by Animals," "Experiments for Altering the Hereditary Nature of Plants," and "The Work and Accomplishments of Lysenko."[27]

Earlier that same year, in Nanjing, Michurin advocates used the lecture hall of the venerable Science Society of China to stage a more confrontational "symposium" on "Michurin Theory and Morgan Theory." Most of those on the speakers' program were already Michurin converts or contemplating conversion. Among the defenses of Michurin biology were remarks to the effect that "the Chinese cannot go by American characterizations of Lysenko or their description of the politicization of Soviet science," and "a Russian friend told me that all the talk about dismissals from the Soviet Academy of Sciences was untrue." Far from any technical discourse, speakers and audience noted the failure of Morganism, unlike Lysenkoism, to create a unity of head, hand, and labor, or theory and production. Lysenko, they enthused, had liberated us from the class-room.[28]

At this symposium, there was definitely an odd man out. Zhu Xi, the distinguished embryologist and student of evolutionary theory, ignored the above sloganizing and tore into Lysenko's notions of speciation and his evaluation of Darwin.[29] These were very substantial issues for which Lysenko soon would be relentlessly attacked by his enemies in the Soviet Union. At this little symposium, however, it was clear that Zhu Xi was outnumbered and that hardly anyone present cared to hear what he had to say, even though he did not dismiss Michurin biology out of hand. He had been educated in France and was comfortably steeped in the neo-Lamarckist tradition; but he was intractably opposed to the mixture of politics and science, and demanded the right to take issue with Lysenko's theories where he found them arbitrary and ill formulated. It is entirely possible that Zhu Xi had mistaken this event to be a real forum for discussion, when in reality it was a ritual for the Lysenkoite takeover of this prestigious organization.[30]

## Soviet-Chinese Tension over Lysenkoism

In 1950, Zhu Xi was a member of the CAS Experimental Biology Institute; and he became its director three years later. Apparently, his research field and this position put him out of the political reach of Luo Tianyu. Consequently, Zhu Xi continued his public criticism of Michurin biology and Lysenko for years to come. The CCP responded to Zhu Xi the way they did with other senior scientists, by letting the Soviets deal with him. For example, he led a 1953 zoologists' delegation sent to the Soviet Union to see how biology was done there. In spite of the Lysenkoite proselytizing, he reported that he had got the Soviets to agree with him that it was inappropriate to apply Michurin biology indiscriminately in China, where concrete conditions often differed from the Soviet Union.[31]

That same year, Zhu Xi became perhaps the earliest example of a Chinese biologist directly criticized by a Soviet Lysenkoite. In response to his recently published article on variation in the evolutionary process, Zhu was paid a visit by I. E. Glushchenko, who was lecturing and consulting in China.[32] Glushchenko, the director of the Soviet Academy of Sciences Institute of Genetics, took offense at the "Morganist" implications in the piece, which he had read in translation. Zhu Xi's published summary of their talk is laced with irony, and he makes the Soviet biologist appear patronizing, preachy, and trite. For example, Glushchenko faulted Zhu for failing to express an awareness of Michurin fundamentals, and warned him that no matter how good Chinese experimental technique might be, conclusions would continue to be lacking unless Chinese studied the dialectical materialism of Engels, Stalin, and Chairman Mao (especially Mao's essay "On Contradiction"). Zhu Xi concluded his report by expressing his gratitude for the attention and "friendly criticism" of his Soviet colleague.[33]

Zhu Xi's piece is among the earliest examples of an emerging minor genre—lightly veiled criticism of Soviet arrogance and the CCP's "learn from the Soviets" policy. Zhu had occasion to develop the genre further that same year when another piece of his research was attacked by Soviet biologists, this time in print.[34] It appears that Zhu Xi had determined that he could provoke Chinese resentment against the Soviet advisers and the CCP policy simply by publicizing details of his encounters, without explicit editorial comment.[35]

Other senior Chinese scientists were proselytized by Soviet Lysenkoites invited to tour China and help with the teaching of Michurin biology. The earliest visitor was V. N. Stoletov, Soviet minister of higher education and chair of the Department of Genetics at Moscow University. He is described in David Joravsky's classic study of Lysenkoism as "one of the original and most virulent Lysenkoites."[36] Stoletov was invited by Beijing Agricultural University to lecture there in December 1949.[37] Close on Stoletov's heels came N. I. Nuzhdin,[38] a specialist in genetics and evolutionary theory who not only stayed for a lengthy lecture tour of the country, he also made house calls.[39]

Tan Jiazhen reports that when Nuzhdin's tour took him south, he visited Zhejiang University, specifically to convince Tan of the merits of Michurin biology and to ask Tan to convert. At this time, Luo Tianyu's campaign was largely confined to the northeast and to agricultural science institutions. Zhejiang was therefore not as yet troubled by Lysenkoism, and Tan was teaching classical genetics without interference. He was therefore quite comfortable telling Nuzhdin politely to go peddle his papers elsewhere.[40] When Tan moved to his new position at Shanghai's Fudan University, in 1952, Lysenkoism had moved into the south and was there to greet him with a vengeance. Meanwhile, Nuzhdin's lectures were translated into Chinese, anthologized, and distributed widely, making it one of the earliest textbooks on Michurin biology. Many more Lysenkoite visitors would follow him, and many more such textbooks would be published.

## Disseminating Michurin Biology: Personnel

Michurinism, during its first years in China, created a strong demand for science technical writers and translators with some biology training. A small industry soon developed to disseminate it to every level of society and to help it become the standard for research and for every level of science education. Publication outlets were likewise needed for the flood of literature that followed; few publishers escaped co-optation. And so there were new employment opportunities, especially for those who could not compete with biologists already trained and working in education and research before 1949, and for those whose progress in biology was ended because they were involved in areas that were effectively banned. The two illustrations that follow are of quietly influential Michurinists who, unlike many of Luo Tianyu's close followers, had substantial education in biology. Their belief in Michurin biology aside, both Mi Jingjiu and Huang Zongzhen were opportunists who early on sensed which way the wind was blowing and set their sails accordingly.

## Mi Jingjiu

One of the most prolific translators and elegant writers working for Luo Tianyu was Mi Jingjiu, whose career provides some insight into what might be called professional Lysenkoism. Unlike others associated closely with Luo Tianyu and the northeast agricultural science institutions, Mi was not a researcher or teacher in the ordinary sense. Instead, from 1950 to 1956, he was an interpreter for visiting Russian biology experts, a translator of Russian language texts, and a publicist of Michurin biology through publications and public lectures. [41]

In 1950, Mi Jingjiu was a twenty-five-year-old graduate student in plant breeding and genetics at Beijing Agricultural University when he decided to drop out of the program and go to work for a visiting Soviet botanist and Darwin expert with whom he had just completed a course. The professor, who was on the faculty of a Soviet provincial normal college, had just come to China by way of a teaching stint in Hungary. She was so impressed with Mi's work that she asked the university to permit him to help her further develop the course and act as her interpreter. Mi had been studying Russian for over two years by then, and he had childhood familiarity with the language which his father had learned as part of his job as a state official in China's far northeast. The instructor had been relying on interpreters (one from Yugoslavia, one from Harbin) who knew little biology. Mi, on the other had, had received a solid science education since youth. At the university, as a student in the plant breeding and genetics program, he had recently studied genetics with a Cornell-educated professor, and population genetics with Li Jingzhun. For advanced plant physiology, he worked with Tang Peisong, who was his research thesis adviser.

Mi felt compelled to join the Lysenkoites because of the pressures being applied on all students and faculty at the university. Further, it was suddenly evident to him that a graduate degree based on an education laced with genetics was a ticket to nowhere. Mi was also seduced by Lysenko's and Michurin's claims about transforming wheat and potatoes, impressive claims that were for him inseparable from the awesome image of Soviet authority created by the CCP.

Mi Jingjiu's work in Michurin biology started with his translations of his Soviet professor's lectures for publication as a college textbook in 1952.[42] Then he began to translate Soviet Lysenkoite writings for the prestigious and powerful Science Press, publisher of major biology periodicals and monographs. At this same time, Mi was appointed to be a peripatetic lecturer in the Beijing area, where he introduced Michurin biology to science popularization organizations, and, more importantly, to groups of CCP cadre. Mi himself was not a member of the CCP. By 1953, Mi became a lecturer at Beijing Agricultural University and head of the school's Teaching and Research Group on Darwinism. Michurin biology's approach to Darwin will be detailed below in a separate discussion.

Under his own byline, Mi Jingjiu's lucid prose can be seen throughout the 1950s in publications aimed at the science community. He claims, however, that Luo Tianyu regularly signed his own name to translation and writing actually done by Mi. These pieces appeared in publications aimed at popular audiences. In any case, Mi enjoyed a busy career promoting Michurin biology until the late 1950s, when Lysenkoism was losing its party-supported authority.

*Huang Zongzhen*

Huang Zongzhen[43] influenced Michurin biology in the early 1950s through his position as an editor at Science Press (Kexue chuban she). He joined the press at the time it was taking over the publications of the Chinese Academy of Sciences and becoming the central clearinghouse for most of the country's science periodical literature.[44] Like Mi Jingjiu, Huang was trained in plant physiology, but his education began in Japan at Hokkaido Imperial University. Then, during the war, he continued his studies in southwest China at Zhejiang University, under the direction of Lo Zonglo.[45] Thanks to his mentor, Huang became a research assistant in plant physiology at the Academia Sinica's Institute of Experimental Biology in Chunqing. After the war, he continued an association with the institute when it moved back to Shanghai.

In 1949, Huang became a believer in Michurin biology and joined the CCP. It is not clear how these two actions may have influenced one another, but Huang is quite clear that the Party chose him for the Science Press because he was a Michurinist. Among his assignments was the editorial staff of *Biology* (Shengwuxue tongbao) and *Botany* (Zhiwuxue bao), serious journals with a large audience among scientists and lay readers. These publications were distributed nationally and were consulted for institutional and educational information as well as for the latest in research news.

Within his assigned publications, Huang had the authority himself to determine what was published, and he determined that a steady stream of Michurinist articles appear. As soon as this material began to appear, however, he began to be lobbied. First, he was visited discreetly by his former teacher Lo Zonglo, then by Tang Peisong and others—all asking him not to promote the expansion of Michurin biology. Huang was forced to concede that they sometimes correctly pointed out the "low level" of Michurinist scholarship; but he would not relent and, like Mi Jingjiu, continued to play an active role in the promotion of Michurin biology throughout the 1950s.

## Disseminating Michurin Biology: Media

The publicity/education campaign for Michurin biology largely owes its

130                                    *Chapter 4*

early and prominent success to taking advantage of the "Learn from the Soviets" and the "Science Popularization" campaigns. Technically, that success was also dependent on the CCP's generous funding of dozens of science periodicals, books, and pamphlets. This resulted in a broad spectrum of outlets for campaign-related literature and was quickly exploited by Luo Tianyu and other Michurinists.

In the pro-Soviet campaign, Michurin biology rarely occupied the position of a mere subset of Soviet scientific culture. Campaign literature commonly used Michurin biology to illustrate the character and prowess of Soviet science in general. Soviet scientists were inevitably and prominently represented by Michurin and Lysenko.[46] And of course in agricultural science journals or biology journals of any type, Soviet science was Michurin biology. Chinese Michurinists had got the tail to wag the dog.

The Science Press contributed materially to this identification of Soviet science with Michurinism, for example, by launching the journal *Soviet Agricultural Science* (Sulian nongyeh kexue) in February 1950. It was a monthly devoted completely to translations from Russian and was edited by Luo Tianyu and colleagues. In the late 1950s and 1960s, when Lysenkoism had lost its monopoly in China and much of its power in the Soviet Union, this journal continued to feature the most extreme Lysenkoites.[47]

Translations of Lysenkoite literature from the Soviet Union comprised the bulk of all Michurin biology publications in China. The Science Press regularly provided space in all of its growing number of periodicals for translations as well as for pieces originating in China. Other presses contributed their fair share of space as well, especially for translations of Lysenko's and Michurin's writing.[48] Lysenko's 1948 Lenin Academy speech, for example, was translated a number of times and circulated in many editions.[49] And there were Soviet Michurin biology textbooks in every field, at every educational level.[50] Biology in China was quickly swamped by this literature.


## Science Popularization Campaign

The CCP's science popularization campaign was motivated by beliefs that science must displace mass superstition and ignorance, and that the people must be able to use science to reach the CCP's ambitious goals for economic construction. By January 1950, every old and every new popular science periodical appears to have been brought into the campaign in lockstep. Their messages are always the same: Science is good because it is necessary for accomplishing what China needs to become prosperous and powerful; the Soviet Union is the world's leader in science and will help China become scientific; learn from the Soviet Union.

This campaign relied on popular science journals begun in the 1930s, under the aegis of the Guomindang, as well as on recently established ones. Childrens' magazines like *Science for the People* (Kexue dazong) and *Popular Science*

*Monthly* (Kexue huabao) had been around for years before they suddenly began featuring simple stories about how the Soviet people loved the Chinese people and were happy to share their science. Well-illustrated articles in these publications introduced topics like "Lysenko—Interesting Man, Interesting Story," and "The Michurin Line Corrects Genetics and Evolution."[51]

Mi Jingjiu and his colleagues were contributors to a new journal aimed at farmers, or perhaps at CCP cadre assigned to teach farmers new tricks. *People's Agriculture* (Dazhong nongyeh) was begun in Beijing in August of 1948. Its main concerns were cultivation techniques and plant varieties. By 1950, it was suffused with short, simply written pieces that broke Michurin biology down into digestible bits. There are short biographies and photos of Soviet agrobiologists; question-and-answer columns (e.g., "Can acquired characters be inherited?"); and outlines of Michurin theory. Occasionally there are translations of folksy wisdom from Michurin, and personable messages from Lysenko himself.[52]

If all this were not enough, the Chinese were additionally treated to a feature-length film biography of Michurin. It apparently opened in a Beijing cinema sometime in 1950[53] and then circulated among Michurin Study Societies throughout the country.[54] The film, simply entitled "Michurin," was in color, had a musical score by Shostakovitch, and for the Chinese audience, it was dubbed in Chinese.[55] According to an article in the *New York Times*, the film was released in the Soviet Union on January 2, 1949, after an unprecedented publicity campaign that was determined "to have the film seen by the maximum audience, especially in the rural areas." The *Times* notes in particular that the film contains one episode "in which an American businessman tries to get Michurin to go to the United States. The American offers to transport him there and set him up in business with all his plant stock. Michurin is shown indignantly refusing."[56]

## Disseminating Michurin Biology: Issues and Ideas

Michurin's scene with the American businessman is of course meant to connote his loyalty both to socialism and to his native land. In the early 1950s, the American businessman could additionally have provoked Chinese audiences to other sentiments as well. The Korean War was the source of a massive patriotism campaign galvanized by virulent anti-Americanism. At the same time, Chinese Michurinism had identified "Mendel-Morganism" with the United States not only because of the central role of American genetics, but also because virtually all of its Chinese practitioners had been educated there. Taking their cue perhaps from Lysenko's persecution of Soviet geneticists, Chinese Lysenkoites attacked "Mendel-Morganists" with the most extreme of accusations, labeling them not merely bourgeois, but disloyal, counterrevolutionary, and even racist and fascist. This labeling came to be known in China as "capping"—a term derived from the public political ritual in which an accused is forced to wear a dunce cap with his/her particular transgressions summarized on the cap in a word or phrase.

In this kind of activity and in most of their early campaign literature, the Chinese Lysenkoites were basically going by the book, and using material like Lysenko's 1948 speech as a template for what they said and how they said it. During the early 1950s, it would be no exaggeration to say that the Lysenkoite's campaign in China introduced virtually every topic and issue found in the template. Much of this was done of course through translations; but Chinese Lysenkoites were making choices about what to translate and how strongly to emphasize particular topics. It is important then to note a number of topics that the Chinese Lysenkoites initially chose to emphasize and some issues that were immediately raised in response to their campaign.

## Genetics and Eugenics

Between Lysenkoites' scientific claims and their "capping" of other scientists, it is impossible to determine which was more shocking and provocative to Chinese audiences. Ultimately it was capping that had to be reigned in and controlled by the CCP, if only because capping and the sanctions that were expected to go along with it were the prerogative of the Party. At the heart of this capping was a logic (or illogic) that concluded that geneticists were fascists, racists, and enemies of the working class because geneticists were directly or indirectly responsible for the development and criminal application of eugenics.

Here, the term eugenics is used in its negative meaning, that is, to remove "undesirable" elements of the population through control of reproduction or through their removal altogether. Well before the advent of Lysenkoism in China, Chinese Marxists in the 1930s had explored in detail the evils of eugenics and its manifestations among thinkers from Malthus to Galton. And a small but intense literature had developed among the few intellectuals who systematically advocated eugenics for China and those who opposed it. Chinese geneticists were virtually absent from this discourse, and it likewise lacked contributions from anyone who could reasonably be called a trained and practicing biologist.[57]

The first published discussion of eugenics in China's Lysenkoist era was from the pen of Zhou Jianren, who was introduced above as a pioneer science journalist in the 1920s. His 1948 polemic against eugenics was written when he joined the CCP (which up to this time he had kept at arm's length).[58] His book mentions neither Lysenko nor Morgan, nor is eugenics associated with genetics. While Darwin and Mendel are discussed it is only to show how their thought was co-opted and distorted by eugenicists, like the American and European Social Darwinists and Nazi Fascists who rationalized racial and class domination, and imperialism. In his prewar writing Zhou employed no Marxist terminology or analysis; this book uses Marxist class rhetoric to provide theoretical foundation for its main argument. Zhou gives genetics itself little attention here, and he suggests that "genetics and inheritance theory" are so complex and in such an unresolved state that it would be best to defer on their discussion. He does however venture a clear exposition of how eugenics arbitrarily attributes various "human flaws" (lack of intelligence, vagrancy, drunkenness, etc.) to hereditary nature, and he notes that the new science of genetics is not in a position to verify these eugenic assertions. Zhou's cool and balanced presentation here contrasts sharply with his overheated and gratuitous harangue of United States' racist discrimination against "Jews, Chinese, and Blacks." His description of the problem even contains a reference to the song "Strange Fruit" (made famous by the jazz singer Billy Holiday), which describes a lynching; and Zhou's text is accompanied by the photo of an actual lynching.[59]

Just within a few months after the publication of Zhou's book, little primers on Michurin biology began to appear with increasingly strident rhetoric on the subject of eugenics and genetics. A popular Chinese-authored textbook specifically links eugenics and its evils to "Weismann-Morgan" genetics, from which only Michurin and Lysenko can rescue the world. These "two schools" are nicely pigeonholed and labeled (idealism versus materialism), and set in Manichaean opposition. Further, here it is argued more fully that science, social class, and politics are inescapably related; that social classes abuse science for their own power and economic ends. Oddly enough, a good portion of this book accurately summarizes what Mendel and Morgan actually thought about heredity. Materials taken from American textbooks, for example, illustrate chromosomal interchange and crossing-over, and there is even a sample of gene mapping. In future textbooks, however, or in any other discussion of genetics, such detail never again appears during the Lysenko era in China. Apparently, the Lysenkoites felt that a complete blackout of information about genetics was the best way to achieve its obliteration and their own victory.[60]

Arguments against genetics were taken about as far as they could go by Luo Tianyu's colleagues when they created a primer out of a few Soviet-authored essays, each more vituperative than the other. Weissman and Mendel, through their belief in the "immutable gene," are accused of creating the belief in a hierarchy of races with immutable human qualities. American genetics is found guilty by association with the very worst of the world's racism. Morgan's laboratory and gene theory are somehow related to fascist racism, and to American antiblack racism, slavery, and antimiscegenation law. For good measure, the "Mendel genetics community" of America is accused of sympathy with Hitler.[61]

### The More Positive Approach

Parallel to the vilification and capping of Mendel-Morganism, another body of Lysenkoite literature addressed itself only to the cultivation techniques of Michurin biology and their accomplishments in China so far. Repeatedly, the techniques and heredity theory of grafting ("vegetative hybridization") and vernalization were explained in every possible forum. These techniques were described promotionally as "easy to understand, easy to use," in an effort to get agricultural experts and party cadre to adopt and teach them. And the techniques were touted as the outcome of melding ideas and practice, as called for in dialectical materialism.

Soon, reports were being published from all over China announcing the splendid results that simple farmers were able to achieve by using Michurin's methods. A ten-year-old student produces championship melons after learning Michurin theory.[62] A seasoned peasant regenerates his fruit trees by using special grafting procedures.[63] A Mongol shepherdess successfully follows Lysenko's breeding guidelines and is declared "a hero of livestock raising."[64] And a northeastern county increases grain production by organizing a mass seed selection movement.[65]

This kind of literature served two crucial functions for the furtherance of Michurin biology in China and the ban of Mendel-Morgan genetics. It presumably demonstrated that the former worked in practice, while the latter was bogged down in useless theory. And presumably it demonstrated further that Michurin biology could increase agricultural production on the cheap, with virtually no investment of the precious capital needed for forthcoming industrial development in China.

### Find a Chinese in This Picture

There is a singular feature shared by all Michurin campaign literature. Virtually never is there any reference, let alone positive reference, to a Chinese scientist, to science done by Chinese, nor to Chinese scientific institutions—ancient, modern, or otherwise. True, the Chinese Michurinists had begun to publish the results of their own research and experiments. Those reports, however, were normally published in technical journals, and not in the campaign literature discussed here. Even in technical journals, however, the Chinese new biologists groveled before the images of Soviet authority. A reader of the popular science literature thus could be understood for thinking that Soviet science was filling an utter void in China. Moreover, Chinese students who used translated Soviet textbooks were confronted with lessons illustrated with Soviet fauna, flora, geography, and regular references to "our" great Soviet nation, "our" beloved Soviet land, and "our" Soviet heroes.

Chinese interested in secondary school biology education were quick to question the wisdom of such dependency on Soviet materials. They called attention to the similarity with an earlier era of Chinese biology education, when students used Western textbooks that created similar problems. Analogous solutions were proposed as well: Educators argued that the Chinese should write (or rewrite) their own textbooks because flora, fauna, diseases, and environments are all "local matters." Soviet experience should be cited, if called for, but a means should be provided for Chinese to learn about their own immediate natural world.[66] Before this kind of criticism could be aired with impunity and become part of a public discourse, Michurin biology in China would have to undergo a considerable diminution in its political status. Meanwhile, its status was about to be enhanced by the CCP, and dissent to be discouraged even further.

# Notes

1. This sketch of Lysenkoism is based on T. D. Lysenko's "The Situation in Biological Sciences," his July 31, 1948, opening speech to the Lenin Academy of Agricultural Sciences, in *The Situation in Biological Science: Proceedings of the Lenin Academy of Agricultural Sciences of the USSR* (Moscow: Foreign Languges Publishing House, 1949), 11-50. And it also draws on major standard analyses: David Joravsky, *The Lysenko Affair* (Cambridge, Mass.: Harvard University Press, 1970), 202-9; Loren R. Graham, *Science and Philosophy in the Soviet Union* (New York: Knopf, 1971), 219-37.

2. For the historical context of the transformist tradition, see Douglas R. Weiner, "The Roots of 'Michurinsim': Transformist Biology and Acclimatization as Currents in the Russian Life Sciences," *Annals of Science* 41 (1985): 243-60.

3. Lysenko, "The Situation in Biological Sciences," 41.

4. See Graham's discussion in *Science and Philosophy*, 223.

5. Graham, *Science and Philosophy*, 209. And see the chilling description of Prezent's career and ideas in Douglas R. Weiner, *Models of Nature: Ecology, Conservation, and Cultural Revolution in Soviet Russia* (Pittsburgh, Pa.: University of Pittsburgh Press, 1988); and Weiner's *Little Corner of Freedom: Russian Nature Protection from Stalin to Gorbachev* (Berkeley, Calif.: University of California Press, 1999).

6. For Stalin's editing, see Kirill O. Rossianov, "Stalin As Lysenko's Editor: Reshaping Political Discourse in Soviet Science," *Russian History*, no. 1 (1994): 49-63. Thanks to Max Okenfuss for the reference.

7. Lu Shougeng, "Zui jin Eguo faming zhe zhiwu shengchang zu duanfa" (Recent developments of vernalization in Russia), *Zhonghua nongxue huibao*, (June 1934): 51-78.

8. Tang Peisong, *Green Thraldom*, 74-75.

9. Chen Jinhua, "Xiao-mo qiubo-xing yu chunbo-xing wenti" (The question of autumn and spring wheats), *Nong bao*, (June 15, 1948): 1-7.

10. Agricultural education and research network: See Chen Fengtong, *Zeran kexueh* (Natural Science), no. 1 (October 1951): 388-89. And interview with Li Jingzhun, January 1985.

11. Beijing Agricultural University: Beijing daxue nongyeh daxue. It is known familiarly as Bei-nong-da. North China Agriculture College was Huabei nongyeh daxue.

12. Wu Zhongxian was born in 1911, near Hankou. His secondary education in China, strong in science and mathematics, was from American and British missionary schools, including St. Paul's in Anwei, Wesley in Wuchang, and Boone Central (the Anglican Yale-in-China). For college, he attended Huazhong University and Shanghai College. In 1935 he won a Boxer fellowship for graduate work at Edinburgh in genetics and animal breeding. After his 1937 doctorate under F. A. E. Crew, he did two years of postgraduate study in chemistry at Cambridge with the Nobel Laureate James Flack Norris (1871-1940). Back in China, between 1939-1948 Wu taught courses in biometry, population genetics, cytogenetics, and animal and veterinary science. In turn, he was on the faculty of Northwest Agricultural College, Northwest University, and National Central University—all in Chengdu. After the war, he became head of animal sciences at Beijing University. (Biographical material from interview with Wu Zhongxian, Beijing, May 31, 1986.)

13. Interviews: Tang Peisong, Beijing, May 19, 1986; Li Jingzhun, 1985. Li recalls that this rescue effort was pointedly undertaken under the direction of Qian Zongrui, CCP minister of higher education. Qian negotiated directly with Li and the school's dean.

14. Interviews: Li Jingzhun, Tang Peisong, Wu Zhongxian.

15. Interviews: Li Jingzhun; Tang Peisong; Wu Zhongxian.

16. See *Journal of Heredity*, no. 90, 1950.

17. Interview, Li Jingzhun. And see Eliot B. Spiess, "Ching Chun Li, Courageous Scholar of Population Genetics," *American Journal of Medical Genetics*, no. 16 (1983): 603-30. H. J. Muller rose to international prominence in genetics after studying with Morgan in the 1920s. An ardent communist at the time, he sojourned to the Soviet Union to realize his revolutionary values. Both Stalin and Lysenko were among the sources of his disappointment, and he returned the United States a strident opponent of both. He received the Nobel Prize in 1947 for his radiation genetics.

18. Interview, Wu Zhongxian.

19. Interview, Tang Peisong.

20. Interview, Tang Peisong. Tang's father Tang Hualong (1874-1918) had a long and distinguished political career, which included prominence as an anti-Manchu revolutionary in 1911.

21. Reporting to the CAS was the Yichuan zaipei shi (Genetics and plant cultivation laboratory). Reporting to the Ministry of Agriculture were the Huabei (North China) and Dongbei (Northeast) nongyeh kexue yanjiu so (Agricultural science research institutes), and the Zhongyang nongyehbu Michulin nongyeh zhiwu xuanzong ji liangzong jiangxiban (Central Ministry of Agriculture, Michurin agro-biology seed selection and improvement study group).

22. For example, see the post-1949 progress reports by Zu Deming, "Liangnian lai Huabei nongyeh kexue di jinchan" (The advance of agricultural science in North China during the past two years), *Zeran kexue yuehkan* (Natural science monthly), no. 6 (November 1951): 483-85; and "Shinian lai Michulin" (Michurin biology in the past decade), *Nongyeh xuebao*, no. 5 (October 1959): 330-43. And Liang Zhenglan, "Michulin xueshuo zai zhongguo di tuiguang" (Popularization of Michurin theory in China), *Kexue dazhong* (Science for the masses), no. 12 (1954): 445-46; and "Shinian lai Michulin yichuanxue zai wo guo di chengjiu" (China's achievements in Michurin genetics during the past decade), SXTB, (October 2, 1959): 462-67. Zu and Liang became number one and two leaders of Michurinism when Luo Tianyu was demoted in 1952.

23. Description of inauguration, KXTB, no. 1 (May 1950): 32.

24. The *Michulin xuehui huikan* was probably initiated at the beginning of 1951. I have only found (in the Hoover Institution Chinese collection) issue September 15, 1951, no. 3. This 108-page issue was published in Beijing by the CAS Institute of Genetics and Plant Cultivation. Luo Tianyu is prominently listed as editor and as president of the society.

25. No. 3 (1951): 106-7. This membership information suggests approximately five thousand members. There is no way to corroborate any of this. Whenever and wherever possible, the society advertised its existence and growth: For example see announcements of new Shanghai branch with 100 members, in *Jiefang ribao* reprinted in KXTB, no. 3 (1950): 194; and Zhejiang University branch of more than 300 members in *Zhejiang ribao* reprinted in KXTB, no. 5 (1950): 343.

26. This organization had its first meeting in 1946 and second one in 1949, according to the 1950 conference program reported in KXTB, no. 5 (1950): 339-41.

27. Conference program, 1950.

28. "Michulin xueshuo yu Morgan xueshuo. Zuotanhui jilu" (Michurin theory and Morgan theory. Symposium record), *Kexue*, no. 3 (1950): 69-78.

29. Zhu Xi (Yu-wen), 1899-1962. In Paris, 1920-1925, he participated in the anarchist Li Shiceng's work-study program. From 1925-1932, at Montpellier University, he studied embryology under J. E. Bataillon, then became his chief collaborator. Zhu became a devotee of I. I. Mechnikov's embryology and his critique of evolutionary theory. From 1932, he was associated with National Zhongshan University, the Beijing Academy of Sciences, and various private research organizations. After 1950, in addition to his CAS position, he was actively involved in national politics representing Shanghai. (See obituary by Tong Dizhou, KXTB, no. 10 (1962): 27-34; *Zhongguo dangdai kexuejia zhuan* (Beijing, 1983), 76-90; ZXS, 177-99. And see autobiographical data in Zhu Xi, *Shengwu di jinhua* (Evolution) (Beijing: Science Press, 1958), 1-2 of foreword.

30. By the mid-1950s, in the science community the term "symposium" (zuotan hui) had taken on the meaning of a conference primarily concerned with the announcement or formulation of policy related to science or scientists. See below for example of the important 1956 Qingdao Genetics Symposium (Qingdao zuotan hui).

31. See Zhu Xi's trip report in KXTB, no. 10 (1953): 23-30, and comments on the trip in *Guomin ribao*, Beijing, May 25, 1956 (SCMP, 1309:7).

32. David Joravsky describes Glushchenko as the chief international salesman for Lysenkoism. His specialty was "vegetative hybridization," that is, grafting. See *The Lysenko Affair*, 224-26.

33. Zhu Xi, "Ji yu Glushchenko jiaoshou di tanhua" (Record of a talk with Professor Glushchenko), KXTB, no. 1 (1953): 80-81.

34. See Zhu Xi's research report on the fate of "excess" sperm that enters an egg cell in KXTB, no. 4 (1953): 70-73.

35. Corroboration of Zhu Xi's attitude and tactic comes in 1956, when he directly criticizes CCP "indiscriminate use of Soviet science" and advises the Chinese and the Soviets to use Anglo-American science. (See *Guomin ribao*, Beijing, May 25, 1956, and SCMP, 1309:7. And also see discussion of the "Hundred Flowers, Hundred Schools" policy, in text below.)

36. David Joravsky, *The Lysenko Affair*, 159.

37. See report on practical results of Michurin biology in its first year, KXTB, no. 11 (1951): 1195-97.

38. From 1935, Nuzhdin was on the staff of the Genetics Institute of the Soviet Academy of Sciences. Joravsky characterizes him as a properly trained geneticist who became a turncoat and cynical opportunist. Because he became a Lysenkoite, Nuzhdin was one of the four professional geneticists, out of the thirty-five staff members of the institute, who kept their jobs after the coup of 1940 (Joravsky, *The Lysenko Affair*, 221n, and passim).

39. Nuzhdin lecture tour: See Li Jige, KXTB, no. 3 (1951): 145-49; and KXTB, no. 11 (1951): 1195-97; and survey article dated December 31, 1955, in *Beijing nongyehdaxue xuebao*, no. 1 (1956).

40. Interview, Tan Jiazhen. But compare Tan's 1952 "self-criticism" wherein he says that during his visit with Nuzhdin, they talked and argued; and the Soviet biologist helped him to greater understanding not by pitching Michurin biology, but by revealing the faults of the old genetics. See "Pipan wo dui Michulin shengwukexueh di zhuowu kanfa" (Criticizing my mistaken attitude about Michurin biology), KXTB, no. 8 (1952): 562-63. See further discussion of this self-criticism below in text.

41. Mi Jingjiu was born in 1925 in Beijing. His father was university educated and a Republican government official, directing the Northeastern Forestry Bureau. His mother

was uneducated and made a career raising his five brothers and two sisters. At least four of the brothers were college graduates. Among their professions were respectively: lawyer and CCP cadre; psychology teacher in a normal college; botanist; and engineer. Both sisters had university educations and became CCP cadre. Between 1944 and 1950, Mi's higher education came respectively from Furen and Beijing University Agricultural Colleges.

This information and the following biographical narrative is primarily based on an interview with Mi Jingjiu in Beijing, June 12, 1986. Some details are corroborated in interviews with Li Jingzhun, Tang Peisong, and Wu Zhongxian. Also see transcript of July 1986 interview with Mi conducted by the Beijing University Center for the Study of Science and Society. The transcript provides some interesting details, but basically restates the information Mi provided in the June interview. See *Yichuanxue yu baijia zhengming: 1956 Qingdao zuotanhui yanjiu* (Genetics and the "Let a Hundred Schools Contend" [Policy]) (Beijing: Beijing University Press, 1996), 50-57.

42. See Mi Jingjiu et al., eds., *Darwin zhui jiben yuanli* (Introduction to Darwinism) (Beijing: Beijing Agricultural University and CAS Genetics and Plant Cultivation Laboratory, 1952.) See section of text below on Darwinism for discussion of this and other texts.

43. Huang Zongzhen: this biographical narrative is principally based on an interview with him, coconducted with William Haas, in Beijing, May 26, 1986. There is a sixty-four-page Chinese-language transcript of the taped interview. Huang continued to work his way up at the Science Press until his retirement as a senior editor in the mid-1980s.

44. For sketches of its imperialistic expansion and general history, see *Kexue chubanshe sanshi nian* (Thirty years of the Science Press) (Beijing: Science Press, 1984). This is an anthology of essays by its staff, including Huang Zongzhen.

45. Lo Zonglo (b. 1898) at this time was a colleague of Tan Jiazhen at Zhejiang University's wartime campus. Lo's graduate training in physiology was also from Hokkaido Imperial University; and this connection brought Huang together with him. In May 1957, Lo became director of the CAS Institute of Plant Physiology.

46. Early on, a favored format was the anthology of Soviet biographies: For example see Chen Wei, trans., *Sulian gujin mingren xiaochuan* (Short biographies of historical and recent famous Soviet people) (Dalian: Sino-Soviet Friendship Association, November 1948). Michurin, but not Lysenko, gets a biography. Both get one in Wen Ji, trans., *Sulian kexuehjia* (Soviet scientists) (Beijing: n.p., 1950); and Li Hua et al., eds., *Sulian nongren xiaochuan* (Short biographies of Soviet agriculturalists) (Shanghai: n.p., 1953).

47. There is no evidence to suggest that Huang Zongzhen personally brokered the deal to publish this journal. From 1959, the journal changed its name to *Nongyeh kexue: yiwen banyuehkan* (Agricultural science: semi-monthly translation journal).

48. The May 1950 issue of KXTB, no. 5 (1950): 25, lists sixteen titles by Lysenko that had already been translated into Chinese and were available for purchase. Li Jingzhun (discussed above in text) hoped to appease Luo Tianyu by translating Lysenko's "Heredity and its variability." Using Dobzhansky's 1946 English translation, Li published what he claims was a best-seller: *Yichuan ji qi bianyi* (Beijing: Xinhua, January 1950, and Guangzhou, June 1950. Preface dated July 1949).

49. The first published translation of Lysenko's 1948 "The Situation in Biological Science," appears to be the August 1949 *Shengwuxue di zhuangkuang,* trans. Li Ho et al., foreword by Zhou Jianren (Beijing: Tianxia Press, 1949). Another translation, most likely from an English version, appeared in October 1950: *Lun Shengwu kexue di xianzhuang,*

ed. Zhou Jianren, trans. Cao Ifeng and Liu Zhun. A long explanatory afterword is meant to make the material accessible to a popular audience.

50. The Shanghai newspaper, *Dagong bao* was already reporting on November 21, 1949, that "the reactionary theories on heredity propounded by Mendel and Morgan have already been deleted from the biology textbooks for senior high schools, . . . and the progressive theory of Michurin should simultaneously be inserted to take the place of the discarded ones." (Cited in S. B. Thomas, "Recent Educational Policy in Communist China" (dated February 20, 1950), in Michael Lindsay, ed., *Notes on Educational Problems in Communist China 1941-1947* (Seattle, Wash.: Institute of Pacific Relations, 1950), 184. And for a discussion of the problem of using translations of Soviet biology textbooks in Chinese secondary education, see *Zhongxue shengwu jiaoxue di gaizao* (Reforming middle-school biology instruction) (Shanghai, n.p., February 1951).

51. *Kexue huabao* (English subtitle: Popular Science Monthly) was edited and published by the Science Society of China. The society was organized in the 1930s to bring together professional "applied scientists" and engineers who were interested in publicizing the practical uses of science. The topics mentioned appeared throughout its issues beginning in January 1950 through February 1952.

52. Available copies of *Peoples' Agriculture* cite the Peoples' Agriculture Society as its sponsor. No other affiliation is noted. It was published in Beijing by Commercial Press. Description here is based on the contents of issues from 1950, no. 6 through 1952, no. 9.

53. The presence of the film in Beijing is evidenced by a Soviet photo of the cinema where it was playing. The cinema's marquee displays a picture of Michurin, large Chinese characters sounding out his name, and a blurb saying "great reformer of the natural world." See reproduction of photo in *Journal of Heredity*, no. 2 (1951): 54.

54. See Michurin Society journal, *Zhongguo Michulin xuehui huikan*, no. 3 (1951) for notice that the film had been shown by the society's branch in Guangxi Province.

55. While I have not been able to find a print of the film, the sound track has in part been reproduced on compact disk: The "Suite from Michurin" is on *Shostakovich: Film Music*, RCA, 1990. The film's director was Alexander Dovzhenko (1894-1956) who also wrote the script. (See: Internet Movie Data Base us.imdb.com.) The distributor in the United States was Artkino Pictures which changed the title to "Life in Bloom."

56. *New York Times*, January 3, 1949 (byline: UP, Moscow, January 2, 1949). The article notes that an extraordinary 1,500 prints were made of the film to insure a huge audience. (Thanks to the Film Archives Division of the Museum of Modern Art for locating this article.)

57. Pan Guangdan (1898-1967), by far China's most active promoter of eugenics, "positive" genetics, had some biology training but never actually worked in the field. Pan stayed in the PRC and recanted his eugenic ideas in 1951-1952. (See BDRC.2: 61-63.) The 1930s discussion of eugenics among intellectuals is epitomized in the Marxist theorist Yeh Qing's critical review of the literature in "Ping yushengxue yu huanjing lun di lunzheng" (Debate over eugenics and theory of environment), *Ershi shiji*, no. 1 (1931): 57-124. Chen Zhen was the only geneticist publicly to entertain the idea of applying modern genetics to human improvement, but his thoughts on the subject were brief and sketchy. (See *Kexue*, no. 6 (1923): 629-36.)

Notwithstanding this lack of eugenic publication from within the professional biology community, eugenic and racial thought did suffuse every corner of Chinese culture from at least the end of the nineteenth century. See Frank Dikotter's discussions in *The

*Chapter 4*

*Discourse of Race in Modern China* (Stanford: Stanford University Press, 1992); and his *Imperfect Conceptions: Medical Knowledge, Birth Defects, and Eugenics in China* (London: Hurst and Co. for Columbia University Press, 1998). And also see my review of the latter in *China Review International*, 7.1 (Spring 2000): 33-36.

58. Zhou Jianren, *Lun yushengxue yu zhongzu qishi* (Eugenics and racial discrimination) (Shanghai, February 1950. Preface dated 1948). Except for the Marxist rhetorical ornamentation, the argument here is the same one he developed in the early 1920s in *Jinhualun yu shanzhongxue* (Evolution and eugenics) (Shanghai 1923).

59. *Lun yushengxue*, 55-61.

60. Chu Qi, *Yichuanxue di Michulin luxian* (The Michurin line on genetics) (Shanghai, May 1950; September 1952, 4th ed.; January 1954, 4th printing).

61. *Michulin xueshuo jieshao* (Introduction to Michurin theory), edited by Northeast Agricultural Science Research Institute, (Beijing, April 1950).

62. KHTB, no. 5 (1950): 344.

63. KHTB, no. 6 (1951): 631-34.

64. Shen Yencheng, ed., *Michulin xueshuo zai chumu jie yingyong* (Application of Michurin theory in livestock) (Nanjing, n.p., 1951).

65. Liang Zhenglan, *Kexue dazhong*, no. 12 (1954): 445-46.

66. See for example, Lu Yudao's essay in the anthology *Zhongxue shengwu jiaoxue di gaizao* (Reform of middle-school biology) (Shanghai, n.p., 1951), 42-51.

# 5

# Lysenkoism as Official Party Doctrine, 1952-1956

Paradoxically, the leader of China's Lysenkoites, Luo Tianyu, lost his position and publicly suffered "thought reform" at the same moment in 1952 that the CCP elevated the practice of Michurin biology to a formal and discrete party policy. Because of problems blamed on Luo, the Party no longer assumed that Michurinism would as a matter of course eventuate in hegemony under the "pro-Soviet" policy. When Luo Tianyu was replaced by some of his former lieutenants, they gained both the authority of party "policy" and the reach of the CCP bureaucracy to further their control over the biological sciences and to squelch dissent. This chapter uses the fate of evolutionary theory to illustrate the nature of their control and extent of their repressiveness. Not until the Lysenkoist hegemony begins to unravel in the Soviet Union does this authority of the Chinese Lysenkoites come to be challenged openly and eventually dissolved.

## Attacking Luo Tianyu, Defending Michurin Biology

In 1951-1952, when the CCP scrutinized Luo's embrace of Michurin biology, it came up with some conclusions that not only cost Luo his job, but also required him later to be subjected to excoriating criticism and "thought reform" before a national audience. Complaints about Luo's tough management style were being lodged with the CCP from early 1950, after he had bullied Li Jingzhun out of the university and out of China. This made for very bad press at a time when the CCP was trying to recruit scientists into the Party and encourage Chinese scientists still overseas to come home.[1]

Prominent biologists soon began using their personal connections with the CCP to lobby against Lysenkoism in general and Luo Tianyu in particular. Tan Jiazhen encouraged his university president, Zhu Kezhen, to raise questions within party circles. Zhu was a party member, head of the CAS Planning Bureau, and quite unsympathetic to Lysenkoism.[2] Tang Peisong repeatedly dunned his contacts similarly.[3]

One account of Luo's loss of party confidence claims that by June of 1950 mounting criticism forced him to defend himself in a letter to Liu Shaoqi, the CCP's second in command. That letter was forwarded to Mao himself, who returned it with the comments: "This letter reveals that the school is not managed well. . . . This comrade's thought is greatly flawed." And based on other information Mao had received, he added, "We must thoroughly investigate this school's leadership . . . [and] create appropriate management." By the end of the year, a Party Central Committee investigation completed its evaluation of Luo Tianyu. Consequently, in March 1951, Luo was fired from his job as dean of Beijing Agricultural University; but his troubles were not over yet.[4]

Firing Luo Tianyu did not end the Party's problems with Lysenkoism. Anti-Lysenkoist lobbying continued and critics made their cases in high places. For example, Tan Jiazhen spoke out against the spread of Michurin biology in May of 1952, at a conference on the reform of higher education sponsored by the Ministry of Education and the Ministry of Agriculture. The former ministry displayed some support for Tan, while the latter continued its support for Lysenkoism.[5]

By 1952, the CCP apparently concluded that the conflict between Michurin biology and "Mendel-Morgan" genetics had to be formally settled, under party auspices, once and for all. This was the year before China's First Five Year Plan was to be launched and the conflict was a potential source of trouble in a number of areas. First, the Party continued to campaign for membership from the scientific community and to recruit the support of that community for the plan. Additionally, since the plan was dependent on a massive amount of Soviet aid and expertise, the CCP could not permit the further expansion of anti-Soviet sentiment that the imposition of Michurin biology had produced.

Last but not least, the CCP wanted to resolve the conflict within the biological sciences because it had convinced itself that Michurin biology did work, that it could increase agricultural production significantly. Such an increase would be an important contribution to the Five Year Plan, which followed a "Stalinist model" that concentrated development on the urban, heavy industrial sector and expected to pay for this development largely out of "surplus" agricultural production.

The Party thought it best to resolve the conflict in biology and cover its posterior through conventional bureaucratic techniques. It began by scapegoating Luo Tianyu. Making a spectacle of his "errors" while giving Michurin biology the imprimatur of party policy, the CCP made itself look decisive, in control, and not at all responsible for the conflict. The conflict was attributed to Luo Tianyu's failure to follow party directives and his abuse of power, not to the content of Michurin biology (and not to the pro-Soviet policy, which was never mentioned in the proceedings).

The Central Committee worked all this out first through a "criticism meeting of Luo Tianyu's many mistakes" in April 1952; and then in a "Symposium on Work within the Biological Sciences" held from April to June of 1952. Organizations participating in the symposium were Zhu Kezhen's CAS Planning Bureau, and analogous bureaus in education and culture, and in public health. Among the dozens of participants were Luo Tianyu and all of his former lieutenants (some of whom had already inherited his leadership of Michurin biology). There were no representatives of Mendel-Morgan biology; but there were some sympathetic administrators, like Qian Zhongshu, the minister of higher education. The detailed conclusions reached by these two meetings monopolized the front page of the CCP's main national newspaper, the *People's Daily*, on June 29, 1952.[6]

## The Case against Luo Tianyu

*Sectarianism*

High on the *People's Daily* list of Luo Tianyu's transgressions was "sectarianism," a concept formulated during the 1942-1944 Rectification campaign. It refers to cadre who use their party status to create factions or to create divisiveness between the Party and nonparty intellectuals. In the vernacular, he was found guilty of "closed-doorism"—shutting out the very scientists he had been commissioned to reform. Instead of reforming them, however, "he [wound up] ingratiating himself to the backward masses." Such demagoguery and grandstanding were part of Luo's "individualistic heroic" style, which rendered him incapable of rallying together nonparty intellectuals. Consequently, the Party accused him of sabotaging its policy toward intellectuals.

Luo further alienated nonparty scientists by acting the part of a "scholar-lord" whose main weapon was libel ("capping"). Luo was admonished that a good Marxist-Leninist takes different approaches to politics and to science; it is therefore thoroughly inappropriate to generalize and say that all of the "old biology" is idealist, reactionary, in service to the bourgeoisie, or fascist. Instead, "we ought to say that the struggles between Morganism and Michurinism are expressions in science of two world views. If some parts of the old biology are demonstrated to be false science, if some conclusions of the old genetics are based on fascist principles . . . then these must be reformed."

The Party drew two conclusions from this: First, "if one says that the old genetics is reactionary, it does not follow that the practitioners of the old genetics are ipso facto members of the politically reactionary element." And second, the CCP would oppose any further use of Michurinism as a weapon in a sectarian battle.

### Empiricism and Dogmatism

The second set of findings against Luo Tianyu addressed his scientific limitations, his "empiricism" and "dogmatism," and his consequent failure to understand Michurin biology. These transgressions are evident, it was charged, in Luo's narrow focus on the minutiae of practical productive experience and his subsequent neglect of systematic agronomic theory and laboratory work. For corroboration, Luo's own words were cited from the first issue of the Michurin Society journal, where he had deprecated formal science education by writing that "the fields are a laboratory; nature is a classroom." The Party's rejoinder here expressed profound concern that "if one separates rational from empirical knowledge, and sets them in opposition—stressing one over the other—then both will be wrong."

In that same Michurin society publication, the Party found further evidence of Luo's naïve understanding of science. He argued there against the value of formal science and the mathematization of natural laws. He lauded the spontaneous and intuitive understanding of the masses. They get directly to the laws of nature, he wrote; they learn via ploughs and fields; and if they do not, they die. Those who approach nature through books and study, however, never reach this kind of knowledge. Therefore, Luo concluded, the masses are the true materialists; intellectuals are merely idealists. The Party dismissed Luo's "leftist" hostility to science, and insisted that—whatever understanding of nature the masses might have—they must also learn from formal science.

The Party determined that Luo Tianyu's limited understanding of science in general made it impossible for him to grasp the essentials of Michurin biology. It followed, then, that he had not been in a position to show the Mendel-Morganists the limitations of the "old biology" and the superiority of the new. A proper presentation of Michurin biology and a proper understanding of its accomplishments would surely make believers out of biologists who had thus far resisted.

### Michurin Biology

Alongside its *People's Daily* criticism of Luo Tianyu, the Party declared, in no uncertain terms, its complete support of Michurin biology and its rejection of Morganism: "In the new China, Morgan is not desired; Michurin is." And the latter received this support because "the greatest aspect of Michurin biology is in the area of promoting agricultural production." The "proven accomplishments" of Michurin, Lysenko, and Soviet agriculture put to shame the Morganists and their useless fruit fly experiments, their reliance on statistics, pure lines, and obscure theoretical issues. Only Michurin biology recognizes and has been able to exploit the central truth of the biological world—the unity of organism and environment. On the front page of *People's Daily*, the Party renewed its campaign for Michurin biology with Lysenko's grandiose declaration that this constituted a revolution in biology, moving it from its previous descriptive and passive phase to a new and aggressive one that would result in the transformation of nature.

## The New Rules of the Game

What then were the new rules for biologists that the Party ritualistically promulgated on the front page of *People's Daily*? In retrospect, it is clear that party cadre and the scientific community understood that "Mendel-Morganism" was completely banned from research, publication, and education. It was likewise understood that a biologist could not be penalized for refusing to acknowledge the validity of Michurin biology or avoiding it in her/his work. One could be sanctioned, however, for violating the ban or challenging Michurinism. The main issue here was that only the Party itself could judge a violation of the policy and impose sanctions. If an organization (e.g., a university) or an individual (e.g., Luo Tianyu) engaged in either it would be usurping party authority and, in effect, be guilty of vigilantism.

The convoluted and coded language of the *People's Daily* policy statement conveys an offer from the Party that the scientific community could not refuse. The Party acknowledged the validity of formal science education and research, and of "theory." The professional view of science that had been attacked by Luo Tianyu and the 1942 Rectification campaign was permitted to express itself again. And, crucially, scientists (biologists at least) were not to be accused of political crimes because of their scientific beliefs. In return, the scientific community would accept the imposition of Michurin biology, without any further fuss.

## Tan Jiazhen's "Self-Criticism"

In July 1952 Tan Jiazhen was required to publish a response to the *People's Daily* policy statements in the form of a "self-criticsm" article for the widely circulated journal, *Science* (Kexue tongbao).[7] Apparently, he was supposed to acknowledge the new policy and, at least implicitly, agree to comply with it. Since Tan does not comment on the origins of this article, one must surmise that perhaps the Party wanted this done out of a sense of political symmetry—we give you Luo Tianyu, you give us Tan Jiazhen. Then again, perhaps it was thought that a conciliatory act from Tan could act as a coda, after which everybody could get on to other things.

Tan perhaps agreed to write the piece as his quid pro quo; however, he was intent on making his true feelings known. As a result, the "self criticism" boldly states his criticism of Lysenko, the Soviet Union, and the transgression of academic freedom. For example, Tan confesses that because "reactionary American and British geneticists slandered Lysenko as a renegade against science," and as a "crackpot," Tan had earlier concluded that Lysenko was a "half-baked scientist who used science to make political capital." But he now realizes that was incorrect. And from those same sources, Tan continues, he had learned "to worship blindly the false theories of academic freedom and the separation of science from politics." But he now realizes that too was wrong. And finally, Tan says that he had earlier come to believe that there was no longer any academic freedom in the Soviet Union, and that geneticists were actually imprisoned there. But now, Tan concludes, he knows that "this was just capitalistic propaganda; no one is jailed in an independent people's nation just because their ideas differ."[8]

Tan's ironic "self-criticism" may have avoided the censor because it went on to criticize Morgan genetics with many of the appropriate Lysenkoist clichés. Nevertheless, when it came to his personal scientific beliefs, he held his ground with straightforward statements: he refused to deny the validity of gene theory; he refused to concede that environment is the basic force in heredity.[9]

The 1952 policy arrangement governed the biological sciences successfully until 1956. Biologists like Tan Jiazhen, who refused to convert to Michurinism, found themselves doing no research at all and teaching general biology courses in which sensitive topics were simply avoided. If they continued to lobby against the arrangement, as did Tan, they did so with great care, through their patrons in the higher echelons of the Party's science bureaucracy.

## Lysenkoism and Michurin Biology at Flood Tide, 1952-1955

Under protection of the 1952 party policy, Michurin biology continued to spread to every corner of China, to every level of education, obliterating the teaching and practice of the "old biology." Michurin biology had no trouble recruiting new promoters, including Western-educated converts with training in genetics. Visiting Soviet experts, nevertheless, continued to tour the China circuit and command large, compliant audiences; and translations of Soviet biology texts continued to dominate Michurin biology publications. With one notable exception, these lectures and translations, however, soon had nothing new to say, and became repetitious in the extreme. That exception was the translation of an ongoing debate on Darwinism in the Soviet Union, a debate that challenged Lysenko and turned out to be a harbinger of his later political problems. Meanwhile, the CCP tried to keep Michurin biology in the national spotlight by sponsoring an endless series of ceremonial conferences and celebrations, like the Michurin centenary, in 1955. Occasionally, Michurin biology was kept in the spotlight as a result of Party-approved public criticism and censorship of scholars who appeared to challenge doctrine, or who appeared to criticize a Chinese or Soviet Michurinist. All of these characteristics were dramatically at play in the treatment of Darwin and evolutionary theory.

## The Trouble with Darwin

Lysenko's 1948 Lenin Academy speech outlines the approach to Darwin that Michurin biologists would follow in China with a standardized, formulaic format.[10] Common to all this literature is the fact that it is actually not about Darwin and evolutionary theory at all. Instead, it is a device for rehashing the fundamentals of Michurin biology from a somewhat novel point of departure. It is really about Michurin's self-styled "Creative Darwinism," which was claimed to embody the accurate understanding of evolution toward which Darwin and a handful of other thinkers had been pointing. Even in book-length discussions, Darwin and his "legitimate" precursors and successors never make it past the first chapter, where they received a breezy and largely negative appraisal. The few contributions to evolutionary theory which are grudgingly attributed to the Darwinians are treated virtually as signs and prophecies of Michurin's coming and the truth of his vision of evolution. Lamarck, St. Hilare, Darwin, K. F. Rulé, Mechnikov—all are depicted in this scheme as precursors of Michurin.

Most of the discussion of Darwin is dedicated to his errors, beginning with his influence from Malthus' population theory. Depicted as a product of his own capitalist society and times, Darwin could not but see nature in terms of the violent social competition and conflict that informed Malthus' law. Michurin biology acknowledges that competition and struggle do occur *between* species, but concludes that Darwin was dead wrong when he suggested that there was struggle within a species or that such struggle could contribute to natural selection and then to evolution. This critique usually concludes by noting that just as Michurin biology rejects the idea of intraspecies struggle it also rejects the idea of intraspecies "mutual aid." Environment thus remains the sole force that drives organic change and evolution.

In this standard format, Darwin is further criticized for muddling the role of environment by creating a false distinction between "definite" and "indefinite" variations. Darwin thought the former occurred when "all or nearly all of the offspring of individuals exposed to certain conditions during several generations are modified in the same manner." The latter showed itself in "the endless slight peculiarities which distinguish the individuals of the same species, and which cannot be accounted for by inheritance from either parent or from some remote ancestor."[11] For Michurin biology, there are only "definite" variations, since any and all variations are a function of an organism's inner response to outer environmental factors.

Next, Darwin is decried for his insistence that evolution occurs gradually, through the accumulation of small changes over very long periods of time. Michurin biology, again, attributes Darwin's confusion here to his failure to grasp fully the fundamental role of environment and the heredity mechanism which reshapes organisms in response to the environment. This error of Darwin is further related to his passive, observational attitude toward nature. Had he thought more like an interventionist and been concerned with practical changes of nature, he might have recognized the potential power of humans to alter nature ("artificial selection") and even to create new species suddenly.

Finally, this standard critique of Darwin faults him for not exploring more fully his correct inclination to give primacy to the role of environment in evolution, and to recognize the inheritance of acquired characters. He is not faulted for what little he contributed to the discussion of a hereditary substance and mechanism. Having left a serious lacuna here, however, Darwin is said to have opened the door to the "neo-Darwinians" with their dangerous ideas about immutable genes and chance variation.

## Michurin Biology Co-opts Darwinism

The earliest full-scale assault on Darwinism by Chinese Lysenkoites came in the form of the 1952 college textbook, *Basic Principles of Darwinism*. It was stitched together from the lectures of a number of visiting Soviet experts, and from a variety of Marxist sources. Six editor-translators are listed, all of them associated with the northeast China agricultural science institutions that remained the base of Michurin biology. This book was supposed to serve as the introduction to Darwinism throughout the entire national higher education system, so it was scrutinized carefully by the education bureaucracy.

The CAS found the first edition problematic, so they called in the science journalist, Zhou Jianren, to advise them on a second draft. In turn, that draft was examined at a CAS advisory conference. In spite of these efforts, however, the revised edition was still, obviously, the product of a committee—one very concerned about staying safely within the bounds of Lysenkoist doctrine. It is a cut-and-paste job, fragmented by scores of citations from Marx, Engels, Lenin, Stalin, as well as Lysenko and Michurin. There is no straightforward narrative summarizing what Darwin said or what evolution is; and the book's scant structure is all polemical, following the standard format described above. A third edition followed with little improvement, but the book continued to be used and to be listed in the 1955 Ministry of Education curriculum guide for the teaching of Darwinism.[12]

This textbook warrants attention here not only because of its role in biology education, but also because it was at the center of the first controversy to erupt after the announcement of the 1952 party policy on Michurin biology. The source of the controversy was a critical review of the textbook's first edition. It was written by Zhang Congping, an obscure science journalist, who was himself writing a series of Darwinism primers, and who clearly considered himself to be a Marxist and a devoted student of dialectical materialism. His review challenged Lysenko's and Prezent's analysis of Darwin by drawing upon his knowledge of dialectical materialist literature, especially Engel's work; and he made the mistake of suggesting that there were "idealist errors" in the textbook's critique of Darwin.[13] For the next three years, even though the textbook had to be twice revised, Zhang was hammered by the establishment, apparently determined to use his case to illustrate where the boundaries of criticism lay.

Zhou Jianren, in a self-indulgent and arrogant piece, was the first to respond in print to Zhang's review.[14] Zhou himself had earlier given the book a very favorable review, and he was the editorial advisor for its revision; he clearly felt personally compromised by Zhang's criticism. By this time, Zhou had become one of the most prominent interpreters and apologists for "Creative Darwinism," and was constantly in demand to write prefaces, book chapters, and journal articles on the subject for student and nonspecialist audiences. Part of his popularity was no doubt due to his prewar career as a science journalist. The rest was clearly a result of his famous family name and the fact that he was an important trophy both for the CCP and Michurin biology. His late brother, the writer Lu Xun (Zhou Shuren), had been canonized by Mao Zedong during the war years; thus, when Zhou joined the Party in 1948, he himself had accrued considerable value. The CCP used him on scores of committees and at as many ceremonial functions; they even made him governor of his home province for a while.[15] Zhou Jianren's publications on Creative Darwinism served it, in effect, as an endorsement.

Zhou's response to Zhang was tame compared to what followed. Zhang's 1954 middle school primer on Darwinism was relentlessly attacked over the next two years. The CAS published a reprimand of the book's errors; and increasingly longer review articles tore it to shreds, guaranteeing that it would not be used in the classroom, and that it would be pointless for Zhang to write the college-level textbook that was his next project. It was probably cold comfort to him that his alleged intellectual mistakes were never linked to issues of social class or politics, and that the Party did not formally proscribe his book, as they would soon be doing to others.[16]

### Fang Zongxi: A Darwinist for All Seasons

At the same time that *Basic Principles of Darwinism* was first published, another important textbook appeared. Aimed at senior middle-school students, *Foundations of Darwinism* was not the center of any controversy and its articulate, coherent text required no rewriting. Along with the former title, it was the only textbook to be listed in the Michurin biology curriculum guide issued by the Ministry of Education in 1955. By that time, it had reached three printings and seventy-five thousand copies just for distribution in the northeastern part of China. Just as *Basic Principles* was meant to serve as the nationwide standard for college-level biology, *Foundations* served as the model for middle school. The two of them must have served well to obscure evolutionary theory and Darwin from the nation's students throughout the 1950s.[17]

Though the content of this successful textbook was derived from a Soviet middle-school original, its editor, Fang Zhongxi, skillfully avoided the clumsiness that characterized most of the many such publications in China. Fang brought to this important assignment only a bit of experience writing science texts for secondary education; but he was one of the best-trained biologists in China. His background bears further scrutiny for at least two reasons: It provides further insight into what kinds of scientists were recruited to Michurin biology;

and it introduces a biologist who became—along with Mi Jingjiu and Zhou Jian-ren—one of China's most influential purveyors of Michurin's Creative Darwin-ism.

Fang Zongxi was born in 1912 in Fujian, China's southeastern coastal province. He received his B.A. in biology from Amoy University in 1934 and an M.A. in 1936, at a time when the biology department was under the direction of Chen Zeying, the outstanding geneticist first trained at Yanjing University and then under T. H. Morgan at Columbia. Fang received a solid education in ge-netics, evolution, and—the local specialty—marine biology. Until the war drove him out of Amoy, Fang remained at the university as a graduate assistant, dis-tinguishing himself by translating Morgan's *Scientific Basis of Evolution*.[18]

Between 1938 and the end of the war, Fang led a precarious existence avoiding the Japanese. First he fled to Indonesia where he taught in a Chinese secondary school and wrote his first science textbook. Then, from 1941 to 1945, he took refuge in Malaya, harbored by the local hill people and working as a vegetable farmer. In 1945, he lived in Singapore, earning his keep by teaching and running the library in an overseas-Chinese middle school. In his spare time, he developed his literary skills writing fiction and poetry, and he began a corre-spondence with the renowned biologist J. B. S. Haldane, head of the genetics program at University College, London. In response to Fang's request, Haldane eventually arranged a fellowship for him; and Fang's Singapore friends arranged for travel and living expenses. He left for England in 1947.

Fang Zongxi spent two years getting his doctorate in genetics under Hal-dane's direction. This put him in London at a time when British left-wing scien-tists, including Haldane, were in considerable turmoil over the development of Lysenkoism in the Soviet Union. Initially, Haldane himself was equivocal about Lysenkoism; but by the end of 1948 he was volubly critical.[19] There is no record of how Fang was affected by these events; but his actions make it clear that he was not deterred from returning to China by the Communist victory or the pros-pect of Lysenkoism. In contrast to Tan Jiazhen's experience, Fang did all possi-ble to return, and was even forced to find a way around the British government's embargo on travel to China from England.

Back home in China at the end of 1949, Fang Zongxi found himself in a dilemma similar to that of Mi Jingjiu around the same time. Since a solid train-ing in genetics was no qualification for a teaching position, he accepted a writ-ing job from the powerful Central Publications Administration whose influential director Fang had met in Singapore. Fang was placed with People's Education Press (Renmin jiaoyu) as biology editor—Michurin biology, that is.[20]

Fang's attitude toward Michurin biology at this time is not on record; so there is no way to know if he had a sudden epiphany and became a believer, or whether economic and political pressures on him dictated this course as a matter of survival. In any case, he did his job very well. Fang worked at the press for the next three years, publishing the *Foundations of Darwinism*, along with three other textbooks on various aspects of Michurin biology. In 1953, he was invited to join the faculty of Shandong University, which obviously was not put off either by Fang's training in genetics or his prominence as an educator in Michurin biology. The university was building a specialty in marine biology and that may have been the source of its interest in him.[21]

There is no record of any research Fang Zongxi might have done during the 1950s. He did, however, publish five more biology textbooks and many prominent articles from 1953 to 1963. These publications display an impressive intellectual flexibility on his part. Each time there was a shift in party line on genetics and evolutionary theory, he produced appropriate textbooks and journal articles. Thus, by the late 1950s, when the Party had abrogated the Michurin monopoly, Fang wrote material that gave equal time and weight to Michurinism and Mendel-Morganism, per current policy. By the early 1960s, as Michurin biology became marginalized, he began to publish impressive explanations and defenses of classical genetics, gene theory, and Darwinism. From 1963, his educational publications mention Michurin biology only in passing, if at all; and his research publications are free of Michurin biology and are clearly informed by the education he received before 1949.[22]

*From Apes to Engels*

Unfortunately for the Chinese writing specifically about the evolution of humans, Michurin biology had concluded that Engels had already stated the complete truth of the matter, and nothing more really needed to be said. Nevertheless, the CCP was apparently taken with the notion that "the Chinese"—in the form of *Sinanthropus pekinensus*—were already making their mark on human progress back in the Pleistocene. Consequently, the Party encouraged public education to deal with the subject of the descent of man, and it likewise encouraged the renewal of digging for the remains of early Homo sapiens.

Things started off well enough. In October 1950, a twenty-seven-day exhibition entitled "Ape to Man" was held in Shanghai as part of the Party's push to popularize science. Spectators reportedly numbered 418,700 individuals.[23] That same year, the famous archaeologist Pei Wenzhong responded to the Party's call with a little secondary school primer on "A History of the Development of Nature."[24] In 1929, Pei had been head of the team that unearthed the remains of "Peking man" at Zhoukoutian; and he went on to direct similarly fabulous discoveries. His primer was a version of his recent lectures and newspaper articles outlining the Michurin biology position on human evolution. Pei had joined the CCP the previous year and undergone an intensive period of training and study that culminated in the first of his public lectures and his declaration that his recent reading of Engels' *Dialectics of Nature* had converted him to dialectical materialism and a new perspective on his work.[25]

Pei Wenzhong's little primer was the first Chinese endorsement and model for Michurin biology's treatment of human evolution. That treatment is virtually identical to the formulaic outline of general evolution discussed above. That is to say, it too is a Trojan horse for Michurinist doctrine on the essential role of environment in organic change and for the inheritance of acquired characters.

Only the introduction of Engel's classic postulate—that "labor creates man"—adds a new and potentially creative dimension.

Though Pei Wenzhong's primer gives it little space, he does declare that the central issue of his text is the one argued by Engels in his essay "The Part Played by Labor in the Transition from Ape to Man." Pei's treatment of this thirteen-page essay from the 1870s is deferential and uncritical; and he brings nothing of his own great experience to its presentation, nor does he bring in anything—corroborative or critical—from the past half century of paleo-archaeology and paleontology.

Pei's approach to human evolution was repeated in more extended expositions of human evolution during the 1950s. Typically, however, narratives merely consisted of larger assemblages of basic paleontological data, sometimes accompanied by no evolutionary analysis, sometimes by a rehash of Engels' essay, but no other analysis from any other source.[26] In 1950-1951, when a physical anthropologist named Liu Xian deviated from this pattern, he became the center of national controversy and a target of party reprisals that lasted for four years.

*The Liu Xian Affair*

Liu's book *A History of the Development of Apes to Humans* appeared at the end of 1950, when he was on the faculty of Fudan University. His book is a respectful but critical dialogue with Engels' essay "The Part Played by Labor, etc." referred to above. He took issue with Engels on the basis of data made available in more recent times and he suggested that one could arrive at a useful determination of when to designate a hominid a "human" following a conceptual scheme that was at least as reasonable as Engels' argument.

Liu argued that Engels was not aware that early Homo sapiens only differed quantitatively from advanced apes and for the most part only in superficial characteristics. Advanced apes are capable of thought and prescience, according to Liu, and some have demonstrated the intelligence of a four- to five-year-old human. Among the basic things they lack to become humans is a larger brain and the physical organs for speech. Liu said that a hominid qualifies for the "human" designation because of a bigger brain, thought, and speech—not because of "labor" or other social arrangements. Liu concluded that Engels had the cultural cart before the physical horse, and implicitly, he rejected Engels' devotion to a Lamarckist evolutionary scheme based on "need"—that is, Engels' belief that apes developed bigger brains and the organs of speech because they needed to communicate better if they were to sustain and expand superior social organization and work patterns.

The second printing of Liu Xian's book in May of 1951 was followed in June by a brief criticism in *People's Daily*. Shortly thereafter, the book's publisher was required by the CCP to recall all copies and was prohibited from further printing or distributing it. The censorship, however did not end Liu's problems. His book continued to be subjected to damning reviews for some time; then, the criticism subsided and he enjoyed a hiatus from public scrutiny until

April of 1955. At that time, the criticism began again, with near-hysterical ve-hemence. The editors of *Science* (Kexue tongbao) invited him to write a recantation of his views, but instead he wrote a defense, which they refused to print. Liu was then brought before a CAS symposium devoted to exposing his errors. The following year, the Fudan University Michurin Biology Study Group and biology department subjected Liu to further and extended harassment. His ordeal did not end until the implementation of the liberalizing "Hundred Schools" policy at the end of 1956.[27]

Why was Liu's case revived so dramatically in 1955, four years after his book was censored? At that time, Michurin biology seemed to have been eminently successful at extending and strengthening its hegemony and seemed not to be again in need of making a public example of Liu Xian. Moreover, that year the CCP and the CPSU were jointly in the process of sponsoring a gaudy celebration of Michurin's centenary and taking stock of all of the accomplishments claimed by Lysenko and Michurin biology. There was however, a shadow cast over the entire proceedings by an ongoing Soviet debate on Lysenko's evolutionary theory, especially his doctrines on the origin of species.

*Soviet Speciation Debates*

The controversy, known as the "speciation debates," began in the Soviet Union in 1952, when some of Lysenko's most adamant foes began to publish pointed and highly technical criticism of his ideas on evolution. The *Botanical Journal* was the main forum for this intellectual assault in the Soviet Union and the base for a formidable alliance among some of the country's most distinguished biologists. Though the debates continued in the Soviet Union, in China they were not reported in any form for two years.[28] It was October 1954 before they began to be translated in a series of CAS pamphlets containing representative selections; it was not until the appearance of the December 1954 issue of *Science* that the Chinese periodical press finally acknowledged and discussed the speciation debates and made available translated samples to a nationwide audience.[29]

Chinese enemies of Michurin biology immediately recognized the debates for their political potential: if the Soviets themselves were openly challenging Lysenko, how could Chinese be faulted for reporting that fact? This message was clear in *Science*'s introduction to the debates in the form of an eloquent statement from one of the Soviet participants. He pled for a free and open debate, and argued that this was no parochial issue, but rather was of vital interest to scientists throughout Eastern Europe where the debate was also in progress. The statement argued that the *Botanical Journal* did not create the controversy, but rather that it was a reflection of the academic community's widespread concerns and their disagreement with Lysenko. Further, the statement criticized the Soviet Communist Party for coercing Lysenko's critics while permitting him to prevail even though he represented a minority of the community.[30]

In the April 1955 issue of *Science*, a provocative editorial complained that the journal's December 1954 introduction of the debates was very little and very late coverage for such an important story. Just so, the editorial continued, the CAS-sponsored translations of the debates was a long time in coming; and meanwhile, since the Soviet debates continue, why has the journal failed to follow up its original coverage of the story?[31] A question at least as intriguing is what changed politically within this important journal and within its sponsoring institution, the CAS, to permit the revelation. While no answer is available, the fact is that the coverage did continue, not only in *Science*, but in other prominent periodicals as well, and it did not stop until two years later, when Michurin biology had finally lost its hegemony in China.

It was in the context of the Soviet debates that the renewal of Liu Xian's persecution must be understood. Chinese Lysenkoites and their party patrons were obviously threatened by the revelations of attacks on Lysenko, and their first reaction was to send out a warning to anyone in China who might want to use the debates as a license for dissent. If there was any doubt that this was the motivation for attacking Liu Xian, it would have been dissolved by the subsequent attack on veteran botanist Hu Xiansu the following month. Earlier in the year, Hu had published a general botany textbook in whose preface he had written the following:

> After their initial publication, Lysenko's new ideas about species in biology were in fashion for a while . . . due to the support of the political authorities. Later on, he was widely criticized in Soviet botanical circles. . . . In the history of modern biology, special attention will be paid to this controversy. Our Chinese biologists, especially plant taxonomists, must understand this fully, otherwise they will be led astray.[32]

In response to these words, and apparently not to anything else in the text, Soviet advisers working in the Chinese Ministry of Higher Education accused Hu Xiansu of being critical of the Soviet government. The CCP responded in turn by confiscating and destroying all copies of the book, then having Hu Xiansu reprimanded from the podium at the ongoing Michurin centenary festivities. The following month, November 1955, a *People's Daily* article completed the attack, showing uncharacteristic delicacy by excoriating his book but not mentioning his name.[33]

### *The Michurin Centenary*

The October 1955 Michurin centenary celebration was intended primarily to accentuate the positive, but it could not ignore the speciation debates. For example, Tong Dizhou, director of the CAS Institute of Experimental Biology, delivered one of the opening speeches to the centenary conference, and, while he devoted most of his time to a perfunctory list of Michurin biology's accomplishments in China, he also heatedly denied that Lysenkoism required the support of the government, the CCP, and the CPSU; he rejected out of hand the recent criticism of Lysenko's evolutionary theories.[34] More typically, the centenary was responsible for some very tedious speeches extolling Michurin's virtues and the publication of some very fancy collections of his works; but its main goal was to sum up the good that Michurin biology had done for China and to take stock of its spread throughout the Chinese educational system.[35]

Dozens of speeches and journal articles listed the accomplishments of Michurin biology in China's agriculture, based on the standard techniques: vernalization, wide crosses, close planting, and vegetative hybridization (grafting). The wide crosses and grafting claimed credit not merely for improved varieties, but for new "species" as well. None of these reports, however, took notice of those speciation debates that argued Lysenko had never substantiated even one of his many claims to the creation of a new species. Nevertheless, all of the centenary reports did continue to take the usual potshots at Mendel-Morgan genetics and its alleged inability to make any positive contribution to agriculture.[36]

The centenary celebration made much of education in Michurin biology, especially so-called popular education for the masses. For the latter purpose, Fang Zongxi crafted a centenary primer on Michurinism, and Zu Deming—Luo Tianyu's successor—wrote one for a slightly more literate audience.[37] Chinese Lysenkoites took considerable pride in their contributions to the Party's campaign for science popularization, and by some accounts the campaign was quite effective. This was evidenced by graded materials in bookstores, children seen reading popular science publications, public science exhibitions, and extensive organization devoted to designing new educational materials and techniques.[38] Michurin biology made special efforts to emulate the celebrated Soviet educator, Ilin (1895-1952), who wrote stories about science for young children.[39]

## Lysenkoism on the Defensive

In spite of the generally upbeat mood of the centenary, hard-core Chinese Lysenkoites grew more apprehensive as the coverage of the speciation debates continued unabated for the rest of 1955. In January 1956, this was expressed dramatically by Mi Jingjiu when he published a lengthy and emotional article, devoted to a defense of Lysenko's speciation theories, and an affirmation of Lysekoism's continuing strength in China. The bulk of the piece is devoted to a summary of the criticisms (by Ivanov, Sukachev, Turbin et al.) made of Lysenko's speciation theories in the debates, and then to a point-by-point rebuttal of the criticisms in the words of Lysenko's major Soviet defenders, such as Nuzhdin. Mi says that his purpose is to provide an objective summary of the debate issues, implying that such a summary will put to rest any doubts that may have been raised by the translations thus far made available to China.

Nevertheless, Mi concludes this article by saying that he is aware that the debates have shaken Chinese confidence in Lysenko and raised many questions. Some people, he admits, are ready to abandon Lysenko, some want an open public debate and criticism of Lysenko, along the lines that prominent Chinese intellectuals were previously dealt with. Mi assures his readers, however, that

Lysenko's theories have not been overthrown and that his contributions have not been wiped out.[40]

Mi Jingjiu's unique article is an exercise in damage control. His nervously defensive tone could have easily led a reader to wonder just how much longer he thought Lysenkoism would last in China. Events already unfolding at home and around the world were in fact contributing to a shorter tenure.

Among the earliest of those events, in late 1955, was the introduction to China of Hans Stubbe's devastating research on Lysenkso's vegetative hybridization theory. Stubbe, the president of the East German Academy of Agricultural Science, had subjected Lysenko's theories and claims to rigorous testing between 1949 and 1952; and in his conclusive 1954 summation, he completely refuted Lysenko's assertions. He wrote that "investigations of the problem of vegetative hybridization of plants, carried out on a large quantity of material and in the course of a prolonged period of time, gave me no evidence of the existence of this phenomenon."[41] An abstract of these findings was published in the Soviet *Botanical Journal* in 1955; they may have been seen by the Chinese before Stubbe himself visited China in December of that year and presented his conclusions in detail to audiences at Beijing Agricultural University.[42]

Stubbe was not content to scrutinize Lysenko's work from afar. First, he sent his pupil, H. Bohme to study this same issue with N. V. Turbin in the Soviet Union,[43] and then Stubbe himself went to visit with Lysenko.[44] Both experiences convinced him all the more of Lysenko's incompetence, a conviction that he conveyed so strongly in China that it was passed on to the CCP Central Committee and eventually to Chairman Mao, along with information about other recent developments in what was now commonly called the "genetics question."[45]

## Developments in the Soviet Union

Between February and the end of April 1956, extraordinary events in the Soviet Union acted as catalysts for CCP evaluation of the genetics question. In February, the CPSU held its Twentieth Party Congress, where Khruschev delivered his secret speech on the crimes of Stalin. Chinese representatives to the congress immediately learned of the speech and conveyed its contents back to Beijing, where it was strongly criticized both for its content and for Khruschev's failure to consult with China beforehand. Nonetheless, the crimes of Stalin—as outlined in the speech—apparently encouraged Mao in the following months to make "policy proposals designed to prevent the abuses of Stalin's Russia from being duplicated in China."[46]

The second event, though less seismic, was of more immediate relevance to the genetics question. Again in February, and thanks again to Chairman Khruschev, there was held an all-Soviet Union conference on the production of hybrid corn seed. Up to this time, Lysenko, for characteristically obscure reasons, had condemned the use of the hybrid techniques in spite of their productive results in North America. Up to 1952 or so, he had banned hybridization; but the ban was then quietly lifted in the Soviet Union while it remained in effect in China. The Soviet conference reviewed the subject in detail and established a national program for the production of hybrid corn. Chinese "Morganists" saw this, like the speciation debates, as more Soviet-based leverage against Lysenkoism.[47]

Where the genetics question was concerned, the most significant Soviet development was Lysenko's loss of the Academy of Agricultural Sciences presidency. In April, he was forced to step down in response to vociferous opposition.[48] By coincidence, news of that event was conveyed to the Chinese quickly and in person by N. V. Tsitsin, a Soviet biologist who had been invited to help China plan a new twelve-year science development program. Not only did he bring the surprising news of Lysenko's apparent downfall, but he shocked his Chinese colleagues as well with his sharp public criticism of Lysenko. Tsitsin had once been a rival of Lysenko for leadership of "agrobiology," but he had defected from those ranks by the early 1950s and begun an active research program to disprove Lysenko's theory of vegetative hybridization.[49]

## The "Hundred Flowers, Hundred Schools" Policy and the Retreat of Lysenkoism

While the Lysenkoites were preoccupied with the Michurin centenary and Hans Stubbe, the CCP concerned itself with fundamental issues of economic development for the second five-year plan, to begin just a year hence. Fiscal problems and concerns about dependency on the Soviet Union drove the Party to consider seriously how China might more independently continue with its fast-paced, costly, industrial development. Early on in Party discussions, there emerged a recognition that—whatever the approach of the next development plan—the key to greater autonomy and productivity was China's scientific and technological elite. Participants in the discussions asked if China's current stock of experts was being treated so as to get the most from their experience. How could new expertise be trained in sufficient numbers? How much longer would China need to go on being dependent on the Soviet Union; and was a more critical attitude to be exercised toward Soviet models and advice? At the same time that these questions were being formulated and addressed, another set of overlapping issues was emerging, largely through the ruminations of Chairman Mao; he was particularly concerned with the relationship of technical expertise and political correctness, and the relationship of the CCP to the technical experts.

Throughout the last quarter of 1955, the Party had sponsored a series of national meetings to begin to evaluate the condition of the technical intellectual community and to consider a long-range plan for scientific development. So-called Democratic Parties—that is, non-CCP intellectual organizations—were called together to consider the "unity and reform of intellectuals." Based on these preliminaries, in January 1956 the CCP called a full-scale national meeting to address the problem of the intellectuals. The results of its work were summarized by Premier Zhou Enlai, who had been in charge of the proceedings.

Everything Zhou had to say affected the genetics question, though it was never mentioned by name; some issues, like party sectarianism and emulation of the Soviet Union, were especially pertinent. Zhou said that "certain unreasonable features in our present employment and treatment of intellectuals and, in particular, certain sectarian attitudes among some of our comrades towards intellectuals outside the Party, have to some extent handicapped us in bringing the existing powers of the intelligentsia into full play."[50] With utmost tact he expressed concern that "between some intellectuals and our Party, there still exists a certain state of estrangement" that needs to be ended and reversed so that intellectuals, especially technical workers, are motivated to apply for Party membership and so that the Party is inclined to admit them.[51]

Zhou said plainly that China could not indefinitely rely on Soviet experts. He added that "on the question of the study of the Soviet Union, . . . in the past there had . . . been such defects as undue haste, arbitrary learning, and mechanical application. Some comrades even arbitrarily rejected the achievements of the capitalist countries in science and technique. These defects should henceforth be avoided."[52]

His summation listed a number of practical steps to be taken to facilitate the work of technical intellectuals, including better working conditions, fewer administrative and political distractions, and incentives for productive work. For the specific development of science, he announced the formation of an agency to develop a twelve-year science plan.[53] And for "problems of a political nature affecting intellectuals" he reaffirmed the leading role of the Central Committee's Propaganda Department, which was now given the mandate to concentrate on the genetics question and find a means of resolving it.[54]

The head of the Propaganda Department was Lu Dingyi,[55] the department's pointman for the genetics question was Yu Guangyuan, who worked in the department's Science Division, and was also assigned to the new twelve-year science plan.[56] Immediately following the January conference on intellectuals, Yu Guangyuan was given the responsibility for planning a March symposium, specifically devoted to the genetics question; but due to the personnel needs of formulating the twelve-year science plan, that symposium was postponed until August.[57] There can be no doubt that the extra preparation time made it possible, before the symposium, for a set of strong principles to be formulated for resolving the genetics question. Nor is there any question that extraordinary events occurring in the Soviet Union made it possible for these principles to be presented to the August symposium more decisively than they could have at an earlier moment.

These new principles were intentionally framed broadly enough to affect the entire science community and its relationship with the Chinese Communist Party. The symposium, likewise, was meant to act as a model device for resolving conflict within the scientific community and between that community and the Party. Principles and symposium were explicitly meant to illustrate the intent of the emerging "Hundred Flowers, Hundred Schools" policy, which is taken up in the next chapter.

## Notes

1. Interviews: Tan Jiazhen; Tang Peisong. Both Tan and Tang agree that the Li Jingzhun affair had strong repercussions on Luo's power. This evaluation is supported by Li Peishan, "Qingdao ichuanxue zuotan hui," 6-7.

2. Interview, Tan Jiazhen.

3. Interview, Tang Peisong.

4. Jiang Shihe, "'Michulin xueshuo' zai Zhongguo 1949-1956: Sulian di yingxiang" (Michurinism in China: Soviet influence), ZBT, no. 1 (1990): 20.

5. Interview, Tan Jiazhen. Also see, Tan Jiazhen, "Pipan wo dui Michulin shengwu kexue zuowu kanfa" (A criticism of my mistaken views of Michurin biology), KXTB, no. 8 (1952): 562-63.

6. The criticism meeting was reported by Liang Xi, "Wo duiyu Luo Tianyu tongzhi suo fancuowu di ganxiang" (My impressions of the political errors of Luo); and the symposium was reported in "Wei jianchi shengwu kexue di Michulin fangxiang er douzheng" (Carry on the struggle in biology for the Michurin line). Both in RMRB (June 29, 1952): 1. This summary is drawn from both pieces.

7. Tan, "Pipan wo dui Michulin."

8. Tan, "Pipan wo dui Michulin."

9. Tan, "Pipan wo dui Michulin."

10. All of the titles discussed in this section follow this format very closely. The format also reflects the lectures of Mi Jingjiu's Soviet instructor, A. V. Dubrovina. See her *Darwin zhuyi* (Darwinism), ed., trans. Mi Jingjiu (Shanghai: CAS, 1953).

11. *Origin of Species*, (New York: Modern Library, n.d.), chapter one, 15-16. Based on the 1860, 2nd edition. And see translation of I. I. Prezent's discussion of the "error," in Mi Jingjiu's translation, *Zeran kexue*, no. 1 (1952): 55-60. For further such discussions, see the influential senior middle school textbook by Fang Zhongxi, *Darwin zhuyi jichu* (Foundations of Darwinism), 1st ed. (Shanghai: Science Press, 1952); and Zhou Jianren, *Shiwu xuebao*, no. 10 (1953): 390-92.

12. Textbook: *Darwin zhuyi jiben yuanli* (Basic Principles of Darwinism), ed. CAS Genetics and Seed Selection Research Bureau and Beijing Agricultural University (Shanghai: Commercial Press, 1952, 2nd ed. 1953, 3rd ed. 1954). Mi Jingjiu is among the listed editors. Problems with first edition: see preface to second edition, dated August 1953. Curriculum guide: See bibliography in *Darwin Zhuyi* (Beijing: Ministry of Education, October 1955), 11.

13. Zhang's review essay in *Shiwuxue tongbao* (Zoology), no. 3 (1953). Zhang had published at least two books by this time: He was editor and translator of *Sulian di xin yichuanxue* (Soviet's new genetics) (Beijing, 1950). This is a unique anthology of European leftist observations on Lysenkoism, with pieces by Jeanne Levy, Marcel Prenant, LeGros Clark, Bernhard J. Stern, Harold Thomas, and J. B. S. Haldane. And see Zhang's primer, *Dongwu di jinhua* (Zoological evolution) (Shanghai, n.p., 1951).

14. Zhou's response (along with others): *Shiwuxue tongbao*, no. 10 (1953): 390-97.

15. For Zhou Jianren's use by the CCP, see biography and evaluation in BDCC, 1: 202-4.

16. Zhang's 1954 middle-school primer was *Darwin xueshuo jichu* (Foundations of Darwinian theory). I could not find a copy of this book and know it only from the reviews, such as: Song Zhenneng, KXTB, no. 1 (1954): 72-75; Huang Zuojieh, "Zai ping Zhang Zongping 'Darwin xueshuo jichu,'" *Yichuanxue jikan*, no. 1 (1956): 92-114.

17. Textbook: *Darwin zhuyi jichu* is based on a 1942 Soviet textbook translated by the PRC's Ministry of Education under the title of *Darwin zhuyi jichu guoben*.

18. Fang's biography and bibliography is based on ZXS: 453-60.

19. For J. B. S. Haldane (1893-1964), British scientists, Marxism and Lysenkoism, see Gary Wersky, *The Visible College* (New York: Holt, Rinehart and Winston, 1978). Haldane was active in the Communist Party of Great Britain from 1942 to 1950.

20. The CPA director was Hu Yuzhi (b. 1896), former editor in chief of the famous journal *Dongfang zazhi* during the 1920s. During the 1940s he took refuge in Singapore and elsewhere in Southeast Asia. After 1949, he was a prominent leader of the Democratic League and poo-bah of the PRC publishing world. The CCP liked to cite him as an important example of nonparty intellectuals who chose to work with the Party. BDCC, 1: 385-87.

21. The job was given to him by Tong Dizhou, dean of the School of Natural Sciences and deputy director of the CAS Institute of Experimental Biology. See DSSI: 6. Tong's prominent work in experimental embryology will be discussed in the latter part of the text below.

22. For example, see Fang's flexible progression in: *Michulin xueshuo* (Michurin theory) (Beijing: China Youth Press, September 1955); "Mendel-Morganism and Darwinism" (no Michurin biology involved), *Shiwuxue tongbao*, no. 4 (1957): 1-7; "On the concept of genes—critically accept the Morganist theory of genes," *Guangming ribao*, Beijing, April 30, 1961 (trans. in *China Background*, 659: 1-5); and finally, a straight-ahead Darwinist, neo-Darwinist summary that mentions Michurinism only in passing, in *Shengwu di jinhua* (Evolution of life) (Beijing: Science Press, 1964). And for the beginnings of Fang's research: "The breeding of a new variety of Haidai (Laminaria Japonica Aresch.)," in English, *Scientia Sinica*, no. 7. (1963): 1011-18. Original in *Acta Botanica Sinica*, no. 3 (1963): 197-209.

23. Exhibition: reported by Pei Shizhang, KXTB, no. 2 (1953): 9.

24. Pei Wenzhong, *Zeran fazhan jianshi* (Beijing, 1950). "Afterword" dated December 1949. The text was drawn from lectures he delivered at a secondary school. These in turn were recycled as lectures at an education conference, and then as newspaper articles.

25. Pei, *Zeran fazhan*, preface; and also see BDRC, 2: 67-69.

26. For example, see the anthropologist Lin Yaohua's *Zong yuan dao ren di yanjiu* (The study of ape to human) (Beijing, 1951). Also see "Yu yuanshi shengwu dao ren so jingguo di jige daguan" (From the origins of life to humans—connections) by Yang Zhongjian, one of China's leading paleontologists. This latter piece is purely descriptive of the data and uses no dialectical materialist jargon. See *Zeran kexueh*, no. 3 (1951): 175-78.

27. The *People's Daily* criticism is in *Renmin ribao*, June 17, 1951. For a review of his case and sample criticisms, see KXTB, no. 4 (1955): 74-81, and no. 11 (1955): 101-2. For the CAS symposium, see *Shiwuxue tongbao*, no. 11 (1955): 5. Also see Liu's exposé of his ordeal in *Guangming ribao*, May 1, 1956, discussed in Roderick MacFarquhar, *The Hundred Flowers* (New York: Columbia University Press, 1960), 90-91.

28. For the Soviet speciation debates, see Loren R. Graham, *Science and Philosophy in the Soviet Union*, 239-41. V. N Sukachev, N. D. Ivanov, and N. V. Turbin were among the principal critics translated in the Chinese survey of the debates.

29. The CAS translation series was *Sulian Guanyu wuzhong yu wuzhong xingcheng wenti di taolun* (Soviet debate on species and their origin) (Beijing: Science Press), vol. 1 (October 1954) to vol. 21 (June 1957). And see KXTB, no. 12 (1954): 26-37 for initiating periodical discussion.

30. KXTB, no. 12 (1954): 29. The statement came from *Botanical Journal*, February 1954.

31. For editorial, see byline "Zhongming," KXTB, no. 4 (1955): 83-84.

32. It will be recalled from chapter 2 that Hu Xiansu was the Harvard-trained botanist who helped to create the biology department of National Central University in the 1920s. Hu's book was *Zhiwu fenlei xue jianpian* (A short course on plant taxonomy) (Beijing: Gaojiao Publishers, 1955). For this citation and the general description of Hu's persecution, see Li Peishan et al., "Qingdao zuotanhui di lishi beijing ho jiben jingxian" (The Qingdao conference of 1956 on genetics: The historical background and fundamental experiences), ZBT, no. 4 (1985). Translated in *Chinese Studies in the History and Philosophy of Science and Technology*, ed. Fan Dainian and Robert S. Cohen (Boston: Kluwer, 1996), 41-54.

33. RMRB, November 1, 1955.

34. Tong Dizhou, October 28, 1955, centenary speech, KXTB, no. 11 (1955): 21-23.

35. Fancy collected works: *Michulin xuanji*, comp. North China Agricultural Research Institute (Beijing, 1956); and Liu Fulin, trans., *Michulin yibainian jinian wenji* (Beijing, 1957).

36. Agricultural contributions: Special Michurin issue of *Shiwuxue tongbao*, no. 10 (1955): 1-28; and Guan Xianghuan, survey article in *Nongyeh xuebao* (Acta agricultura sinica), no. 4 (1955): 373-98.

37. Fang Zhongxi, *Michulin xueshuo* (Beijing: Youth Press, 1955); and Zu Deming et al., *Michulin shengwuxue tongsu jianghua* (Beijing: Association for the Popularization of Science and Technology Knowledge, 1955).

38. As early as 1953, Joseph Needham reported on the enthusiastic popular science campaign and its acceptance by the populous. See his "Chinese Science Revisited (2)," *Nature*, no. 4346 (1953): 283-84. And see reports by the Association for the Dissemination of Science and Technology Knowledge in NCNA Tientsin, July 10, 1953 (SCMP/608: 32); and Bejing, *Guangming ribao*, September 3, 1953 (SCMP/656: 26-27). The campaign was even stronger by 1956 according to reports in NCNA. Peking, October 31, 1956 (SCMP/1403: 18; and also 1404: 14; 1412: 4-18).

39. See Gao Shiji, ed., *Zeran kexue tongsuhua wenti* (On the question of popularizing natural science) (Beijing, 1955). This is an anthology of articles celebrating Ilin.

40. Mi Jingjiu, "Guanyu wuzhong ji wuzhong xingcheng wenti di taolun" (On the speciation debates), *Yichuanxue jikan* (Genetics), no. 1 (1956): 54-70.

41. For Stubbe's work, see L. I. Blacher, *The Problem of the Inheritance of Acquired Characters*, ed., trans. F. B. Churchill (Washington D.C.: Smithsonian Institution and National Science Foundation, 1982), 237-38. Original Russian edition (Moscow: Nauka, 1971). Stubbe's summation of his research: "Uber die vegetative Hybridisierung von Pflanzen Versuche an Tomaten-mutanten," *Kulturpflanze*, no. 2 (1954): 183-236. Stubbe conducted his research at the Genetics Institute, Martin Luther University, Halle-Wittenberg. Blacher's summary of Stubbe's work notes that he reached his conclusion "On the basis of 2,455 grafts which he performed on different varieties of tomato to clar-

ify the mutual influence of the stock and scion, Stubbe concluded, 'not one instance produced evidence of the transfer of genetically conditioned traits from one of the graft components to the other in the graft generation.' He studied the traits of 15,560 progeny of 351 grafted plants and observed no influence whatsoever of the scion on the stock. Segregation of traits in the offspring of heterozygous graft components and in ungrafted control plants took place in accord with normal monohybrid crossings" (Blacher, 237).

42. Abstract of Stubbe's research: translated by D. V. Lebedev, in *Botanical Journal*, no. 40 (1955): 603-4. For Stubbe's visit to Beijing Agricultural University, Interview, Wu Zhongxian.

43. H. Bohme: Interview with Raisa Berg, St. Louis, November 1984. Dr. Berg was on the biology faculty at Leningrad University at the time that Bohme came to work with Turbin. The latter had once been a strong Lysenkoite but by 1954 had turned against Lysenko. In the Soviet Union, Bohme researched the same issue as Stubbe, came to the same conclusions, and, according to Berg, announced those conclusions in the Soviet Union even before Stubbe. For Bohme's work, see Blacher, 238. Bohme concluded in 1954 that "directed changes do not occur in grafted plants, grafts do not affect the segregation of hybrid traits, and surely they do not produce anything resembling a vegetative hybrid" ("Untersuchungen zum Problem der genetischen Bedeutung von Pfropfungent zwischen genotypisch verschiedenen Pflanzen," *Z. Pflanzenzucht*, no. 33 (1954): 367-418).

44. Stubbe's visit: Yu Guangyuan, "Zai 1956 Qingdao yichuan xuehui shang di jianghua" (Unofficial materials on the Qingdao genetics symposium), ZBT, no. 5 (1980): 6.

45. Information about Stubbe to CCP and Mao: Gong Yuzhi, "Fazhan kexue di biyu zhi lu" (Develop the necessary road for science), *Guangming ribao* (December 28, 1983): 2.

46. Roderick MacFarquhar, *The Origins of the Cultural Revolution*, vol. 1 (New York: Columbia University Press, 1974), 48.

47. Hybrid corn: See Joravsky, *Lysenko Affair*, chapter 9; and Graham, *Science and Philosophy in the Soviet Union*, 243. And see Chinese evaluation in Laurence Schneider, ed., *Lysenkoism in China: Proceedings of the 1956 Qingdao Genetics Symposium*, trans. Laurence Schneider and Qin Shihzhen (Armonk, N.Y.: M. E. Sharpe, 1986), 16.

48. See description in Graham, *Science and Philosophy in the Soviet Union*, 240-41.

49. Tsitsin: See Joravsky, *Lysenko Affair*, 81-83, 160, 400. And also see Blacher, *Problem of the Inheritance of Acquired Characters*, 237. For the Chinese response: Li Peishan, "Qingdao zuotanhui," 44; and Schneider, *Lysenkoism in China*, passim.

50. Zhou: "On the Question of the Intellectuals," January 14, 1956, in *Communist China 1955-1959: Policy Documents with Analysis*, ed., trans. R. R. Bowie and John K. Fairbank (Cambridge, Mass.: Harvard University Press, 1962), 129.

51. Zhou, 136, 138.

52. Zhou, 137.

53. The most extensive discussion of the twelve-year science plan is in Xu Liangying and Fan Dainian, *Science and Socialist Construction* (Armonk, N.Y.: M. E. Sharpe, 1982). This is a translation of *Kexue he wo guo shehuizhuyi jianshe* (Beijing: Renmin, 1957). The discussion also has much to say about the genetics question and the Double Hundred policy. And also see organizational details for the plan in NCNA, March 14, 1956 (SCMP, 1251: 7) and December 29, 1956 (SCMP, 1453: 8).

54. Gong Yuzhi, "Fazhan kexue."

55. Lu Dingyi (b. 1901): A party member since 1924, he was associated with CCP publishing, propaganda work, and education policy. He spent most of his long-lived career at the center of CCP power, and was at his pinnacle when he delivered his famous Hundred Flowers speech in 1956. BDCC, 2: 661-65.

56. Yu Guangyuan (b. 1915) has eluded biographers, in spite of his significant role in party-intellectual relations. He was educated at Qinghua University during the 1930s, and received a strong foundation in social theory as well as mathematics and the philosophy of science. After 1949, he was a prolific contributor to *Xuexi* (Study) the major party intellectual journal that later was called *Hongqi*. He was a sophisticated student of dialectical materialism and a translator of Engels' *Dialectics of Nature*. He also wrote knowledgeably on probability theory and statistics in natural science research, and was discreetly critical of Lysenko's attitude on these subjects. After the Cultural Revolution, he lent his clout to the creation of China's most significant group of intellectuals devoted to the history and philosophy of science, and science policy. Their work can be seen in the journal *Zeran bianzhengfa tongxun* (Dialectics of Nature journal), to be discussed in the text, below.

57. Genetics symposium: See Gong Yuzhi, "Fazhan kexueh"; also see Tong Dizhou, opening speech at Qingdao Genetics Symposium, August 1956, in Schneider, *Lysenkoism in China*, 1.

# 6

# The "Double Hundred" Policy and the Restoration of Genetics, 1956-1961

By 1956, the "genetics question" had become framed by the forthcoming second five-year plan and twelve-year science plan as well as the Hundred Flowers, Hundred Schools policy. It might have been lost among these complex initiatives, but instead it gained considerable prominence because party leaders chose to use the resolution of the genetics question as a model for future relations between the Party and the science community. In the following analysis, the genetics question will initially be seen as a reference point and illustration in some prominent policy statements. Then, when the "Double Hundred" policy invites scientists to vent their concerns publicly, the genetics question becomes a source of unanticipated rage against the Party, Lysenkoism, and the Soviet Union. The genetics question most fully assumes its role as model when it comes to be resolved in a set of carefully crafted principles and a well-orchestrated political ritual that remove the Party's ban on genetics and abrogate its sanction of the Lysenkoist monopoly. The science community, however, had little time to consider the implications of this apparent victory or to take advantage of it, for it was quickly followed by a massive purge of the intellectual community in the anti-rightist campaign and then by the Great Leap Forward. Only during the

efforts to recover from the Great Leap disasters did biologists find significant
and sustained opportunities to rebuild their community.

## Initiating the "Double Hundred" Policy

At the end of April and in early May 1956, Chairman Mao delivered two major
policy speeches that further enabled Lu Dingyi's Propaganda Department to
resolve the genetics question. In "The Ten Great Relationships," Mao had
something further to say about China's dependency on other countries: "We
mustn't copy everything indiscriminately and transplant mechanically. Naturally, we mustn't pick up their shortcomings and weak points." He said China
had adopted weaknesses from the Soviet Union in particular: "While [Chinese]
were swelling with pride over what they had picked up [from socialist countries
and the Soviets], it was already being discarded in those countries; as a result,
they had to do a somersault like the [legendary] Monkey Sun Wugong."[1]

On the second of May, Mao presented his "Hundred Flowers, Hundred
Schools" speech, but it was never published. Experts are convinced that the substance of that speech was accurately reflected in Lu Dingyi's detailed May 26
address, which became the main guide for the Double Hundred campaign and
for the resolution of the genetics question.[2] Lu Dingyi's speech can be thought
of as an elaboration of certain aspects of Zhou Enlai's list of recommendations
for the treatment of intellectuals—especially in the areas of party-intellectual
relations and the influence of the Soviet Union. It is based on the same premises:
that intellectual creativity and innovation require the open interaction of view-
points, ideas, and opinions; that disagreement and debate are potentially engines
of advancement; and that the advancement of the Chinese revolution in the area
of economic development is dependent on the expertise and creativity of Chinese intellectuals. Thus, the Double Hundred's relatively liberal policy toward
intellectuals is derived from bluntly utilitarian criteria; the policy has nothing to
do with concepts like academic freedom or individual rights. As Maurice Meisner sums it up, "intellectual repression was becoming an economic liability."[3]

At the outset of his presentation, Lu Dingyi underlines this utilitarianism
when he distinguishes between the arts and literature (symbolized in the Hundred Flowers) and the natural sciences (the Hundred Schools). The latter are
privileged in this discussion by being declared not to have a class character; that
is, they are not part of the cultural superstructure, where art and literature reside;
but rather they are a fundamental component of the relations of production, they
are part of the process of production.

> [The natural sciences, including medicine] have their own laws of develop-
> ment. The only way they tie up with social institutions is that under a bad social
> system they make rather slow progress, and under a better one they progress
> fairly rapidly. . . . It is, therefore, wrong to label a particular theory in medicine,

biology or any other branch of natural science "feudal," "capitalist," "socialist," "proletarian" or "bourgeois."[4]

Lu concludes from this that it is "therefore wrong, for instance, to say that 'Michurin's theory is socialist,' or that 'Mendel's and Morgan's principles of heredity are capitalist.' We must not believe such stuff." He then connects this kind of error with the questions of party/scientist and Soviet/Chinese relations: "Some people make this sort of mistake because they are sectarian. Others do it unconsciously by trying to emphasize, but not in the proper way, that one ought to learn from the latest scientific achievements in the Soviet Union."[5]

By the same token, Lu expresses intolerance for those scientists who would not consider Michurin biology because they are politically opposed to the Soviet Union or because they belong to another school of thought. In effect he is saying: Don't throw the baby out with the bathwater; welcome useful science wherever it comes from and separate political from scientific issues.

Lu reemphasizes Zhou Enlai's charge to the Party to develop better relations with intellectuals. He likewise urges intellectuals to give more considered attention to the Party and to what it stands for; but he clearly expects the initiative to come from party cadre, whom he accuses of arrogance and hostile behavior toward intellectuals and a doctrinaire approach to academic studies, artistic and literary work, and scientific research. Lu expects the Party to encourage free and open discussion among the intellectuals and between them and the Party—without recourse to any special administrative measures; and he further expects that everyone will respect the right of minorities to be heard.

The Double Hundred speech concluded with another call for a balanced approach to learning from abroad: "We must have our national pride, but we must not become national nihilists." Lu warns that "It is utterly wrong not to learn from the Soviet Union. Nevertheless, in learning from the Soviet Union we must not mechanically copy everything in the Soviet Union in a doctrinaire way. We must make what we have learned fit our actual conditions."[6]

## Expressions of Dissent under the Double Hundred Umbrella

Well before Lu Dingyi's speech was delivered, party cadre had been charged to solicit intellectuals to air their concerns, criticisms, and suggestions, but there was resistance on the part of some party leadership to carry this out. For different reasons, there was some reluctance on the part of some intellectuals to go public. Nevertheless, from the beginning of 1956, scientists and literary intellectuals were already expressing in personal terms the kinds of issues that Zhou Enlai, Mao, and Lu Dingyi spoke about abstractly.[7]

For example, in the May 1 issue of the *Guangming Daily* news, Tan Jiazhen and Liu Xian, both from Fudan University in Shanghai, wrote candidly about their problems. Liu wrote a stunning exposé of his ordeals since the publication of his book on human evolution. Tan, who had been discreetly silent since his 1952 "self-criticism," wrote as chair of his biology department that

> many top class intellectuals are tied down with administrative duties and very few of them are actually doing any research. This is an enormous waste. . . . Here I think it is necessary to repeat: there is a difference between science and politics. A scientist should not separate himself from politics; but sciences themselves—particularly the natural sciences are classless.[8]

Many other general complaints were published at this time about mechanically copying the Soviets, the overwhelming presence of Soviet textbooks, and the proscription against using anything from American science and culture.[9] By the end of May, scientists were speaking out directly to the issue of Lysenkoism. For example, a Beijing Agricultural University professor with a doctorate from Cornell wrote of his support for the new openness policy because of his fearful experience with Lysenkoism. In the Beijing *Guangming Daily*, he wrote that those faculty who were critical of Lysenko had not spoken out before for fear of reprisal and being charged with undermining confidence in Soviet science.[10]

### The Case of Bao Wenkui

All of these revelations are important for an appreciation of the collective experience of Chinese scientists during the 1950s, but a few stand out because of the indisputable stature of the scientists in question and the extraordinary impact their stories had on party leaders and the public. The experience of the plant geneticist Bao Wenkui is a major case in point. Bao was discussed earlier, as a precocious graduate assistant at the wartime National Rice and Wheat Improvement Institute in Sichuan Province. His outstanding contributions in polyploid research at that time were subsequently rewarded with a fellowship to Caltech, where he received his doctorate in genetics and biochemistry in 1950.[11]

In 1949, Bao Wenkui had familiarized himself with Lysenko by reading *The Situation in Biological Science*, but he never expected to have to deal with Lysenkoism in China. He felt a strong obligation to work for China and rejected possible alternatives to returning home. Taiwan was not appealing to him, even though his former patron, Li Xianwen, had moved there; and he turned down an offer from his Caltech professors to stay in the United States.

Bao had a good position waiting for him in Chengdu University in the fall of 1950. He found the government very helpful to returning overseas students, and the university and party officials quite open-minded about his work, allowing him to teach and do research in his choice of subjects. The central and provincial governments provided him with generous research funding. All of this continued for two years without a hint of Lysenkoism in Sichuan Province. Although in 1952, the Soviet Lysenkoite I. I. Ivanov came through Chengdu on a national lecture tour, long-term Lysenkoite advisers did not arrive until the next year. Meanwhile, party cadre assigned to Bao's research institute revealed themselves to be completely unfamiliar with polyploid research, but saw no reason why Bao should not continue with it. That he did, prolifically and with great success, until the winter of 1954, when provincial authorities met to assess biological research and realized that Lysenko had condemned polyploid research years earlier. Though nothing else had changed, authorities at the institute immediately destroyed all of his experimental plots, and with them, years of research. He of course was ordered not to resume his work.[12]

In early 1955, Bao Wenkui wrote to the Ministry of Agriculture about this incident; in June, the ministry, in an unprecedented act, ordered his institute to permit him to resume his research. On September 8, 1956, *People's Daily* featured his experience as an example of how the Double Hundred policy sets things right. The paper gave Bao three-quarters of a page to tell the entire story in detail.[13] Before the year was over, Bao had been appointed to the CAS and to a full professorship at Beijing Agricultural University. He accepted both appointments and continued to work in both institutions until his retirement in the 1980s. Perhaps because of its appearance in the nation's most important political newspaper, Bao's story became a near-legendary and oft-told tale.[14]

In August of 1956, personal testimony for the Hundred Schools policy had spread from the national papers to the science journals. *Science* began an extensive series of forums for which dozens of scientists wrote paragraph- to page-length statements on what was good about the policy and what problems it would help to solve. Among the most common sources of outrage was the Soviet Union's direct or indirect interference with the heretofore commonly accepted scientific techniques for determining facts and laws, and its arbitrary and dogmatic imposition of alleged scientific truths.

For this forum, an editor of the Science Press condemned the role that the press was forced to play in Soviet dominance of science in China. He wrote that since its 1950 inception to mid-1956, Science Press had published 330 science monograph titles, but only 11 came from outside of the Communist bloc. He said the press had to smuggle translations of European or American publications into China under the camouflage of phony Russian title pages. He revealed how Science Press was censored and tightly controlled where the speciation debates were concerned; as a result, the press was permitted to translate nothing that was critical of the Lysenkoites. This editor was convinced that general deference to Soviet science in combination specifically with Lysenkoism were responsible for a chilling effect throughout China's scientific community. His local remedies were to enlarge the editorial independence of Science Press; expand its scope beyond the current "one field, one school" approach; and begin providing China's science community with translations from all schools of Soviet science, as well as from Europe and America.[15]

## Resolving the Genetics Question

In the thick of this public outpouring of criticism, Lu Dingyi scheduled an elaborate, two-week-long symposium devoted to the genetics question. The genesis of this event can be traced back to Mao, earlier in the year, when he was apprised of the blow struck against Lysenko by East German scientists. Mao instructed Lu's Central Propaganda Committee to research the question of Lysenkoism and genetics with the aid of the CAS and all other relevant government agencies and persons.[16] The symposium was one of the outcomes of that research, and Lu Dingyi invested it with particular importance because it would be the first science conference since the promulgation of the Hundred Schools policy. He felt that, being first, the symposium should be a model for developing and implementing the policy.[17] The best way to accomplish this, Lu told Yu Guangyuan, would be to bring together representatives from both "schools" for "continuous discussion and heated talk," and "free, unrestrained debate."[18] To facilitate all this, the symposium was held in the northeastern resort city of Qingdao, from August 10-25, 1956.

*Preparing for the Qingdao Genetics Symposium*

Yu Guangyuan asked the CAS and the Ministry of Higher Education to supervise the organization of the symposium. From a representative cross section of biology fields and the two schools, they invited fifty-three scientists to participate in symposium presentations and discussions; forty-three were able to attend. They also invited another seventy-three observers, representing the Party, educational and research institutions, and the ministries. Notably, the strongly pro-Lysenkoist Ministry of Agriculture was not involved in organizing the event though it was invited to send a representative. The symposium organizers invited a few leading figures from each side to write up short introductory presentations for each session, and others were invited to prepare short commentaries; the rest of the discussion was expected to be extemporized. Additionally, the organizers commissioned a group of scholars to prepare for the symposium participants a detailed chronicle of the career of Lysenko and the development of Lysenkoism in the Soviet Union. This resulted in a sixty-seven-page narrative that was by far China's most detailed, accurate, and neutral introduction to the topics.[19]

Before the symposium began, Lu Dingyi and Yu Guangyuan had distilled a set of broad principles from the Hundred Schools policy specifically for the genetics question. They asked the participants to keep these in mind as they carried out their discussions and tried to come to an understanding of what steps should be taken next. The first principle was that the CCP would not make decisions about the specific meaning and application of the Hundred Schools policy; such decisions would be left to the scientific community itself. Second, science orthodoxies and monopolies were impermissible: The CCP would not do what the Soviet Communist Party had done when it "created the Lysenko faction and gave it its privileged place." The CCP would not make rulings on scientific controversies; and the ideas of minorities would be protected. Third, political attacks disguised as philosophical labels must cease. And last, only scientists should decide if and when philosophy is of any aid to science. The symposium was asked to consider how these principles might be implemented in the area of biology through reforms of education and research.[20] Not on this occasion, nor on any other did the Party acknowledge that its 1952 policy on Michurin biology was the source of the problems that Lu and Ding were trying to fix; or that the CCP had already done what the CPSPU did when it created the Lysenko faction.

### New Party Guidelines for Biology

These "principles" fundamentally resolved the key problems caused by the 1952 party endorsement of a monopoly for Michurin biology and a ban on "Morgan genetics." Before the symposium began, the principles were already party decisions. Lysenkoites came to the symposium knowing that their privileged position was gone and their hegemony was over; Morganists came with the knowledge that they could resume their teaching and research according to their own lights, and that doing so could no longer be grounds for libelous attack and party repression. In fact, Lu Dingyi and the Party had reached a decision to abrogate the 1952 policy even before the symposium plans were in place. Wu Zongxian reports that the Party informed the Morganists "to return to their work" in education and research in the early summer of 1956. The Morganists met to consider this welcome development, however, Wu and others argued that it would be a mistake to resume their work without a specific and clearly articulated statement from the highest party level that abolished "capping" and its attendant party-sponsored penalties. According to Wu, Lu Dingyi eventually brought back assurances from Zhou Enlai that the abolition of capping would be among the principles in place when they resumed their work—presumably after the Qingdao Symposium.[21]

If the basic issues were already resolved and no longer up for discussion, then what role was the symposium supposed to play? Why did the Party insist on holding it? One motivation was perhaps to promulgate the principles and to resolve the genetics question in an appropriate ritual attended by representatives of all the main constituencies: the Party, the CAS, the ministries, the Lysenkoites, the Morganists. Part of this ritual was the symposium's open discussion of Western genetics—the first since 1949—and the unprecedented debate between Lysenkoites and Morganists. Lu Dingyi suggested another reason for the symposium when he told the participants that the clash of ideas was healthy for the advancement of science and that he hoped the clashes at the symposium would lead each side to adopt the strong points of the other; hinting vaguely at some kind of synthesis to come about eventually.[22]

The most formal statement of the symposium's purpose was made by its chair, Tong Dizhou, in his opening speech, delivered less than a year after his Michurin centenary defense of Lysenkoism and denunciation of its Chinese critics.[23] The symposium was called, Tong said, "to overthrow the one-sided and narrow point of view so that genetic science in China may develop within the twelve-year [plan] to reach international standards." Tong stated that among the reasons for genetics' lack of development in China was the failure to borrow intelligently from the Soviet Union and to study scientific development in other countries. As a result, China mistakenly relied on Lysenko while "Morgan's theories were only criticized without study; . . . [and consequently] a segment of our scientific workers not only did not dare to rely on Morgan's theories to advance their research work, but further, they did not dare introduce to anyone the details and current advances of Morganist theories."

Tong's speech bemoaned the "mental barrier" that was dividing China's biological sciences because there has never been a thorough study of each school's theories and accomplishments before coming to conclusions and determining truth or falsity; and because scholarly dialogue had not developed between the schools and therefore there had been no opportunity for mutual understanding and study. The opening speech went on to set three goals for the symposium, which could be achieved, Tong said, only if all participants would lose their inhibitions and speak out boldly, safe in the knowledge that no one was going to be putting nasty labels on them. The first goal was to achieve a comprehensive analysis of the recently developing international conditions of genetics. Herein, he requested that there be no arguments about which theories were better, Morgan's or Lysenko's, but rather, analytical discussions should be advanced on the basis of concrete facts and recent accomplishments.

The second goal was to discuss how genetics should be taught, and to achieve a unified viewpoint that could be proposed and considered for adoption by the ministries of education. "At present," Tong said, "there is great discomfort about teaching genetics in universities. Previously, Morgan's theories were criticized; one dared not introduce them. Now, Lysenko's theories are criticized, and all the more one does not know where to turn."

The last goal was to achieve a unified point of view on the goals of genetic research. "In the past, the majority of work was carried out in compliance with Lysenko's theories. There was virtually complete prohibition of research using Morgan's theories." Now, Tong concluded, proposals for genetics research needed to be included in the twelve-year science plan.

## The Qingdao Genetics Symposium, August 1956

The symposium was not organized to address Tong Dizhou's three goals. Instead, its fifteen days were divided up equally among four ponderous topics: the material basis of heredity; heredity and environment; heredity and individual development; heredity and evolution. On the last day, just a few hours were allotted to a discussion of genetics in education. At the beginning of each of the five main units, a representative of Morganism or Michurinism gave a detailed presentation introducing his school's position on the main topic; this was followed by a number of prepared responses to the presentation; then discussion was opened up to all. The initiative was clearly in the hands of the Morganists to describe recent international developments in genetics and the current state-of-the-field in each topic area. The Michurinists either rehashed their very familiar positions on the key topics or took a defensive posture as they dealt with the unfamiliar need to respond to questions—many pointed questions—and to cope with a barrage of new information and data.

Political roles were in plain evidence and affected the tone and rhythm of the symposium. The Morganists were led by Tan Jiazhen, who gave the symbolically important first presentation on the first day. He succinctly outlined the main historical developments of genetics from the nineteenth century through the 1940s. Typical of his demeanor throughout the symposium, he was controlled, factual, and unprovocative. Always politic, he nevertheless showed himself willing and able to engage in intelligent discourse with the Michurinists; he left it to his colleagues to engage in heated rhetoric and to throw difficult questions. Some of those colleagues had been Tan's pre-1949 students, recently returned from the United States with fresh doctorates in genetics. They played a decidedly more feisty and uncompromising role to Tan's statesmanship; and they were a major source of intelligence about the development of genetics since 1949.[24]

The leader of the Michurinists was Zu Deming,[25] who had inherited the position from Luo Tianyu in 1952. He matched Tan's courtesy and diplomatic air, and he obviously saw to it that his colleagues curb their libelous inclinations, in keeping with the latest party fiat. Throughout the symposium, Zu and his colleagues were careful to conclude their comments with a conspicuous salute to the wisdom of the Hundred Schools policy and a fervent endorsement of Lu Dingyi's admonition that the two schools cooperate and learn from each other's strong points.[26] Most notably, Zu Deming and his colleagues made efforts to distance themselves from Lysenko without undermining the authority of Michurin biology in China. They requested that, in their analysis and criticism, the Morganists "should not lump together Academician Lysenko and Michurin"; and that Lysenko's mistakes "not be taken as the mistakes of the entire Michurin school."[27]

*Reaching for the Symposium's Goals*

Did the symposium accomplish the goals set out for it by Tong Dizhou's opening speech? The transcript of the proceedings shows that—given the limited time frame and the rather chaotic structure of the symposium—the Morganists were able to present useful, if not "comprehensive," state-of-the-art summaries of international developments in genetics. By the end of the first day of the first session, for example, Tan Jiazhen and two of his former students had made presentations that concisely indicated how the science of genetics had come to draw upon the methods and findings of other areas of biology like biochemistry, biophysics, and molecular biology in general. They outlined the implications of identifying genes with DNA and RNA; and illustrated the kinds of research, in microbiology for example, that the work of Watson and Crick suggested for the future. For the topic of heredity and evolution, the veteran entomologist Chen

Shixiang had been delegated the responsibility for the main survey. He provided a sophisticated introduction to the "evolutionary synthesis," and to the insights of population genetics.

Notwithstanding these well-crafted presentations, however, the proceedings' transcript indicates that in each session of the symposium, the goal of understanding the current international state of genetics was quickly forgotten and replaced by a free-for-all debate about the fine points of Michurin biology. Whatever value this may have had, it clearly diminished the symposium's ability to deal with the current state of international genetics. As a result, the symposium was not able to respond to its second goal, that is, to recommend to the twelve-year science planning committee the direction that genetics research should take. Nevertheless, Yu Guangyuan—behind the scenes—did broker an institutional recommendation from the symposium to the twelve-year planning committee. On behalf of the symposium, he proposed that the CAS establish an Institute of Genetics, thereby recognizing the importance that this science had acquired. The new institute could assume the responsibility of guiding the future direction of genetics research.[28]

Nor was the symposium able to address adequately the third goal, making a recommendation for the teaching of genetics. In the few hours given to discussion on this subject there was strong agreement that biology textbooks in China were wretchedly bad, largely due to the Education Ministry's insistence on the use of Soviet materials. There was also agreement that the biology educational curriculum must open itself up to material from wherever good science is done. Chen Zhen's famous general biology textbook was recalled with fondness and nostalgia and suggested as an antidote to the current crop of translated Soviet materials. All in all, the discussion here expressed enormous frustration and a sense of intimidation by the immense task of retraining teachers, writing and translating new materials, and trying to figure out when or whether to teach only Morganism, only Michurinism, or both.

The education ministries were impatient for policy guidelines on these latter questions, and so Yu Guangyuan made sure that some decisions were made "in the corridors" if not at a formal symposium session.[29] As a result, Tong Dizhou was able to announce in his closing speech that it had been decided that all colleges and universities would be expected forthwith to teach both Morganist and Michurin biology in genetics courses, and that the Ministry of Higher Education would be obliged to facilitate the creation of the new curriculum. The Ministry of (secondary) Education would be left to determine on its own what kinds of curriculum changes to make within secondary education; it was clear, however, that it was expected to integrate Morganism into the curriculum soon.[30]

### Critiquing Lysenkoism as Science

If the symposium was less productive than hoped for in the realm of policy, it obviously provided ample opportunity for Morganists to express some of their fundamental frustrations and dilemmas in dealing with Lysenkoism as "science." The proceedings' transcript records a series of piecemeal observations on methodology and "language" that may have revealed as much to the symposium audience about Lysenkoism as all the arguments about speciation and heredity combined.

The issue that seems to have triggered these observations was Lu Dingyi's expressed hope that the two schools, through dialogue, could find some middle ground or generate some kind of synthesis. In the symposium, when a follower of Michurin biology speculated on areas of "compromise" between the two schools, one of Tan Jiazhen's protégés shot back that

> there is no reason to compromise for the sake of compromise or for the sake of teaching. Whether or not [these schools] will compromise with each other depends on whether there is a common language. This is the key point. The point is not whether grafting [for example] is successful, or whether the grafting is to be done at all. It is whether there is a common language. If people have to use an interpreter to communicate with each other, it is very difficult to compromise.[31]

In their symposium comments, critics of Lysenkoism constantly struggled with the exclusiveness and arbitrariness of its language. Wu Zhongxian, for example, took issue with a Lysenkoite's defense of his school's introduction of new scientific language and his claim that this should not be a source of difficulty: "If we know the work of others," said the Lysenkoite, "we understand their language; but if we try to force others to use our language, then we have problems." Wu responded:

> If you are speaking about a Chinese who must communicate with a foreign country, then naturally he must understand the other people's language. But in regard to science, this is not appropriate. No matter which particular science, science has only one language. In physics, one cannot say there is one language for Newton, one for Einstein. Genetics is the same. Science has only one language. We cannot take the views of one person or one minority as another scientific language.[32]

In yet another discussion, Li Jingxiong, a veteran plant geneticist, observed that "a controversy between two schools must first have a common language." If it is otherwise, he said, "it is like the husband saying he is right and the wife saying she is right, and it is very difficult to talk. I believe that—no matter which faction's methods, results, or explanations—if the other faction disagrees, it must use the same language to debate."[33] Li argued that essential elements in the universal language of science are the very controls and statistical methods arbitrarily denounced by Lysenkoism. He suggested that the unilateral imposition of Lysenko's scientific language is the counterpart of Lysenkoite dismissal of a hundred years' development of biological research, data, and theory.

Li Ruqi, Tan Jiazhen's mentor, chose not to say much at the symposium, but this was one area where he did speak out, quite emotionally. He confessed to being utterly confused by Lysenkoite "double-talk"—especially their conceptions of "facts." "What we call scientific facts," Li said, "are facts that are obtained with relatively reliable materials and methods, requiring a minimum of controls, and necessarily treated statistically. This is the common language of our debates. If we want to understand the work of others, we must adopt this kind of language. When we adduce facts, we can then argue their validity."[34]

Lysenkoite double-talk, Li Ruqi said, was in part a result of their idiosyncratic approach to the facts, and it was also a symptom of their perverse logic. "When one is doing scientific research," he reminded the symposium, "at the outset one has to have an hypothesis, and this most certainly has to be based on the foundation of the scientific facts established by predecessors. After the hypothesis, then experimentation; and then analysis of the data, synthesis, and a conclusion. This is a step-by-step systematic process. This is what we call logic." However, the Lysenkoites exercise yet another kind of logic, Li observed: "First you have the conclusion, then you get the facts; then you take your conclusion and absolutize it, formulize it, and vulgarize it."[35]

These observations on the basics of scientific method apparently did have some resonance at the symposium with influential party intellectuals. For example, Yu Guangyuan stepped out of his ordinarily neutral public persona to express his contempt for the oft-repeated Lysenkoite disparagement of statistical techniques in science. In one of his between-session talks to the symposium, Yu made it clear that he found intolerable Lysenko's notion that "chance is the enemy of science," and that he personally would defend the validity and essential importance of statistical analysis to good science.[36]

## The Soviets, the CCP, and the "Genetics Question"

However much the Morganists may have appreciated the support of party officials, like Lu Dingyi, they nonetheless faulted the Party for forcing them to submit to Soviet "colonial science." At the Qingdao Symposium, no one publicly interpreted the Party's change of policy toward genetics to be a result of the admonitions of scientifically sophisticated cadre, like Yu Guangyuan. On the contrary, the common understanding was that the CCP only lifted the ban on genetics because the Soviets had changed their policy on Lysenkoism. The CCP was accused of merely continuing to ape the CPSU, incapable of acting on its own initiative and unwilling to trust the advice of its own science community.

While this cynicism is understandable, a more accurate interpretation would probably be that the changing position of Lysenkoism in the Soviet Union enabled spokespersons for the Chinese science community to get a hearing in higher echelons of the Party. Moreover, the new CCP policy toward genetics—in retrospect at least—clearly was part of a much more comprehensive evaluation of Chinese-Soviet relations, an ongoing evaluation that was increasingly more concerned about China's dependency, and increasingly disappointed with the benefits of the relationship.

Ultimately, imitation of the Soviets is an unconvincing explanation for the initiation and termination of the CCP's support of Lysenkoism. In the light of the CCP's future relationships with the science community, it becomes apparent that many of the most egregious attitudes of Lysenkoism toward science and scientists were renewed and implemented when the CCP was in an anti-Soviet mood, and when Lysenkoism was fading from significance or altogether marginalized in China. That is to say, these were occasions when CCP's science policy could not be explained as slavish imitation of the Soviet Union. No one blamed the Soviets for the harsh treatment of science during the Great Leap Forward (1958-1959) and the Cultural Revolution (1966-1976). Yet both of these campaigns reiterated the most provocative of Lysenkoism's principles, such as the class nature of science, the invalidity of "ivory-tower" science education and research, and the sovereignty of "people's science."

In the following discussions, there is a recurrent sense of déjà vu in the reappearance of antiscience policies once associated with Lysenkoism, followed by renewals of the Hundred Schools policy. This pattern strongly suggests that if Soviet Lysenkoism had not existed, the CCP would have invented something like it on its own.

## Third Time's the Charm: The Double Hundred and Genetics after Qingdao

After the Qingdao Symposium, the Double Hundred policy had to be recharged twice during the decade—once in 1957 and again in 1961. It was not until the third time around that genetics began to realize the promises of the policy and the Qingdao Symposium. Before that last point was reached, China's political economy experienced a series of upheavals, beginning with the 1957 antirightest campaign—a massive purge of the intellectual community; then the disastrous Great Leap Forward; and finally the unfriendly divorce of China and the Soviet Union, beginning in 1960 when all Soviet experts left China permanently.

In the wake of the Qingdao Symposium, it appeared that the Double Hundred policy was working well for the biological sciences, and vice versa. Lu Dingyi's Propaganda Department, the CAS, and the press facilitated maximum national exposure of the symposium and the principles and resolutions associated with it. The national New China News Agency relentlessly covered the Qingdao Symposium and its aftermath.[37] Then Yu Guangyuan spoke at the Eighth Party Congress in September 1956, explaining how the genetics question was resolved by the Hundred Schools policy.[38] The CAS was so overwhelmed with requests for information about the symposium that it took the unusual step of publishing and circulating its complete proceedings.[39]

In September and October, high-profile articles on the Qingdao Symposium appeared in *People's Daily*. In the first, Tan Jiazhen, now clearly considered the spokesperson for Morganists, wrote a powerful and uncompromising statement, quite unlike his cautious and diplomatic comments at the symposium. Tan asserted the strength and reliability of Morganist genetics, the fundamental weakness of Michurin biology, and the impossibility of compromise between them. The material basis of heredity, he said, was without doubt chromosomes, genes, and DNA. In nontechnical language, he corrected Lysenkoite misrepresentations of Morganist gene theory, and challenged their simplistic claims for the role of environment in biological variation. Tan concluded that Michurin biology had nothing to offer because it lacked a scientific basis and sufficient explanatory power.[40] Responding to Tan's statement, Zu Deming, leader of the Michurinists, displayed surprise and displeasure at Tan's independent stance and his uncompromising aspersions for Michurin biology. Appealing to Double Hundred values, his only rejoinder was the weak and defensive one that each of the two schools was incomplete and superficial without the work of the other.[41]

In spite of the success of the Qingdao Symposium, party cadre resistance to the Double Hundred policy continued. To some in party leadership, it seemed apparent that the Double Hundred was a weapon that Mao wanted to use for a rectification of the Party. Mao did indeed want a party rectification and recognized that criticism solicited by the Double Hundred could be useful for that purpose; however, for the first half of 1957, he was devoted to convincing the Party that the Double Hundred had its own reasons for being.[42]

In February 1957, Mao reemphasized the Double Hundred policy in his important speech "On Correctly Handling of Contradictions among the People." The most controversial point of this complex statement was that even though social classes and social class conflict had been essentially eliminated from China, there was perhaps always going to be some kind of contradiction or conflict within the new socialist society—among some of its elements, like intellectuals and party, or intellectuals and masses. Such conflicts need not be seriously disruptive, however, if the Party handled them in noncoercive ways. He said that

> Using administrative force to impose one particular genre coercively, or one particular school, while banning another genre or school, is, we believe, detrimental to the growth of art and science. Issues of right and wrong in art and science should be resolved through free discussion in artistic and scientific circles and through the practice of art and science.[43]

Mao took to the road and became a "roving lobbyist" for the Double Hundred campaign until May 1957, when the response of intellectuals reached a crescendo and newspapers and journals were packed with complaints and criticism, much of it stridently aimed at the Party. Where science was concerned, Mao's talks with various cadre groups around the country repeated the same themes: The Party leadership and rank and file lack the number of scientists necessary for the Party to give close, detailed guidance to the science community; therefore, the Party must not impose decisions on science that are best left to scientists.[44]

## Anti-Rightist Campaign, Great Leap Forward

At the beginning of June 1957, the criticism of the Party—encouraged by Mao and the Double Hundred campaign—had quite unexpectedly reached enormous and threatening proportions. Under heavy party pressure, Mao was forced to halt the campaign abruptly; and within months, the Party began attacking "extremists" who had revealed their "dangerous rightist tendencies."[45] By the end of 1957, party backlash accounted for "sending down" (exile to the countryside) thousands of intellectuals, while others were sent to labor camps and jail. The punitive process and results have been described at length and it is only necessary to note here that, while literary intellectuals seem to have taken the brunt of the anti-rightist campaign, many natural scientists were also victimized. It appears, however, that there was no ill consequence for any of the biologists or others who spoke out during the Double Hundred campaign against Lysenkoism and Michurin biology, or against Soviet intrusiveness. The anti-rightist backlash was aimed at those who directly criticized the Party or socialism.[46]

The Great Leap Forward of 1958-1959 presented a different kind of threat to the science community. Where the anti-rightist campaign sought to arrest and quash the development among the intellectuals of any kind of independence from strict party control, the Great Leap was concerned with the relationship of the technical intelligentsia and the masses. The Great Leap Forward campaign occurred at the time that the second five-year plan was supposed to have begun, and it was in preparation for that same occasion that the Double Hundred policy had addressed the needs of scientists whose skills would be essential to the success of the plan.

Mao Zedong and his supporters acknowledged the importance of being able to rely on Chinese instead of Soviet expertise; but they were obviously concerned no less that the Chinese people might surrender their future to a technical elite. It was felt that if the elite were not controlled ideologically and politically they surely would use their position to their own advantage and create social inequality. Therefore, they must be made to embody political correctness along with their expertise, they must become both "red and expert," and they must recognize that "redness" takes priority over expertise. But Mao's populist beliefs led him to a solution beyond creating and controlling technical experts; in the Great Leap Forward he expressed a conviction that the common, working masses—who were "naturally red"—were themselves capable of acquiring and using the technical expertise necessary to drive the Chinese economy to higher levels of output. As Meisner summed it up:

Mao's program, simply put, envisioned the development and application of modern science and technology without professional scientists and technocrats.

Just as the Chinese nation was to become technologically self-reliant and not dependent on other nations, so too were the Chinese people to become self-reliant and not dependent on a technological elite. Technological development was conceived as a mass movement, and one of the great rallying cries of the Great Leap Forward years was the slogan "the masses must make themselves masters of culture and science."[47]

To accomplish all this, a number of strategies were undertaken, chief among them being the mobilization of the peasant masses for science and technology innovation. Scientific experts were obliged to leave their institutes and universities to visit the farms and "learn from the farmers, sum up their invaluable experiences to lead to higher yields."[48] These same "sent down" experts were supposed to organize small local schools that would eliminate the separation of teaching from research and research from production.[49] Obversely, sometimes the hands-on experience of the masses was brought into the academy, as when twenty-one farmers were invited to join the regular research staff of the new CAS Institute of Agricultural Science.[50]

Most importantly, the object of the Great Leap Forward was to encourage and enable the spontaneous, natural scientific and technological creativity of the working masses. The rhetoric of the Great Leap Forward is often reminiscent of anarchist—especially Kropotkin's—celebration of the inherent creativity of the worker on the front line of production. In a May 1958 *People's Daily* article entitled "Science Is No Mystery," this populist belief is straightforwardly expressed in a call for workers and peasants to train themselves as scientists. "To a large extent, inventions come not from experts or scholars," the article asserts, "but from the working people at various times." It continues:

> Most of the important inventions of history, whether in China or in any other country, have come from the oppressed classes, from among those in lower social position, younger in age, less learned, in bad circumstances, and those who even suffered setbacks and discrimination.[51]

The implications of the argument are spelled out this way: "In the old society, the inventiveness of working class people suffered suppression and frustration at the hands of the ruling class." The conclusion was that the people's creativity must be encouraged and their inhibitions eliminated.[52] The Great Leap Forward certainly tried to do that, and soon the news media were bursting with reports of successful mass research campaigns that claimed tens of thousands of projects[53]—all carried on without expert scientists who "blindly worship foreigners and themselves, and do not dare to do things which have not been done abroad."[54]

The Great Leap Forward is well known for its preoccupation with production quotas that aimed to surpass capitalist industrial output; but those fantastic quotas were always paired with the social revolutionary goal of achieving a true communist society in the immediate future (and hence, ahead of the Soviet Union). Mass mobilization for science was the means of achieving the former;

mass organization into the giant, comprehensive communes for the latter. In both instances, there was little place for professional, specialized science based in formal institutions—that is, the kind of science that China had painstakingly begun to develop in earnest, using the model of Soviet science, during the first five-year plan.

The Great Leap Forward ended in disaster, most likely for reasons having little to do with this attempt to reinvent science. Within two years of its initiation, the economy was in shambles, a catastrophic famine was killing tens of millions of people, and the Soviet Union was in the process of withdrawing most of its support for future economic development.

Because of its own fiscal problems, the Soviet Union had, in effect, begun withdrawing its support from China on the eve of the second five-year plan; and some historical analyses suggest that this prospective disappearance of outside financing was an important catalyst for the Great Leap Forward. Be that as it may, the Great Leap Forward, which was denounced by the Soviets, was one among many issues that contributed to the break between the two countries. Technology, that is techno-politics, was especially divisive. Since 1952, for example, the Chinese had correctly assumed that the Soviets would not honor their original promise to share fully atomic energy information and to provide the Chinese with a prototype atomic bomb; so the Chinese set about creating the bomb on their own. In 1957, the Soviets' launching of the space satellite Sputnik made the Chinese aware that there was yet more crucial technology that the Soviets were not going to share; ditto the Soviets' subsequent success with the ICBM. And here again, the Chinese determined to develop these technologies on their own. It is important to recognize then that the Party supported the successful development of these technologies, at enormous expense, throughout all of its various policy campaigns aimed at reforming the technical intellectuals or giving precedence to mass science.

The military bureaucracy was relatively better able to insulate its science and technology needs from the depredations of party campaigns. As a result, the CCP's schizoid approach to science and technology is nowhere more clearly visible than in the late 1950s when the military pursued high technology, big science projects while Mao pushed the Party to promote nativist people's science and deprecated professional, specialized, technical expertise.[55]

Where the future of the Double Hundred policy is concerned, these contradictory approaches to science have great significance. When the policy was reinstated in the early 1960s, it had acquired new meaning because it was responding to phenomena quite different from those in the mid-1950s. The first time around, the policy was aimed squarely at the relationship between the Party and science intellectuals; after the Great Leap Forward, it was asserting and defending the legitimacy of an entire culture of science and its practitioners.

## Post–Great Leap Forward Renewal of Science

In the post-Great Leap Forward wreckage, Mao was forced to step back from center stage and allow the economists and central planners along with the proponents of professional science to resume their leading positions in the Party. Obviously galvanized by the departure of Soviet experts, in January 1961 the Party Central Committee began devising a plan to repair the damage done by the Great Leap Forward and resuscitate natural science. It is impossible to determine the extent of that damage, but it clearly followed some obvious patterns: the entire educational system had been expected to go on a half-work, half-study schedule; students and teachers were expected to get out of the classroom and go down to the countryside to engage in farm labor; little schools run by farmers proliferated and preempted the regular schools; research by institute and university workers was supposed to be replaced by manual labor or group work with farmers and cadre; scientific expertise needed to legitimize itself with "redness" and scientists whom cadre deemed insufficiently red were subject to harassment and sending down.

The Party Central Committee addressed these problems with a positive approach that advocated science education and research, unencumbered by politics, ideology, or manual labor on the front line of production. It thereby set the stage for recovery and future development.[56] In February 1961 the Party sponsored a national science conference in Canton specifically to lay out the steps necessary to get science research and education back on the tracks they were on in 1956-1957. Among other things, the work of the conference enabled the restoration of full-time education, the relegitimization of specialized training and research, and the relative autonomy of teaching staff from party control.[57] The next month, an article in the Party's main policy journal, *Red Flag* (Hong Qi), officially cast the recovery effort in the mold of the Double Hundred policy.[58]

In the spring, the Party, via Lu Dingyi's Propaganda Committee, had begun sponsoring scientific conferences on more specific issues, including genetics. At the April-May Canton Genetics Conference, biologists were assured that it was safe and appropriate to resume their post-Qingdao Symposium efforts to develop genetics according to the principals and guidelines established at that time. Tan Jiazhen reports that the conference did its job well, and this encouraged him and his colleagues to begin a vigorous campaign to rebuild genetics.[59]

## Notes

1. Mao, "On the Ten Major Relationships," section 10, April 25, 1956, in *Selected Works of Mao Zedong*, vol. 5 (Beijing: Foreign Languages Press, 1977). Sun Wugong is the monkey king in the very popular novel *Xi Yuji* (known in translation as *Monkey*), a favorite of Mao.

2. Mao and Lu Double Hundred speeches: See discussions in MacFarquhar, *Origins*, vol. 1, 53; and also see MacFarquhar, "The Secret Speeches of Chairman Mao," in *The*

*Secret Speeches of Chairman Mao: From the Hundred Flowers to the Great Leap Forward*, ed. R. MacFarquhar et al. (Cambridge, Mass.: Harvard University Press, 1989), 6.

3. Meisner, *Mao's China and After, a History of the People's Republic*, 2nd ed. (New York: Free Press, 1986), 171.

4. Lu, "Let a Hundred Flowers Blossom, a Hundred Schools of Thought Contend" (Beijing: Foreign Languages Press, 1958). I am following the annotated presentation in Fairbank and Bowie, *Communist China*, 153, 156-57.

5. Fairbank and Bowie, *Communist China*, 157.

6. Fairbank and Bowie, *Communist China*, 161.

7. For the early outpouring of criticism by intellectuals, see MacFarquhar, *Origins*, 33-34.

8. Tan, cited in MacFarquhar, *Hundred Flowers*, 91; and for similar examples, 112.

9. MacFarquhar, *Hundred Flowers*, 80, 91, 110-11, 128.

10. See Professor Lin Chuanguang, Vegetation Protection Department, Beijing Agricultural University, GMRB, Beijing, May 25, 1956 (SCMP/1309: 9).

11. Bao Wenkui (b. 1916) was born into a merchant family in Ningpo, Zhejiang Province. In 1939, he received his B.S. in biology from National Central University (two years in Nanjing, two years in Chunqing) under the supervision of outstanding faculty such as Bing Zhi, Lo Zhonglo, and Wang Jiaji. From 1939-1947, in Chengdu, he was a graduate assistant under Li Xianwen at the Zichuan University Institute of Agricultural Research (as described in text above). In addition to the research he conducted he taught embryology and genetics. Li's connections made it possible for Bao to attend Caltech, where Bao was Sterling Emerson's first Ph.D. student; he also studied with Jose Reissig in biochemistry. Bao was the first Chinese biologist to be trained to use Neurospora. His dissertation was "Investigations of the thermophobic character in Neurospora crassa, especially of the relationships between temperature and carbohydrate utilization." Sources: Interview, Bao Wenkui, Beijing 1986; and biographies in ZXN, 505-18; ZBT, no. 3 (1979): 85-93.

12. Interview, Bao Wenkui.

13. RMRB, August 25, 1956.

14. For example: Guo Moro, China's minister of culture used Bao's story to illustrate to foreign diplomats the inanity of China's blind worship of Soviet culture (RMRB, December 18, 1956). Xu Liangying's 1957 *Science and Socialist Construction* calls Bao's experience "an extremely isolated case" which is nevertheless retold in his book (81). In 1960, T. H. Chen brought the story to an American audience in his survey of science and politics in China. See *Sciences in Communist China*, ed. S. H. Gould (Washington D.C.: AAAS, 1961), 70. And Bao's story is highlighted by Li Peishan et al., in their 1985 history of Lysenkoism and the Hundred Schools policy in "Qingdao Yichuanxue zuotanhui," 43.

15. Zhao Zhongzhi, "Kexue chuban gongzuo yu 'Baijia zhengming'" (The work of Science Press and the Hundred Schools), KXTB, no. 11 (1956): 48-50.

16. Mao to Zhang Jichun, Central Propaganda Committee, April 18, 1956. Cited in Gong Yuzhi, "Fazhan kexue."

17. Undated letter from Lu Dingyi to Yu Guangyuan, cited in Yu, "Zai 1956 Qingdao Yichuanxue hui," 13.

18. "Zai 1956 Qingdao Yichuanxue hui."

19. See "1935-1956 Sulian shengwuxue jia di san ci lunzheng" (1935-1956 Soviet Biology's Three Controversies), in *Baijia zhengming fazhan kexue di biyu zhi lu*, ed. Li Peishan et al. (Beijing: Commercial Press, 1985), 356-423.

20. Yu Guangyuan, "Zai 1956 Qingdao zuotan hui," 5-13.

21. Interview: Wu Zhongxian.

22. Yu Guangyuan, "Zai 1956 Qingdao zuotan hui," 5-13.

23. See complete translation of Tong's opening speech in Schneider, *Lysenkoism in China*, 1-3.

24. Tan's three students at the symposium: Sheng Zujia and Zi Liji (Caltech Ph.D'.s 1951); Liu Zedong (University of Michigan, Ann Arbor Ph.D. 1953). Sheng and Liu were on the faculty with Tan at Fudan University; they had studied with him in western China during the war.

25. Zu Deming (b. 1905): 1929 B.S. in agricultural science from Hobei Agricultural University; 1936 advanced degree (doctorate?) from genetics and plant improvement program of Tokyo Imperial University Agriculture School, under Takashi Masao; 1936, faculty of Hobei Agricultural University; 1948, joins CCP; 1949, on faculty of North China Agricultural University, and founding member of North China Agricultural Science Research Institute, under the Ministry of Agriculture, publisher of *Agricultural Studies* (Nongyeh xuebao) and *Soviet Agricultural Science* (Sulian nongyeh kexue); 1959-1966, director, CAS Institute of Genetics. (ZXN: 237-47; DSSI: 33, 37.)

26. Zu Deming, in Schneider, *Lysenkoism in China*, 18.

27. Li Jitong, in Schneider, *Lysenkoism in China*, 25.

28. Yu, "Zai 1956 Qingdao zuotan hui," 13.

29. Interview, Yu Guangyuan, Beijing, May 1986. Yu suggested that the most important decisions about and by the symposium were made "in the corridors" and not at the formal sessions recorded in the proceedings.

30. *Baijia zhengming*, 14th session, 300-317, and closing speech, 320.

31. Sheng Zujia, in Schneider, *Lysenkoism in China*, 29-30.

32. Schneider, *Lysenkoism in China*, 25-27.

33. Schneider, *Lysenkoism in China*, 14-15.

34. Schneider, *Lysenkoism in China*, 34.

35. Schneider, *Lysenkoism in China*, 34.

36. Yu, "Zai 1956 Qingdao zuotanhui," 10.

37. NCNA coverage is translated in SCMP/1350-60; 1444: 15-16.

38. Yu Guangyuan's speech, September 26, 1956. Cited in Li Peishan, "Qingdao yichuanxue zuotan hui," 46.

39. The proceedings: *Yichuanxue zuotanhui fayen jilu* (Beijing: Science Press, April 1957). The preface speaks to the demand for the proceedings.

40. Tan Jiazhen, "Yichuanxuejia zai Qingdao Yichuanxue Zuotanhui hou di ping lun" (Geneticists' debates after the Qingdao Symposium), RMRB, September 6, 1956. Reprinted in *Baijia zhengming*, 331-34.

41. Zu Deming, "Ge xuepai gongtong nuli, ba wo guo yichuanxue tuixiang guoyuan shuiping" (Each school cooperatively strive to advance our nation's genetics to the international level), *Renmin ribao*, September 15, 1956. Reprinted in *Baijia zhengming*, 335-39.

42. Meisner argues convincingly that Mao always intended the Double Hundred policy primarily as a means of challenging the Party from the outside and shocking it out of its growing bureaucratic conservatism and routine: "It was more to revitalize the party than because of any desire to liberate some intellectuals from ideological and political discipline—although Mao saw some limited virtues to be derived from a limited degree of intellectual ferment and opposition. . . ." And "Mao now was attempting to turn the

tables: it was the Party that was to be rectified, and the nonparty intelligentsia was the instrument to be used for that purpose" (*Mao's China and After*, 177-78).

43. "On the Correct Handling of Contradictions among the People," 330 (February 27, 1957), translated and discussed in J. K. Leung and M. Kau, *The Writings of Mao Zedong*, 2 (Armonk, N.Y.: M. E. Sharpe, 1992), 307-50.

44. Mao, "Talk at the Forum of Heads of Propaganda, Culture, and Education Departments from Nine Provinces" (March 16, 1957), and "Talk at a Conference of Party Members and Cadres in Tianjin Municipality" (March 17, 1957), in *Secret Speeches*, 208, 288-89. And also see 279-80, 285, 288, 356-57 for further expressions of Mao's concerns about party leadership of science.

45. For CCP internal politics leading to this campaign, see MacFarquhar, *Origins of the Cultural Revolution*, 1, part 4.

46. From August 1957, KXTB began reporting anti-rightist speeches, meetings, and conferences dealing with the natural- and social-science intellectuals. The common issue, vaguely delineated, was failure to comply with the policies of party leadership. No one in the biological sciences was mentioned, nor was there any allusion to the genetics question. See KXTB, no. 15 (1957): 449-69; nos. 20, 22 (1957).

47. Meisner, *Mao's China and After*, 224-25.

48. Tong Dizhou, "A Decade of Biology in China," *Scientia Sinica*, no. 12 (1959): 1429. Original in KXTB, no. 18 (1959): 577-85.

49. For example see "Kiangsi Party Committee Decides on Attack on the Fortress of Science," RMRB, May 6, 1958 (in SCMP/1780: 18-19).

50. RMRB, June 6, 1958, 1 (in NCNA, no. 263, February 6, 1959).

51. Cited in NCNA (English) Peking, May 22, 1958 (SCMP/1780: 4041).

52. Editorial from *Zhongguo qingnian bao* (Chinese youth), cited in SCMP/1780: 4041.

53. For example see "Mass Line Proves Successful in Science Research," NCNA (English) Peking, September 17, 1958 (in SCMP/1859: 24); and "China's Science Enterprise Marches Forward with Epoch-Making Speed," NCNA (English) Peking, September 25, 1958 (SCMP/1879: 16-18). The latter claims that "in Chekiang, more than forty-five thousand people are engaged in science research work in the Chinhua district alone."

54. RMRB, June 30, 1958, 8 (cited in China News Analysis, no. 263, February 6, 1959, 3).

55. See for example John Wilson Lewis and Litai Xue, *China Builds the Bomb* (Stanford, Calif.: Stanford University Press, 1988).

56. See discussion in Merle Goldman, "Party Policies toward the Intellectuals: The Unique Blooming and Contending of 1961-1962," in *Party Leadership and Revolutionary Power in China*, ed. John W. Lewis (Cambridge, Mass.: Cambridge University Press, 1970), 268-303.

57. National Science Conference: Interview with Huang Zongzhen, who attended the conference. And also see discussions in Byung-joon Ahn, "Adjustments in the Great Leap Forward and their Ideological Legacy, 1959-62," in *Ideology and Politics in Contemporary China*, ed. Chalmers Johnson (Seattle, Wash.: University of Washington Press, 1973), 257-300; and Fan Dainian, "Retrospect of the Science Policies of the PRC," paper for the Australian/China Science and Technology Symposium (Beijing, November 29, 1983), 44.

58. Goldman, "Party Policies toward the Intellectuals," 272.

59. Interview, Tan Jiazhen. And see discussion in Pai Chen, "Science Work," in *Communist China 1961*, vol. 1 (Hong Kong: Union Research Institute, 1962), 203.

# 7

# One Step Forward, Two Back: Genetics from the "Double Hundred" through the Cultural Revolution

## Institutional Developments for Genetics

Developments at a number of levels contributed to the expansion of genetics after the Qingdao Symposium and again in the early 1960s, after the Great Leap Forward. Most conspicuously, there was the creation of a new CAS institute, devoted—nominally—to genetics. One of the recommendations of the Qingdao Symposium, brokered by Yu Guangyuan, was that a CAS Institute of Genetics should be established in recognition of the great importance genetics promised to have in the future of a number of "frontier" areas of the biological sciences.[1] The institute was formally established in 1959, the product of consolidating a number of extant northeast China research organizations controlled by longtime leaders of Michurin biology. Like other institutes at this time, the Genetics Institute profited from the abundant resources of the CAS. Its initial but temporary home was in a small building near the Friendship Hotel compound (Yuyi bing-guan), in Beijing's western suburbs. (Appropriately enough, that compound was originally built to house Soviet experts.) In 1964, the institute moved into its own, spacious five-story, custom-built facility, also in the suburbs.[2] It remained there until the late 1980s, when it moved into a much larger and up-to-date facility.

The institute's director was Zu Deming, Luo Tianyu's successor, and its first members were all veteran Michurinists.[3] By the early 1960s, the institute's expanding membership began to include some biologists who had not been involved with Michurinism, and who received training in classical genetics. Most of these newer members had advanced training in the Soviet Union, but a few had been educated in the United States or at Fudan University under Tan Jiazhen.[4] Nevertheless, the CAS Institute of Genetics continued to display the legacy of Michurin biology into the 1980s.

## Fudan University Institute of Genetics

It is not at all clear why the CAS permitted Michurinists to dominate the new Genetics Institute and thereby to acquire the CAS imprimatur as well as scarce resources. Quite possibly, it was part of a deal with Tan Jiazhen and the "Morganists" who established their own Genetics Institute at Fudan University in 1961. Five years earlier, at the Qingdao Symposium, it had been argued that if the proposed CAS Genetics Institute were established in Beijing then it ought to have a "southern branch" as well. The Fudan institute may have been the product of that proposition—a southern (Morganist) counterpart to the northern (Michurinist) CAS institute. Had it not been for the political upheavals of the late 1950s, the Fudan institute might have been established years earlier. In any case, Tan Jiazhen was required to lobby strenuously for his institute, even going so far as to make his case personally, on four different occasions, to Mao Zedong. Tan is convinced that his last audience with Mao, in 1961, did the trick and got him approval for the institute.[5]

The Fudan University Genetics Institute was not a branch of the CAS, but rather it stood independently, funded by the central government (probably by the Ministry of Education) and staffed completely by Fudan faculty with joint appointments. It was meant to be a vehicle for graduate education and for research. In the early 1950s, with the adoption of the Soviet model for organizing science, research was for the most part not encouraged or supported in the universities. By connecting faculty geneticists to state funding, independent of CAS or ministry competition and control, the Fudan Genetics Institute was a way around this barrier to university research. Tan Jiazhen, its director, was able to staff the institute initially with a small but dynamic group of young geneticists who had completed their doctorates in the United States just a few years earlier. Under Tan's direction, this group led the recovery of genetics during the 1960s.

## Developments in the Chinese Academy of Sciences

There were developments in the CAS, outside the Institute of Genetics, that did contribute to the advancement of genetics in the 1960s. The academy continued to refine its areas of specialization in biology, a process that began in 1950 when academicians from the Institute of Medicine were mandated to create an Institute of Biochemistry and Physiology. In 1958, as the CAS budget increased greatly, some members of this institute were given the resources to initiate three separate institutes: Physiology, Biophysics, and Biochemistry. With some overlapping membership, these three developed their respective fields cooperatively, and in turn, they were to work closely with Tan Jiazhen's genetics group at Fudan. The head of the new Biophysics Institute was Bei Shizhang (b. 1903), whose 1928 doctorate in physiology was from Tubingen. During the war, as director of Zhejiang University's Biology Institute, he had worked closely with Tan Jiazhen. In the early 1950s, he had been the head of the CAS Institute of Experimental Biology. [6]

Spinning off the new institutes made it possible for the the the CAS Institute of Biochemistry to achieve greater depth. It was already one of the strongest areas of biology research in the country, and thanks to the formidable leadership of its new director, Wang Yinglai (b. 1907), this institute maintained that status for decades. Wang was a 1941 Cambridge University Ph.D. and had spent the war years as a postdoctorate, doing further research there. From 1950 to 1957, he was director of the CAS Institute of Biochemistry and Physiology. As new director of biochemistry, Wang was fortunate to have recently well trained colleagues like Wang Debao (b. 1918). The latter's study of molecular biology began with a Washington University, St. Louis, M.A. in 1949, continued with a Western Reserve doctorate in 1951, and was completed with Johns Hopkins postdoctoral research from 1951 to 1954. He then returned to China, establishing himself at the CAS as the country's leading expert in nucleotide and DNA research. [7]

## Plant Genetics

In the area of plant genetics, there were institutional developments that helped to break the stranglehold of Michurin biology and to give renewed energy to China's strengths in this field. In March of 1957, the Ministry of Agriculture established the Academy of Agricultural Sciences, independent of the CAS. [8] One of its four research arms, the Institute of Crop Breeding and Cultivation, was put under the directorship of Wang Shou, the outstanding plant geneticist trained in the Cornell-Nanjing program and famous for his breeding of the strain of barley that bears his name. Up to this time, Wang had been working quietly, and safely, in out-of-the-way Shenxi Province. For his first deputy director, he chose the wheat specialist Dai Songen, also a Cornell Ph.D. in plant genetics. Thus, the previously pro-Michurinist Ministry of Agriculture had turned one of its most important research organizations over to a bona fide Morganist plant geneticist. The Michurinist Zu Deming was also named a deputy director of this institute. It is clear, however, that this was a nominal position and that Wang's appointment was responsible for reviving and carrying on the work of Shen Zonghan's pre-1949 National Agriculture Research Bureau and the Rice and Wheat Improvement Institute.[9]

## Building a Biology Community

Fortunately for the development of China's science in general, and genetics in particular, successful efforts were made to bring scientists together, across institutional as well as geographical lines. China's Soviet-style science organization had a decidedly vertical character that offered little opportunity for communication and cooperation between and among ministries and institutes. Some CAS biologists, however, held university appointments as well, if only because there was a serious dearth of senior scientists, with outstanding research experience, to do substantive teaching. As CAS budgets increased, however, there was diminishing incentive for a CAS member to moonlight at a university. The most common way of bringing scientists together of course was the regional or national meeting or conference, of which there were growing numbers in the 1960s, thanks to CAS funding.

In January of 1960, for example, the new CAS Institute of Biochemistry sponsored the first All-China Biochemistry Conference in Shanghai,[10] followed two years later by the first All-China Cytology Conference in Beijing. The latter brought together sixty-seven participants from forty-three different work-units (danwei).[11] In 1964, the Zoological Society of China celebrated its thirtieth anniversary in a week-long meeting in Beijing that included, along with dozens of technical panels and papers, speeches about the history and current status of the profession of zoology in China.[12]

In these same years, there were literally dozens of other national and regional science conferences that contributed importantly to the creation and maintenance of science communities. Between August 1961 and February 1962 alone, there were five genetics research conferences, each held in a different province, on the campus of a university or institute with a substantial biology department. The series stretched from Sun Yatsen University down south in Guangzhou to Shandong University in the northeast. The latter conference claimed an attendance of over three hundred participants from over thirty different work-units. Fudan University's brand new Genetics Institute was central to the organization of this series and it was responsible for compiling and publishing reports on each meeting.[13] This was just the beginning of Fudan's efforts to bring the entire biological science community into the revival of genetics and to integrate genetics into all aspects of biological work.

## Genetics in Education and in the Public Mind

Soon after the Qingdao Symposium, the Morganists began to address the monumental problem of providing fresh teaching material for all levels of education and accessible information about genetics for the broader reading public. Their publications in these areas were slowed down, but not stopped altogether, during the troubles of the late 1950s; however, from the moment the Double Hundred was revived in early 1961, there was an explosion of publishing on all areas of genetics, the product of an extraordinary collaborative effort supported from every corner of the biological sciences. None of this would have been possible, of course, without the cooperation of the press, which not only seems to have given this enterprise carte blanche, but also actively took the initiative by inviting specific kinds of materials or staging forums on specific topics.

Where to begin the remediation for a seven-year information blackout? In 1957, the editors of *Biology*, the main forum for the life sciences, addressed this question by inviting Li Ruqi to write an introductory survey of classical genetics. He did so in an unprecedented thirteen-part, sixty-five-page serial that took the reader from Weismann in the nineteenth century through the Morgan school in the 1930s. He left the discussion of molecular genetics for others more familiar with the subject than himself. The series was meant to provide middle-school and college teachers quickly with a reliable guide for new courses the state was expecting them to teach, as well as a reliable introduction for all those who only knew of classical genetics through the misrepresentations of Lysenkoism. Remarkably, Li's survey was never polemical; neither Lysenko nor Michurin biology was mentioned.[14]

As this series began, the Science Press published Wu Zhongxian's translation of Mendel's principles, in a typically inexpensive paperback edition.[15] Within the next two years, Tan Jiazhen had published a nonpolemical upper middle-school/college primer on Morgan genetics, and T. H. Morgan's *Theory of the Gene* was translated.[16] Of equal importance was the publication of the first dictionary of genetics, the product of a collaboration mainly between Beijing- and Shanghai-based biologists. It contained four thousand entries and became the point of departure for all genetics-related terminology in China for the rest of the century.[17]

In 1959, Morganist genetics got its first major college textbook. It came from the unlikely pen of Fang Zhongxi, who just a few years earlier had authored a very popular Michurin biology textbook. During the Qingdao Symposium, Fang had already begun to shed his Michurin skin and give cautious support to Morganism. In this textbook, his metamorphosis was complete, and he was able, quite unself-consciously, to detail every aspect of genetics from Weismann to DNA with his characteristic directness and lucidity. This three-hundred-page survey draws on contemporary American textbooks and on a score of primary sources from the United States and Great Britain; it also extensively cites the work of geneticists like Tan Jiazhen and Li Ruqi, who are among the many Chinese Morganists he thanks for reading and commenting on his manuscript.[18]

Fang's textbook is not only important for its accurate survey of classical genetics and its contemporary developments, it is also notable for presenting this material in isolation from Michurin biology. In his preface, he curtly tells the reader to go elsewhere if they want information about that school of thought; enough has been written about it already, he says.

In the short afterword to his textbook, written in 1957, Fang confronts some standard issues associated with the "genetics question" in order to put the whole problem into perspective. First, he argues that science in general and genetics in particular have no class nature. He does this by showing that just as Lysenkoism was far from the only approach to cellular genetics in the Soviet Union, Morganist genetics is far from being the product of one country or social class. Additionally, Fang rebuts the claim that Morganist genetics can contribute nothing to "socialist construction" by citing its role in American agriculture and medicine. Finally, in the last two pages of the book, Fang provides an outline of Michurin biology's basic principles, and then a quick patronizing evaluation to suggest which have a relatively solid foundation, which have insufficient evidence, and which have failed to develop adequate proofs. Whether he intended it or not, these last pages prove to be a splendid device for snubbing Michurin biology and undermining its credibility. It might have been still more effective if Fang had explained his own role in promoting Michurin biology or apologized for his role in the suppression of Morganism. He never did either one.

The publication of new textbooks and translations continued unabated into the 1960s in an effort to meet unrelenting demands of education and, apparently, public interest. Most of these publications came out of Fudan University. For example, Sheng Zujia's *The Genetic Code*, a forty-five-page pamphlet, was an upper middle-school level introduction to molecular genetics and DNA studies that was the first monographic treatment of the subject in China, and—as things turned out—the only one before the 1980s.[19] Under Tan Jiazhen's editorship, the Fudan group also published *Genes and Heredity*, a short, deft survey of the development of genetics from Mendel to DNA. It aimed for general audiences or middle-school students, and was more accessible than Fang Zhongxi's comprehensive college text.[20] Neither of these Fudan publications was the least bit polemical; neither mentions Lysenkoism or the "genetics question."

## Darwinism and Evolution

Morganist publication in the area of Darwinism and evolutionary theory was considerably less robust than in genetics as such. Part of the reason for this may have been a desire to avoid this most embattled of subject matter, and to give it time to cool down. It is also likely that there simply were insufficient personnel to carry on a full-blown Morganist campaign here as well as in all the other areas.

In the post-Qingdao period, one of the most important interpreters of evolutionary theory was Chen Shixiang, a French-trained entomologist and taxonomist who was the director of the CAS Institute of Entomology.[21] His presentation at the Qingdao Symposium was a brilliant reintroduction to China of the "evolutionary synthesis," a term that had not been uttered in China since 1949. He cited the work of Ernst Mayr, Julian Huxley, and Th. Dobzhansky in order to summarize the possible roles of geographical, ecological, and reproductive isolation in the development of species; and he carefully laid out the relationships between variation, isolation, and selection.[22]

Over the next two years, Chen published articles in major journals expanding his Qingdao Symposium presentation. In these, he introduced important conceptions and definitions of species that had been absent from Lysenkoist evolutionary discourse in China. He demonstrated the essentialism of Lysenko's definition of species, and showed how similar it was to the old Linnaean scheme. Chen also reviewed the spectrum of literature on the question of gradual and sudden evolution of new species (saltation) and the role that polyploidy plays in the latter. These discussions along with extensive and up-to-date bibliographies made these articles excellent points of departure for teaching and further study of contemporary evolutionary theory, and they were a source of encouragement for other knowledgeable biologists to contribute complementary publications.[23]

Chen Shixiang's efforts were most fully continued by Wu Zhongxian, the Beijing Agricultural University expert in animal breeding and genetics. More than anyone else, Wu clarified how evolution had been reconceptualized in recent decades. He further elaborated Chen's discussion of the new synthesis, and was alone in outlining and explaining in plain language the mathematical principles of natural selection, as expressed in the field of biometrics, and in the mathematical modeling of G. L. Simpson.[24] The efforts of Chen and Wu were supported by Tan Jiazen and his Fudan colleagues with a translation of Th. Dobzhansky's *Genetics and the Origin of Species*.[25] These publications in China are notable both for their competence and their uniqueness. Only a few Chinese biologists devoted themselves to evolutionary theory during this period.

The topic of human evolution was carefully reopened in the early 1960s, but by just a few paleontologists and paleobiologists. Discussion here was refreshingly free from Lysenkoist dogma and mindless deference to Engels; it was wide open to international literature and discourse. Nevertheless, attempts to remedy the damage done earlier to this field had little opportunity to accomplish much before the Cultural Revolution closed down the academic world.[26]

*Morganist Media Blitz*

The Fudan University geneticists were central to a different kind of effort that began early in 1961, as soon as the Party showed signs of renewing the Double Hundred policy and supporting the Morganists pursuit of a satisfactory resolution for the "genetics question." The project in question was a nationwide, high-profile education/public relations campaign, using the muscle of major newspapers and journals. For a two-year period, beginning in March 1961, the project successfully blitzed these publications with carefully crafted articles, eighty-one of which were then anthologized by Fudan University and published in three hefty volumes. By all means this project made sure that appropriate articles were published by major technical journals, but more important to its goals, it also managed to have at least two articles a month published in a major newspaper—*People's Daily, Guangming Daily*, or *Wenhui Daily*.[27]

The Morganist campaign was orchestrated to introduce genetics systematically to a national lay audience, and to do so through a set of discrete topics, each of which addressed a major area of Michurinist criticism. The object was not to criticize Michurin biology, but to focus attention on the actual substance of Morgan genetics and its practical accomplishments—to drown out the objections of Michurin biology in a deluge of information. Just a few of the contributors were given the job of engaging in polemics; most avoided the distractions of debate and focused on the details of their topic. All contributors pointedly acknowledged the importance of the Double Hundred policy to their work and the advancement of genetics. Unlike the Michurinist campaign of the 1950s, every contributor here held a position in the CAS and/or a major university; every one held a doctorate from a major Western biology program; there were no hacks.

Among the main themes of this campaign was that the identity of "genetics"—indeed of experimental biology altogether—had radically changed in recent decades, and that it had spread far beyond those areas typically cited in the clichéd rhetoric of Michurin biology. Therefore, contemporary genetics must be understood as something quite different from Lysenko's "Mendel-Morganist" straw man. Genetics had become the sum of collaborative work carried out by a variety of biochemists, biophysicists, physiologists, etc., all of them using new techniques and technologies nonexistent or not readily available in 1949. The net result of the changes, it was argued, was to provide definitive answers to the standard objections raised earlier by Michurinism.

To questions about material basis of heredity, CAS biochemists like Wang Debao answered that the contemporary understanding of the gene in terms of DNA and protein chemistry could hardly be a more solid demonstration that genes were very "real" things, and that this approach to heredity was solidly grounded in a material base. To questions about the utility of genetics, plant geneticists like Bao Wenkui piled up mounds of data illustrating successes in breeding hybrid double-cross maize, while others wrote about the successful use of genetics in silkworm breeding and forestry. Most notable was the introduction of the subject of human genetics and its use in medicine and public health.

It is remarkable that Fang Zhongxi was selected in this campaign to play the part of polemicist and point man for engaging Michurinist hostility. It is hard to imagine a person more dramatically suited for the job than the man who had recently been Michurin biology's most prolific educational writer. The pieces he wrote for this campaign are blunt confrontations with his erstwhile Michurinist colleagues, and they call for complete capitulation to Morganist genetics. For example, in a Beijing *Guangming Daily* article, Fang stridently tells the Michurinists that "the gene has gradually gone through the transition from a hypothetical and abstract unit to a real chemical substance," so if you are really concerned about controlling the "direction" of heredity, you had better learn to work with this new genetics.[28] A year later in the same newspaper, his article "Positive Influences of Genetics during the Past Few Years" sings the praises of the central interdisciplinary role that genetics has come to play in biology. Here, he dismisses out of hand any question about the material basis of genetics or about the significant and varied contributions that genetics has made to "the national economy and the life of humankind." "The gene theory as expounded by the Morganist school," he confidently concludes, "is fully backed by the facts."[29]

In the technical journals, the Morganist campaign was no less prolific. It was dominated by virtuoso state-of-the-field articles by the CAS biochemists. Wang Debao's encyclopedic surveys were especially notable, beginning with his three-part 1959-1960 series in *Biology* on cytology and cytogenetics, and continuing with his periodic reviews in *Science* of protein chemistry and molecular biology.[30] In 1962, *Science* published the most comprehensive overview of the new approaches to genetics in a forty-page symposium of articles from the CAS biochemists on topics including genetics and enzymes, molecular studies of proteins, RNA and DNA, and Wang Debao's long and typically detailed piece on messenger RNA.[31]

During the period 1961-1965, biological discourse was changed fundamentally, and it became common to see discussions on topics that had been anathema until the Qingdao Symposium—human genetics for example. Liu Zedong was devoted to promoting this field in articles like his 1962 survey of studies of the X and Y chromosomes.[32] More broadly, others surveyed the current state of structural-functional understanding of chromosomal genetics.[33] None of this technical literature was polemical. All of it cited a wealth of up-to-date science literature from the West and Japan, but notably, nothing from the Soviet Union.

*Access to International Scientific Literature*

Where did Chinese biologists obtain post-1949 Western literature on genetics, let alone up-to-date literature from the 1960s? The most important source was the CAS Institute of Scientific and Technical Information, which in 1958 became an independent agency reporting directly to the central Scientific and Technology Commission. This information agency had been collecting all kinds of scientific literature from all over the world since the early 1950s,[34] including comprehensive acquisitions from the West in all areas of biology. Up to the Qingdao Symposium in 1956, the agency locked away any literature perceived to be related to Mendel-Morgan genetics. Just before the symposium, it opened up its hoard and made it accessible for the first time to symposium participants. That is why—when Tan Jiazhen made the opening survey of Western genetics at the symposium—he remarked that his presentation was going to be a bit unfinished because he had not had time to absorb all the literature released to him only a month earlier.[35]

From 1956, the CAS began to spend huge sums of foreign exchange in "a strenuous effort to secure Western scientific books and periodicals, both back and current publications."[36] Where biology is concerned, as evidenced in the Morganist publishing campaign of the 1960s, the effort was successful. During this same period, individual biological research institutes and university libraries would have been able to begin subscriptions to scientific periodical literature previously proscribed under the ban on Morganist biology. The library of the CAS itself—which is distinct from the Institute of Scientific and Technological Information—tripled its holdings in the late 1950s, largely through acquisitions of current literature, especially from the United States.[37]

## Post-Qingdao Research: The Morganists

CAS enthusiasm for updating its library holdings was matched by its expensive program to enhance research facilities and equipment. The overall CAS budget increased threefold between 1953 and 1957, much of it allocated to the 1958-1960 construction of new buildings outfitted with research amenities up to contemporary world standards. Between 1955 and 1960, the total number of CAS research institutes more than doubled (from 47 to 105), and the senior research staff increased similarly (from 428 to 800). In that same period, the total CAS research staff almost tripled (2,483 to 7,000). Nevertheless, as a 1960 survey concluded, there was a serious shortage of senior, experienced research scientists—so much so, that on occasion in the late 1950s budgets went unspent and facilities underutilized.[38]

That shortage is nowhere more evident than in experimental biology in general, and in those various areas concerned with the revival of genetics. For the latter, the burden of remedial work—in education, public relations, and administrative politics—was in itself quite consuming, even before starting to meet the challenges of retooling for genetics research. Thanks in particular to the handful of recently returned young Ph.D's, however, fresh research projects were undertaken in a variety of contemporary areas.

At Fudan University, for example, Tan Jiazhen's Genetics Institute embarked on an extended research project using radiation technology techniques inspired in part by the work of N. P. Dubinin, the distinguished Soviet geneticist who in 1956 had organized the Moscow laboratory of radiation genetics in the Biophysics Institute.[39] The Fudan project was concerned here with issues related to mutation and human heredity. Ultimately it was concerned with the potential ill effects on humans from radiation generated by x-rays and atomic radiation, and would later be followed up in China by epidemiological studies of workers exposed to various forms of radiation. Tan's research used rhesus monkeys as experimental animals to study the effects of different dosages of radiation on chromosome aberration in spermatogenesis. Liu Zedong, who was a major proponent of medical genetics, was Tan's chief collaborator on this project.[40]

## Plant Breeding and Genetics

Plant genetics, which had been an especially embattled area before the Qingdao Symposium, recovered relatively quickly and with great strength, largely due to the convergence of government support and the outstanding leadership of Bao Wenkui and Li Jingxiong. The government was intensely concerned about agricultural production after the Great Leap Forward and the catastrophic famine, and so the 1960s saw the development of centrally coordinated programs for increasing agricultural inputs. At the same time, the government began relying on the expertise and following the advice of leading plant geneticists.

Bao Wenkui's polyploid research had resumed in 1956 and by the early 1960s he had become a major spokesperson for the application of genetics to breeding work.[41] The quality of his work and his outspokenness were early incentives for the government's approval of a significant program for the development of hybrid maize. In a prominent newspaper article, he wrote in 1962 that "we naturally can select breeds without genetics, yet with genetics we can select breeds with greater success and can develop new breeds which have never existed before in the natural state."[42]

Li Jingxiong was one of the many successful plant geneticists trained in the Cornell-Nanjing program. Like Bao Wenkui, he was on the faculty of Beijing Agricultural University, and he was devoted to the once proscribed technique of hybridization. His ongoing 1960s research on hybrid maize was begun after the Qingdao Symposium, under a research grant from the Ministry of Agriculture.[43] While his carefully controlled experiments did not yield any fundamentally new genetic insights, they did represent the successful application of a sophisticated understanding of genetics and breeding techniques, especially those applicable to double-cross hybrids.[44] While of a lesser order of significance, the plant genetics work of Fang Zhongxi is notable, and not only because of his erstwhile role as a promoter of Michurin biology. Under Fang's direction, the CAS Institute of Oceanology engaged in successful long-term breeding studies of important marine algae using standard practices of classical genetics.[45]

## Plant Genetics and the Green Revolution

Economic historians and historians of China's agriculture agree that a "green revolution" occurred in some areas of China during the 1960s, in part due to the work of China's plant geneticists. In Ramon Meyers' definition, such a revolution occurs when "the total output per sown area each year increase[s] more rapidly than the combined inputs of land, labor, and capital." In consequence, Meyers observes, in the localized areas where this occurred, "food grains exceeded population demand, and thus a reserve was created for the support of a rapidly growing animal husbandry industry. For the first time in

China's agrarian history the supply of farm animals, such as hogs and poultry, rose more rapidly than the human population."[46]

There is a strong consensus among economic historians that from 1962-1965 grain production increased significantly, catching up with and then surpassing the most successful years before the collapse in 1959-1960. Most studies of this phenomenon attribute it to the Party's policy of "putting agriculture first," and the subsequent allocation of funds for farm machinery, chemical fertilizers, and irrigation projects. Taken for granted in these studies is the germ plasm to which these improvements were applied. It goes without saying, however, that if one begins with seeds that produce disease-prone plants, the application of more fertilizer and the use of machines for harvesting is likely at best only to produce more of the disease-prone plants; and if one begins with late ripening rice, chemical fertilizer will not shorten its growing season. Clearly one of the key factors in China's green revolution was the availability and continued development of new strains of seeds that overcame debilitating characters and disease vulnerability, or that ripened earlier and permitted more than one crop per year in a given location, or permitted cultivation in a previously unfeasible area of China. Meyers is one of the historians who clearly acknowledges that the "cropping revolution" of the 1960s is attributable first to the fact that "the supply of early-ripening, high-yield, and disease-resistant seeds increased for rice, cotton, and wheat," and second, to the great expansion of irrigation, drainage, and organic fertilizer application in the communes.[47]

Those who have studied this green revolution attribute the systematic use of these seeds to the post-1956 work of geneticists like Bao Wenkui who developed a "rye-wheat" (a backcross between rye and octoploid wheat) for double-cropping in the northwest provinces,[48] and those who developed improved varieties of rice and hybrid corn.[49] Corn occupied a significant and growing role in the Chinese diet over the twentieth century, and China was second only to the United States in corn production.[50] It is significant then that specialists in developing hybrid corn, like Li Jingxiong, were already reporting enormous success with their new varieties in 1957.[51] Over the next year, the area planted with hybrid corn expanded sevenfold, increasing at an ever faster pace in subsequent years.[52] Additionally, plant improvement and genetics work guided by the Cornell-Nanjing program, from the 1920s to the mid-1940s, is also cited as a major source of seeds used systematically in the 1960s.[53] Be that as it may, economic studies of this period suggest that poor record keeping and reporting by the Chinese government make it difficult to measure the specific economic gain brought about by the wider use of improved seed.[54]

A study of China's green revolution suggests that it was the result of a conscious decision of the Party, formalized as policy in September of 1962.[55] The policy was refined the following month at the first of a number of small, high-level meetings between party officials and twenty-six agricultural science specialists. Among those specialists were Bao Wenkui and Li Jingxiong. Half of those scientists at the meeting had a foreign doctorate, five of which came from Cornell.[56]

## Biochemistry

During the 1960s, China's biochemists were heavily engaged in research that typically aimed at understanding the chemical structures and processes that informed the actions of genes. Like their colleagues in plant genetics, their work yielded no results that earned them international attention; but unlike the plant geneticists, their work was seldom of immediate practical use, even though it was being done on a frontier where every step promised to bring a clearer understanding of more fundamental mechanisms of genetics. Their research demonstrates that they were in close touch with international trends and were in a position to act as reliable educators in this area of fundamental importance. Their mastery of concepts and research techniques proved to be of inestimable value to China a decade or so later, when the practical application of their work in biotechnology became a priority of China's science modernization program.

Biochemical research relevant to genetics was being done only in a few places—primarily at the CAS Institute of Biochemistry[57] and the Institute of Experimental Medicine of the Academy of Medical Science.[58] Additionally, some work was being done at the CAS Institute of Experimental Biology, and the biology department of Beijing University. While the former two institutes concentrated on very fundamental issues, the latter two were attempting to create a bridge from this basic work to practical applications in key areas of production, such as the reproductive process of wheat or of silkworms.[59]

## The Fate of Michurin Biology

What had the Michurin biologists been doing while all these efforts were being made to revitalize genetics? In spite of the Qingdao Symposium principles and guidelines, Michurinists were tempted in the late 1950s to test the resolve of the CAS and the Morganists by challenging them from time to time with especially inflammatory public rhetoric or actions. In 1957, for example, they touted their trip to the Soviet Union—sponsored by the Ministry of Agriculture—to visit Lysenko, tour his Lenin Hills Experimental Farm, and hear him lecture.[60] During the Great Leap Forward, they cast themselves as supporters of its policies on mass science while depicting the Morganists as its enemies.[61] For this kind of activity, they were able to count on the virulent zealotry of a small pocket of die-hard Michurinists at Wuhan University, in central China.[62] There is no evidence, however, that there was any other significant support outside of Beijing Agricultural University and the CAS Institute of Genetics. From 1961 and the renewal of the Double Hundred, even those bases were becoming diluted by the growing strength of Morganist genetics; and consequently, Zu Deming had begun to back away from pugnacity to a more cautious and conciliatory position.

In the spirit of Double Hundred, Zu and his colleagues had been invited to contribute to the 1961-1962 Morganist public information campaign, but their response was tepid, defensive, and sometimes contradictory. Mi Jingjiu, for example, conceded the necessity of Michurin biology incorporating the mass of indisputable facts generated by the new genetics, and he continued to argue for the melding of the "two schools."[63] Zu Deming likewise dutifully invoked the "convergence of the two schools," but remained unmoved by any of the newly aired information. His attitude toward the revolutionary developments in genetics was epitomized in a curt, defiant statement: "It is nothing but an inference based on available data that nucleic acid is a transmitter of genetic information."[64]

Where its own research was concerned, throughout the 1960s Michurin biology persisted with the very same programs it had begun a decade earlier, and the very same methods as well. This is clear from its main forum, *Genetics* (Yichuanxue jikan), the journal of the CAS Institute of Genetics,[65] as well as from extensive surveys of its research published in prominent journals.[66] The list of research activities includes vernalization, distant breeding and wide crosses, multiple pollination, grafting (vegetative hybridization), and photoperiodism. Methodology continued to be undisciplined; analysis continued to be based on presumptions like the preeminence of environment in heredity, the inheritance of acquired characters, and the possibility of creating new species rapidly, through special techniques.

Over the course of the 1960s, in spite of the predominance of doctrinal Michurinism at the CAS Institute of Genetics, some important changes did occur there, and these are evident from changes in their journal *Genetics*. Not only were radically new topics introduced in the journal, but articles now began to cite material in languages other than Russian and from Western Europe and the United States. Radiation and human genetics was the most important addition to the research program, and it was carried on by younger workers, recently associated with the institute, and trained outside of the narrow circle of agricultural colleges that had been the breeding ground for Michurin biology. Among these new arrivals were people trained at Beijing and Fudan Unviersities, in the biology departments of Li Ruqi and Tan Jiazhen respectively.

By 1965, a group of six workers at the Institute of Genetics had begun to explore both ends of the problem of radiation and human health. Some conducted a national survey of chromosome aberrations among medical x-ray technicians.[67] The others, using various experimental animals, studied the genetic effects of various kinds and various dosages of irradiation.[68] Published reports from this program indicate highly disciplined methodology and analysis, quite comparable with the work coming out of Fudan University's Institute of Genetics, and with the work in the United States and the Soviet Union that provided the foundation for these studies. Internationally respectable as this work surely was, it nevertheless occupied a small place in the institute's research program, still dominated by the narrow, inflexible concerns of 1940s Lysenkoism. No explanation was offered by the Michurinists for this addition to the institute's program. One can only speculate that it was done as a sop to the CAS in its efforts to promote Hundred Schools ecumenism.

## Michurin Biology's Legacy on the Eve of the Cultural Revolution

If it is difficult to assess the economic effects of bona fide plant genetics in China, it is impossible to determine those of Michurin biology. The former difficulty is largely a function of poor government records and reporting; however, there was no dispute that food plants improved via Mendel-Morgan genetics were actually grown in abundance. The latter impossibility is due to the same problem as well as the largely anecdotal reporting of the Michurinists themselves.

If Michurin biology were responsible for significant increases in crop quality or yields, there would be no way to demonstrate it. Nor is there any way to sustain the credibility of their claims to creating new varieties and so-called species of food plants. Likewise groundless were their theoretical explanations of changes brought about by their work. With these as the general characteristics of fifteen years of effort, and using their own criteria of productiveness, it does not seem too harsh to suggest that resources spent on Michurin biology were scarce resources down the drain. Were Michurin techniques or Michurin-bred plants harmful in some way? Even the most adamant anti-Michurinists do not make this accusation. Perhaps the closest that Michurin biology came to biological harmfulness was in its advocacy of "close planting," a product of their belief that there was no intraspecies competition and therefore "same species" plants would not crowd each other out when bunched together for the purpose of achieving more output per unit of land. This technique, however, proved to be such a dud from the outset that it was abandoned before it had a chance to do much damage.[69]

The great harm of Michurin biology lay rather in opportunity costs. While it had its monopoly from 1949-1956, it prevented or slowed the development of areas of biology that could have made important contributions to agriculture and public health. How much grain failed to be produced in China because of Lysenko's prejudice against hybrid techniques and polyploid research? How much grain failed to be produced because the CCP put their trust into Lysenko's counterfeit science and ignored the demonstrable success of Chinese plant geneticists?

Where scientific development is concerned, the lost opportunities are just as clear. At the Qingdao Symposium, one biologist summed up the ordeal saying that when he began in 1956 to survey the scientific literature denied access to him, he painfully realized how seriously outdated his science had become.[70] The reign of Michurin biology in China denied Chinese geneticists the opportunity to build on their substantial pre-1949 heritage, consigning their field after 1956 to remedial, catch-up work for years. Michurin biology left in its wake a legacy of obfuscation, the intellectual equivalent of a giant oil spill. After the Qingdao

Symposium, many talented biologists had to spend precious time cleaning up the mess when they might have been devoting themselves to productive research. It was not just a matter of teaching legitimate genetics to future generations, it was also a problem—perhaps even more difficult—of what to do about those who had already been schooled solely in Michurin biology, or those who were mandated to be educated in both "schools" of biology. What kinds of confusion must this have caused in the minds of potential biologists?

Finally, the legacy of Michurin biology is a legacy of dogmatism, censorship, and political coercion. The CCP would have provided plenty of this sort of thing even without Michurinism; but the effects of suppressing a whole area of science were so chilling and so egregiously oppressive that they would be remembered for the rest of the twentieth century as the outstanding instance of just how far wrong the CCP could go in its relationship with the science community. The memory of this debacle even survived the catastrophe of the Cultural Revolution.

## The Cultural Revolution and the Scientific Community

The Cultural Revolution's devastation of all aspects of Chinese society has been recounted many times over. Nevertheless, its effects on the science community warrant a brief summary here, if only to suggest the kinds of serious damage that had to be repaired for at least a decade after the formal end of the Cultural Revolution, in 1976. It is no exaggeration to say that the development of natural science education and research was virtually stopped dead in its tracks from the initiation of the Cultural Revolution, in 1966, until after President Nixon's visit to China, in the early 1970s. During the most violent and destructive period of the Cultural Revolution from 1966 to 1970—remembered for rampaging Red Guards—the higher education system was shut down, CAS institutes were closed or seriously disrupted, science journals ceased publication, and many researchers and professors were forcibly "sent down" to labor in the countryside. It was cold comfort for the scientific community to know that they were part of a much larger community of intellectuals, bureaucrats, and even party cadre who were treated in this fashion.

The period from about 1973 to Mao's death in 1976 has also been considered part of the Cultural Revolution, even though repair work on earlier Cultural Revolution damage had already begun. Science journals began to be published again, and some CAS institutes restarted research programs. Exiled researchers and professors began to return to their laboratories and schools, and very gradually colleges and universities began admitting students again.[71]

Still, these same years saw continuing efforts to rectify science according to Cultural Revolution values. In 1974, visiting American plant scientists found that Cultural Revolution educational reforms were still being pursued, but in such a confused manner that the agricultural schools, for example, were in a state of chaos. They were all supposed to have moved to new, spartan rural campuses, but these were still only partially built, at best. And they were to have developed a new three-year curriculum to replace the former four-year one; but the new "production-oriented" curriculum was still being debated. These same visitors also observed that at newly reopened CAS life science institutes and university biology departments, one-third of the staff were required to take turns working away from their laboratories and classrooms, in "the field," where they were to share their skills with the peasants and learn from the peasant's hands-on practical experience. While the visitors acknowledged some potential short-term advantages to agriculture from this arrangement, they doubted that any necessary long-term and coherent research and testing would be possible.[72]

Those who have studied the Cultural Revolution in the larger context of PRC history make the convincing case that the Cultural Revolution grew out of a fundamental set of Mao's principles that he had attempted to implement re-peatedly in the past, perhaps as far back as the 1942 Rectification campaign, but most notably in the Great Leap Forward.[73] Assuredly, Great Leap Forward atti-tudes toward science and technology, discussed earlier in this study, resonate clearly with the Cultural Revolution. For example, the Cultural Revolution did share the Great Leap Forward's assertion of the Chinese people's independence from the science and technology of foreigners on the outside and Chinese ex-perts on the inside. And the Cultural Revolution appears to have expanded on the Great Leap Forward position by attributing all science and technological creativity to the masses, especially to poor rural youth.

There are, however, some distinctly different qualities to Cultural Revolu-tion rhetoric and mood. The Cultural Revolution did not exhibit the Great Leap Forward's frenzied enthusiasm for fantastic production goals to be achieved by way of mobilized masses of workers and peasants. In 1966, the positive, uto-pian, even chiliastic bombast of the Great Leap Forward was replaced with a mood of danger and a sense of threat to the revolutionary accomplishments of China under the Communist Party. No doubt, the Cultural Revolution's sense of an endangered China was heightened by the presence of the American military in Vietnam, and the Soviet military, massed along China's northern border. The Cultural Revolution, however, was preoccupied with China's internal problems. Its goal was to restore the momentum and trajectory of socialist revolution by eliminating the threat of a "bourgeois restoration" by a new social class of bu-reaucrats, managers, and technical experts. This new class was accused of forming an elite and gaining definitive power in China, not through ownership of the means of production, but through its administration.

The new class was accused of taking selfish advantage of its trusted admin-istrative role, of feathering its own nest at the expense of the workers and peas-ants. Evidence of this, it was charged, could be seen in the elevated lifestyles of its members: in their relatively luxurious homes and the schooling for their chil-dren; in the "inheritance" of their lucrative and influential jobs by those same children; in their general isolation from manual work and workers, and their absence from the front lines of production. Members of this new class were to be found widely not only among bureaucrats and intellectuals, but within the Party itself. The same Party that supervised the Great Leap Forward was now accused by Mao of leading the restoration of bourgeois values, and of resuming the development of those aspects of Chinese society and culture that Mao had long mistrusted or downright rejected as incompatible with his radical egalitarian vision. These included the resumption of centralized planning with a decided emphasis on urban industrialization; occupational specialization and elitism; and exclusive formal higher education.

Thus, the problems addressed by the Cultural Revolution, Mao claimed, were not problems in the material substructure, in the relations of production, but rather in the "superstructure"—the realm of politics and culture, values and ideas, art and science. A premise of the Cultural Revolution was that bourgeois values had been corrupting the revolutionary accomplishments of China through the medium of foreign literature, art, and science, or through those who had been influenced by foreign values in the past: intellectuals trained abroad or in missionary schools on Chinese soil, or Chinese universities modeled after foreign schools. The Cultural Revolution took these notions still further by arguing that even residual Chinese culture from the old regime was a potentially corrupting and counterrevolutionary influence. So, "old things" and "old people" also came to be perceived as a threat.

According to the Cultural Revolution, only the working masses, especially the young peasants, were free of taint and to be trusted with the rectification of China's bad elements and with continuing the revolution in the right direction. In the late 1960s, such young peasants, who were bused or marched into the big cities by the millions, were the core of little-red-book-waving Chairman Mao worshipers. They were portrayed as the most immediate victims of bourgeois restoration, for their education and their livelihood suffered most. Many of them were prepared to punish their exploiters and destroy the media of hegemony: publicly to thrash and humiliate intellectuals, especially old ones; to ransack the homes of bureaucrats and cadre; to vandalize museums, libraries, and monuments. And while they engaged in their near-ecstatic antinomianism, Mao, with the aid of the People's Liberation Army, closed the educational and research institutions that stood accused of fostering the growing bourgeois hegemony. For awhile, he even closed down the CCP itself. But when the violent behavior of the young people in the cities was perceived to be spiraling out of control, Mao used the army to restore order and hustle them back to the countryside.

In this state of virtual civil war and unbridled inquisition, natural scientists appear to have been manipulated and targeted for abuse to the same degree as other intellectuals. No special distinctions were made among scientists, and scientific institutions were attacked at all levels. In 1967, for example the general assembly of the CAS was abolished; and in 1969 the Science and Technology Commission met the same fate. Meanwhile, at the local level, faculty at many of the closed colleges and universities were collectively trucked off to work on farms in the hinterland, as laborers and sometimes also as instructors.

Biologists in general and agricultural scientists in particular were suscepti-
ble to this rural exile known as "sending down." For example, the entire Beijing
Agricultural University, the largest of its kind in the nation, was moved to the
countryside because the "revolutionary committee" controlling its fate decided
that it was inappropriate for agricultural science to be taught in a city. It was
relocated over four hundred miles to the west, to an abandoned village in Shenxi
Province, near Yan'an. Its hundreds of faculty and staff "became farmers."[74]
Meanwhile, the university's substantial Beijing campus became a barracks for
the Peoples Liberation Army.[75]

The CAS Institute of Genetics and the biology faculty of Fudan University
experienced similar fates. Their research staff and faculty, however, were not
sent down en masse to the same place, but rather, were broken up and sent to
various work camps in the northwest, leaving their families far behind. This kind
of treatment continued in many cases for two years or more. Reflecting on all
this more than fifteen years after the fact, Tan Jiazhen concluded that the only
good thing about the Cultural Revolution was that some old Lysenkoites were
sent down with the rest of the biologists, and finally got a taste of their own
medicine.[76]

In still other cases, scientists were singled out for public humiliation and
punishment. Tong Dizhou, for example, was in his mid-sixties when his institute's revolutionary committee crowned him with the familiar dunce cap and
marched him through the streets of Beijing. He was an exemplary target: aged,
foreign educated, head of a CAS institute, and an experimental biologist.
Charged with "revisionism," he was beaten and put under house arrest in the
winter of 1966 for nearly two years.[77]

In sum, the Cultural Revolution's harm to science occurred on two levels.
On the institutional level, it was easy enough to see—if difficult to grasp—the
damage of denying education to a whole generation of students, and denying
scientists access to laboratories and up-to-date literature. When finally permitted
to do so, it was impossible for the "lost generation" of students to catch up; and
as it will be seen in the next chapter, in spite of the monumental amount of resources made available to them, the scientists' efforts to "catch up" with the
outside world were, more often than not, Sisyphean.

The Cultural Revolution harmed science on another, less tangible, level as
well. Here, science was deprecated and rejected because of its class nature.
Reminiscent of the Lysenkoist treatment of biology, Cultural Revolution ideology denied the legitimacy of most of the sciences because they were "bourgeois" (due to their Western derivation), or "revisionist" (if influenced by the
Soviet Union). Science thus conceived, was not an element of the substructure
and was divorced from production; scientists thus conceived were not members
of the working class, but pretenders to an elitist, new bourgeoisie—enemies of
the people. And again reminiscent of Lysenkoist thinking, the Cultural Revolution considered the only valid science ultimately to be that practiced by the
masses, spontaneously, as an adjunct to production. It was going to take years
and the pointed efforts of post-Mao party leadership to dissolve the Cultural

Revolution miasma of contempt and distrust, and to rehabilitate the stature of the scientific community.

# Notes

Some of the material in Part II was introduced in my essay "Learning from Russia: Lysenkoism and the Fate of Genetics in China, 1950-1986," in *Science and Technology in Post-Mao China*, eds. Merle Goldman and Denis Simon (Cambridge, Mass.: Harvard University Press, 1989), 45-65.

1. Interview, Yu Guangyuan; and also see Yu, "Zai 1956 Qingdao Zuotan Hui," 13. After the symposium a public relations announcement was made of a "decision" to establish a CAS Genetics Institute. See KXTB, no. 10 (1956): 69-70.

2. Interview, Hu Han, director CAS Institute of Genetics, August 8, 1984.

3. For example: Hu Qide (b. 1921) was on the faculty of Northwest Agricultural University, 1949-1950; Beijing Agricultural University, 1950-1951; CAS Genetics and Seed Selection Research Laboratory, 1951-1952; he attended Harbin Russian Language School, 1952-1954 (CASS: 224).

Li Jijing (b. 1923) studied at Beijing Agricultural University, 1948-1952, and Moscow University, 1953-1957 (CASS: 214). Liang Zhenglan (b. 1921) worked closely with Luo Tianyu from the beginning of the Michurin biology campaign in 1948-1949. Before his appointment to the institute, he was associated with the Northwest Agricultural College and Beijing Agricultural University. He was Zu Deming's second in command (CASS: 209).

4. By 1965, the CAS Genetics Institute had twenty-four members, of varying ranks, and at least that many support staff. Eight had been educated at Beijing Agricultural University, eight at other agricultural schools. Eight members had advanced education in the Soviet Union, either at Leningrad University or at the Timiriazev Agricultural Academy. One member was trained under Tan Jiazhen at Fudan; one under Li Ruqi at Beijing University; and one at Yanjing University. Six of the members were women (CASS: 207-32).

5. Interview, Tan Jiazhen. And also see Tan's "Hold Aloft the Great Banner of Chairman Mao and Raise Rapidly Genetics in China to a Higher Level (sic)," *Yichuan xuebao* (Genetics), no. 4 (1977): 277-82.

6. Bei Shizhang: From the late 1950s Bei was a member of the powerful planning and coordinating State Science and Technology Commission. He was not a CCP member until 1978, when many scientists, including Tan Jiazhen joined the Party. ZKJ, vol. 1 (1982): 31.

7. Wang Yinglai received his B.S. from Nanjing University in 1929 and then was on its faculty until 1937. His Cambridge Ph.D. research was on "Chemical Methods for Estimation of Water-Soluble Vitamins B, C, and Nicotinic Acid." From 1944-1948, he was on the faculty of National Central University. He joined the CCP in 1958 (CASS: 296; ZXS: 331-42).

Wang Debao got his B.S. in chemistry from National Central University in 1940, and was then on the school's faculty in agricultural chemistry until 1943. From 1943-1946, he studied biochemistry at the university's Medical School. His 1949 Washington University M.A. thesis was "Studies on the Metabolism of Pyrimidine Compounds in E-coli." His doctoral dissertation was "Metabolism of Purine and Pyrimidine Compounds by Bacteria" (CASS: 297). His early publications were: "Enzymatic Deamination of Cytosine Nucleotides," *J. Biol. Chem.*, no. 17 (1950); and "Kinases for the Synthesis of Coenzyme A and Triphospho-pyridine Nucleotide," *J. Biol. Chem.*, no. 311 (1954).

8. For the Academy of Agricultural Science (Zhongguo nongyeh kexue yuan), see DSSI, item 110: 110-11; and R. W. Phillips and L. T. C. Kuo, "Agricultural Science," in *Sciences in Communist China*, ed. Sidney Gould (Washington, D.C.: AAAS, 1961), 279-80.

9. DSSI: item 140: 130-31. Apparently, some political force was trying to work out a balance of influence between the Michurinists and Morganists here: Just as Zu Deming was given a deputy directorship of this research institute, Dai Songen was given the same position at Zu's North China Agricultural Research Laboratory, making Dai its first non-Michurinist.

10. Biochemistry conference report in KXTB, no. 5 (1960): 157-58.

11. Cytology conference: KXTB, no. 10 (1962): 46-47, and RMRB September 9, 1962 (SCMP/2832: 7-8).

12. See proceedings of zoology conference: *Zhongguo dongwuxuehui sanshi zhounian xueshu taolun ji* (China Zoological Society thirtieth anniversary conference) (Beijing, 1965). There are 590 abstracts in the collection, most from the July 8-18 Beijing meeting, and some from the society's 1963 Guangzhou meeting.

13. Conferences: August 16-25, Amoy, 46 participants; September, Guangzhou; November, Zhejiang; December, Shandong; February 1962, Jiangsu. Reports published in *Yichuanxue wenti taolun ji* (Anthology of discussions on the genetics question), 3 vols. (Shanghai: Fudan University, 1962).

14. Li Ruqi, "Yichuanxue jiangzuo: yichuanxue di jiben yuanli" (Discussions of genetics: the fundamentals), SXTB, from no. 3 (1957): 42-45 through no. 3 (1958): 42-46.

15. *Zhiwu zajiao di shiyan* (Beijing: Science Press, January 1957). This translation is based on the Gregory Bateson English version, "Mendel's Principles of Heredity," in G. Mendel, *Experiments in Plant Hybridization* (Cambridge: Cambridge University Press, 1913).

16. Tan Jiazhen, *Tan-tan Morgan xuepai yichuan xueshuo* (Shanghai: Science Press, 1958). And Lu Huilin, trans., *Jiyin xueshuo* (Shanghai: Science Press, 1959).

17. *Yichuanxue mingci* (Beijing: Science Press, 1958). Leading editors were the Fudan University genetics faculty, Li Jingxiong at Beinongda, Li Ruqi at Beida, and Fang Zhongxi. The 1966 revised edition changed its title to *Ying-Han yichuanxue mingzi* (English-Chinese genetics dictionary).

18. Fang Zhongxi, *Putong yichuanxue* (Introductory genetics) (Beijing: Science Press, 1959). English index. Subsequent editions in 1961, 1964, 1974. The last edition had 509 pages and listed a total printing of 89,000.

19. Sheng Zuji, *Yichuan mima* (Shanghai, 1960).

20. Tan Jiazhen et al., *Jiyin ho yichuan* (Beijing: Knowledge Books, 1962), 3rd printing, 1964.

21. Chen Shixiang (b. 1905) received a 1928 biology B.S. from Fudan University. In 1931-1932 he had a China Foundation fellowship to study at the Laboratoire d'entomologie, in Paris; in 1934, he received a Ph.D. from the University of Paris. From 1934, he was associated with the CAS Natural History Museum. In 1950, he became head of the entomological division of the CAS Institute of Experimental Biology; in 1953 the Institute of Entomology was established as an independent CAS unit, with Chen as its director. He joined the CCP in 1979. See China Foundation, *Report*, 1931-1932, no. 6; DSSI: 43; ZXS: 296-306.

22. Chen's presentation: *Baijia zhengming,* 253-63.

23. Chen Shixiang, "Wuzhong jiegou yu wuzhong xingcheng" (The formation of species and the origin of species), SXTB, no. 12 (1956): 51-54; "Guanyu wuzhong wenti" (The species question), KXTB, no. 2 (1957): 33-42. And also see Fang Zhongxi, "Mendel-Morganism and Darwinism," SXTB, no. 4 (1957): 1-7.

24. Wu Zhongxian, "Darwin xueshuo di xianshi" (Contemporary views of Darwinian theory), SXTB, no. 2 (1963): 24-33, and no. 5 (1963): 45-53.

25. Tan Jiazhen et al., trans., *Yichuan yu wuzhong qiyuan* (Beijing: Science Press, 1964).

26. Wu Rukang (b. 1916), CAS Institute of Paleobiology, was central to the revival of human evolution discourse. He was a 1949 anatomy Ph.D. from Washington University in St. Louis. See his running discussion beginning with SXTB, no. 6 (1963): 35-41, and KXTB, no. 8 (1964): 701-5. And Wang Dianyuan, KXTB, no. 7 (1965): 626-33.

27. The anthology: Fudan University Genetics Institute, ed., *Yichuanxue wenti taolun ji* (Anthology of the genetics question debates), 3 vols. (Shanghai: Science and Technology Press, 1961-1963). This comprised a total of 81 articles in 653 pages.

28. Fang Zhongxi, "On the Gene Concept—Critically Accept the Morganist Theory on Genes," Beijing, GMRB (April 30, 1961), 8. (Translated in *China Background,* no. 659: 1-5). And also see his "What Is the Morganist Theory on Genetics?" Beijing, RMRB (June 15, 1961), 31.

29. Fang Zhongxi, "Positive Influences of Genetics during the Past Few Years," Beijing, GMRB (May 3, 1962).

30. Wang Debao's surveys, SXTB, no. 12 (1959) through no. 4 (1960); and KXTB, no. 6 (1960): 491-92; and on molecular biology KXTB, no. 9 (1965): 753-59.

31. CAS biochemistry survey: KXTB, no. 9 (1962): 1-41.

32. Liu Zedong, "Renlei xipao yichuanxue di fazhan" (Progress in human cellular genetics), KXTB, no. 12 (1962): 26-32.

33. Chromosome studies: Zheng Guochang (1951 genetics Ph.D., University of Wisconsin), KXTB, no. 2 (1965): 1059-72.

34. Lindbeck, "Organization and Development of Science," 14.

35. Tan, *Baijia zhengming,* 42.

36. Lindbeck, 14.

37. Lindbeck, 14.

38. Lindbeck, 10-12, 16-23.

39. N. P. Dubinin (b. 1907): Originally trained under A. S. Serebrovsky in drosophila genetics. He was a contributor to developing the concept of "genetic drift," and subsequently was appointed head of the genetics division of N. K. Kol'tsov's Institute of Experimental Biology (1932-1948). He was renowned for his studies there of drosophila population genetics, the original source of Tan Jiazhen's acquaintance with his work. Dubinin opposed Lysenko and as a result suffered professionally. See biography by Mark Adams in *The Soviet Union: A Biographical Dictionary,* ed. Archie Brown (London: Weidenfeld and Nicolson, 1990), 82.

40. See "Effect of Different Dosages of Gamma-Radiation on Chromosome Aberration in the Spermatogenesis of Macaca Mulatta," *Acta Biologica Sinica,* no. 4 (1962): 386-97; "Effects of X-Irradiation," *Scientia Sinica,* no. 8 (1964): 1253-64; "Effects of Different Dosages of Gamma-Irradiation on Spermatogonia and Spermatocytes," *Scientia Sinica,* no. 9 (1964): 1447-52.

41. Bao summarized his research in "Women yanjiu dobeiti di qianhou" (The before and after of our study of polyploidy), RMRB (August 26, 1956).

42. Bao, "The Application of Genetics to Breeding Work," Beijing, *Guangming ribao*, October 24-27, 1962 (SCMP/2866, November 26, 1962, 1-10). This is a comprehensive survey of the recent accomplishments of plant genetics in China.

43. Li Jingxiong et al., "On the Studies of Breeding Hybrid Corn in North China: The Evaluation of Varietal Crosses and Hybrids between Inbred Lines" (English title), *Bejing nongyeh daxue bao*, no. 1 (1957): 1-22.

44. See Li Jingxiong hybrid maize research survey articles: *Hongqi* (Redflag), nos. 2, 3 (1962); and RMRB, January 8, 1963.

45. Fang Zhongxi et al., "The Breeding of a New Variety of Hadai (Laminaria Japonica Aresch)," *Scientia Sinica*, no. 7 (1963): 1-17.

46. Ramon F. Meyers, *The Chinese Economy Past and Present* (Belmont, Calif.: Wadsworth, 1980), 209, and "Wheat in China—Past, Present, and Future," *China Quarterly*, no. 74 (1978): 297-333.

47. Meyers, "Wheat in China."

48. Interview, Y. C. Ting, Beijing, August 4, 1984.

49. Carl Riskin, *China's Political Economy* (Oxford: Oxford University Press, 1987), 169 (citing Thomas Wiens, "Agricultural Statistics in the People's Republic of China," in *Quantitative Measures of China's Economic Output*, ed. Alexander Eckstein (Ann Arbor, Mich.: University of Michigan Press, 1980), 88-95.

50. By the period 1930-1939, corn comprised 9 percent of the farm diet in north China, 3 percent in the Yangzi Valley, and 1 percent in the south. After World War II, because of rising wheat and rice prices, corn production and consumption increased greatly. By 1950, 66 percent of the corn crop was used for food, 19 percent for livestock feed, and 9 percent for seed. The northeast and southwest are the largest producers. The largest consumer is the northeast. See T. H. Shen [Shen Zonghan], *Agricultural Resources in China* (Ithaca, N.Y.: Cornell University Press, 1951), 204-5. Corn is eaten boiled and on the cob, or as a fried bread.

51. Li Jingxiong et al., "Studies of Breeding Hybrid Corn in North China."

52. Wu Shaogui, "Utilization of Hybrid Vigor in Increasing Maize Production in the People's Republic of China," *Acta Agriculturae Sinica*, no. 5 (1959): 360-67.

53. Also see Benedict Stavis, *Making Green Revolution: The Policies of Agricultural Development in China* (Ithaca, N.Y.: Cornell University Press, 1974), 83-86.

54. Kang Chao, *Agricultural Production in Communist China, 1949-1965* (Madison, Wisc.: University of Wisconsin Press, 1970), 174. And also see Leslie T. C. Kuo, *The Technical Transformation of Agriculture in Communist China* (New York: Praeger, 1972).

55. Stavis, *Making the Green Revolution*, 95-97. Also see Alva Lewis Erisman, "China: Agricultural Development, 1949-71," in *People's Republic of China: An Economic Assessment*, ed. Joint Economic Committee, Congress of the United States (Washington, D.C., 1972), 130-31.

56. Stavis, *Making the Green Revolution*, 157-72.

57. For illustrations of their work, see Wang Debao, "Keyongxing hetang hesuan di yanjiu" (Solubility of ribose and nucleic acid), *Acta Biochemica et Biophysica Sinica*, serialized 1963-1965; and Xu Yucheng, Wang Debao et al., "Effects of Conditions Commonly Used for the Preparation of RNA on the Activity of Bovine Pancreatic Ribonuclease," *Scientia Sinica*, no. 5 (1965): 757-64.

58. Illustrations of their work: Liu Zhengwu et al., "Studies on the Analysis of Purine and Pyrimidine Bases of Nucleic Acid," *Scientia Sinica*, no. 5 (1963): 673-84; and "Studies on the Soluble RNA of E. coli," *Scientia Sinica* 15.4 (1966): 507-20.

59. For example see: Hu Shixuan, "Morphological and Cytological Observations on the Process of Fertilization of Wheat [Dynamic aspects of nucleic acids in the fertilization process]," *Scientia Sinica*, no. 4 (1964): 923-38; and Lu Jiahong, "RNA in the Silk Gland of Silkworms," *Scientia Sinica*, no. 5 (1966): 683-95.

60. See notice of trip in *Sulian nongyeh kexue*, no. 1 (1958): 1.

61. Liang Zhenglan, "Shinian lai Michurin yichuanxue zai wo guo di chengjiu" (China's achievements in Michurin genetics during the past decade), SXTB, no. 10 (1959): 462-67. Translated in JPRS, 2322/CSO: 3302-N.

62. Interview, Huang Zhongzhen, who indicated Professor Wang Xiangming as the Michurinist ringleader at Wuhan. And see Wang Xiangming's especially nasty tirade against the Morganists in *Wuhan daxue ziran kexue bao* (Wuhan University natural science digest), no. 3 (1959): 107-15; and the hyper-Lysenkoist book *Mendel-Morgan yichuan xuepai pipan* (Critique of the Mendel-Morgan school), Wuhan University Biology Department, trans. (Beijing: Science Press, 1960). This was a set of lectures delivered at the school by visiting Soviet expert H. A. Tikonova.

63. Mi Jingjiu, "Several Questions in the Discussion of Genetics," Beijing, *Guangming ribao*, June 7, 1961.

64. Zu Deming, "Academic Discussion and Practice of Genetics," Beijing, *Guuangming ribao*, June 14, 1961.

65. This journal was begun in August 1956, before the formal establishment of the institute, which took it over in 1959.

66. See Liang Zhenglan, "Shinnian lai," and Tong Dizhou, "A Decade of Biology in China," *Scientia Sinica*, no. 12 (1959): 1437-44.

67. See Du Rofu (b. 1930), "Survey of Genetic Effects of Occupational Chronic Irradiation among Chinese Medical Workers [title in English]," *Yichuanxue jikan*, 1965. Du was trained in the Soviet Union. From 1958-1962, he was a research assistant at the CAS Institute of Atomic Energy; he joined the Institute of Genetics in 1962.

68. See Chen Xiulan (b. 1932), "The Effect of P32 on Frequency of Chromosomal Aberrations of the Germ Cells in Male Mice," December 1965. Ms. Chen was trained at Yanjing and Beijing Universities, then was a research assistant at the CAS Institute of Experimental Biology. She went from research assistant to full member of the Institute of Genetics from 1959-1962. Also see Zhou Xingting (b. 1938), "Comparative Study of Acute and Chronic Co60-Gamma Ray Irradiation on Chromosomal Aberrations in Peripheral Leukocytes of Rabbits Irradiated in Vivo." Zhou received his B.S. from Fudan in human genetics, then joined the CAS Institute of Genetics [Original alternate English titles].

69. See S. D. Richardson, *Forestry in Communist China* (Baltimore: Johns Hopkins, 1966), 143-49; and also Kang Chao, *Agricultural Production in Communist China*, 88.

70. Wu Zhongxian in Schneider, *Lysenkoism in China*, 27.

71. Higher education: During 1966-1969 no students were admitted. Enrollments fell from a high in 1965 of 674,436 to a low of around 5,000 at the beginning of 1970. Forty-two thousand were newly admitted that year. By 1976, total higher education enrollments were still a hundred thousand below their 1965 level (and the total population of China was of course a lot bigger as well). Saich, *China's Science Policy*, 112-13.

72. *Plant Studies*, 1975, 4, 6, 46-47, and passim. Also see similar but more sympathetic descriptions in *China: Science Walks on Two Legs*, ed. Science for the People (New York: Avon, 1974). The latter is a trip report from an American group "politically sympathetic" to China and the Cultural Revolution.

73. For example, the interpretations of Maurice Meisner in *Marxism, Maoism and Utopianism,* and in *Mao's China and After.* Also see Roderick MacFarquhar's interpretation in which the Cultural Revolution is in effect the culmination of the first fifteen years of PRC history: *Origins of the Cultural Revolution,* vol. 1 (Oxford: Oxford University Press, 1974); vol. 2/3 (New York: Columbia University Press, 1983, 1997).

74. Interview, Mi Jingjiu, 1986. Mi was a member of this "sent down" faculty at the time.

75. In 1986, a branch of the local constabulary prominently used some of the campus buildings for offices and barracks.

76. Interview, Tan Jiazhen.

77. Interview, Shi Yingxian, Beijing 1984.

# Part III

## Deng's China, 1976-2000

# 8

# Science Reforms and
# the Recovery of Genetics, 1973-1986

The scientific community attracted an immense amount of party attention and effort over the decades following the Cultural Revolution. During these years, post-Mao leadership (especially the Deng Xiaoping regime) initially set a goal for itself to reestablish the scientific community's pre-Cultural Revolution status, and then to do whatever was necessary to enable the natural sciences in China to reach international levels by the end of the century. Party and state became heatedly devoted to the process of reviving and developing science education and research. Each year saw a growing number of national and regional conferences on these questions, accompanied by a blizzard of policy statements with an ever expanding list of increasingly more penetrating reforms. Every aspect of the scientific community's operation and structure was addressed. Genetics—whether referred to as "molecular biology," "genetic engineering" or "biotechnology"—consistently worked its way up the priority list of science development plans, and was gradually expanded and transformed.

The premise of all this attention was fairly straightforwardly suggested in the mid-1970s, in the ambitious, broad-based economic reform program called the "Four Modernizations," aimed at industry, agriculture, national defense, and science/technology. The latter pair, it was argued, was the engine that would power the development of the other three areas; so it stood to reason that everything necessary had to be done for the recovery of a science and technology moribund in many sectors since the late 1960s. The reform program was only able to gain momentum, however, after closure on the Cultural Revolution was begun, a process that was interwoven with the politics of post-Mao succession, and the legitimization of the Deng Xiaoping regime.

A critical turning point in the reform of science came, sometime in the early 1980s, when evaluation of China's science (along with that of the entire economy) was fundamentally reframed. Basic problems of growth and development were no longer attributed to the Cultural Revolution, in spite of its acknowledged damaging effects. Rather, it was the entire Soviet-derived system of science organization and practice that was subjected to scrutiny and often severe criticism. As a result, by the mid-1980s, the reforms in place or unfolding for the science community were far beyond anything suggested at the outset of the Four Modernizations campaign.

The commercialization of science was the major case in point, closely linked to a gradual imposition of the "individual responsibility system" on science research institutions and programs. Here, science research funding—in effect an entitlement—was shifted away from state budgets, to two other areas: income from contractual services, and competitive grants from new public agencies designed to encourage basic research. These developments in science resonated closely with the trends of broader economic reforms that were opening major sectors of the economy to market forces and internationalization.

Another set of science reforms was devoted to structural changes of the science community for the general purposes of utilizing personnel more rationally and linking research and production as efficiently as possible. Research, long absent from the universities, was promoted and enabled with new funding sources. Persistent efforts were made to facilitate horizontal communication among CAS institutes, and between the CAS and scientists in educational institutions and the ministries. Loosening up the "work-unit system" (danwei) and encouraging the mobility of science workers was an intimate part of this comprehensive project to facilitate creativity and innovation. Aiming at the same goals was a policy to reconnect China's science education and research with the international science community outside of the Communist bloc.

Additionally, scattered efforts were made to inform these institutional developments ideologically and normatively. Deng Xiaoping worked to depoliticize science and technology, to legitimize professional scientists and their work, and ipso facto to put an end to the populist "mass science" approach to science that had characterized the Great Leap Forward and the Cultural Revolution. Thanks to his regime's efforts, by the end of the first decade of science reforms, the old sine qua non of "redness" had all but succumbed to the triumph of expertise. Parallel to this process of redeeming the scientific community in general, the biologists were still at work laying to rest the ghosts of Lysenkoism. Eventually, they found closure by using their experience with Lysenkoism as an ethical and political guide for the future development of science in China.

## Early Signs of Revival in the Science Community

There was talk of science reform as early as 1972, in the wake of the Kissinger and Nixon visits to China. Some of it was inspired by foreign visitors like C. N. Yang, the Nobel laureate in physics, who was among the earliest of a distinguished line of Chinese-American scientists to visit China and offer advice on science development. Yang visited with Mao and argued the necessity of reviving basic scientific research. At the same time, Premier Zhou Enlai began to endorse and foster the repair of Cultural Revolution damage to science as part of a larger concern for "economic revival and modernization." He proposed that once schools and research institutes were reopened, scientific theory must be taught again, and laboratory research should be conducted in research institutes rather than in fields and factories. He further argued that science and technology students should be enrolled in college directly from middle school rather than forced to delay their formal education for factory or agricultural labor. "If class struggle is grasped exclusively at the expense of struggle for production and scientific experiment," Zhou warned, "then the socialist revolution and construction will be weakened."[1]

It was not until January of 1975, at the Fourth National People's Congress, that Zhou was able to introduce a coherent proposal for economic reform—the "Four Modernizations"—in which science and technology played a central role. His report emphasized the importance of basic research and imported technology. It was endorsed by a set of Deng Xiaoping's policy statements that elaborated the requirements of scientific development, especially the restructuring of higher education and regaining support of the intelligentsia for the state and party. Widespread party debate on these issues was temporarily interrupted by the terminal illnesses of both Mao and Zhou, and an all-out political battle between the Deng Xiaoping faction and the Gang of Four, the last major avatars of the Cultural Revolution and the first major hurdle for the Four Modernizations.[2]

The Gang of Four were arrested and incarcerated in October 1976; Mao and Zhou were dead; and Hua Guofeng became Mao's putative heir.[3] He forcefully renewed the Four Modernizations campaign the following year at the Eleventh Party Congress. Deng Xiaoping and his associates, on a platform featuring science reform, resumed their quest for party leadership and achieved it within the next three years

## Appraisals of Genetics Research, 1974-1978

During this same period of the 1970s, scattered efforts at the local level were made to revive science research, publication, and education. Something is known about the state of genetics thanks to the evaluations made by a series of American and European scientists who visited China under the auspices of the Chinese Academy of Sciences. After the Nixon visit, China invited informal diplomacy and cultural exchanges with the United States and other countries. Many of the American visitors were scientists who acted as consultants and no doubt provided grist for science reform proposals. On the American side, these visits were brokered through the Committee on Scholarly Communication with the PRC, and sponsored by the United States National Academy of Sciences, the Social Science Research Council, and the American Council of Learned Societies.

Five trip reports compiled by American scientists between 1974 and 1978 provide a consistent picture of genetics as a once promising field of science now largely stagnant and disorganized. American scientists variously from the fields of plant genetics, chemistry, and biomedicine all called attention to a set of fundamental problems with genetics research. It was repeatedly noted that, even though universities and CAS institutes had begun to reopen and operate, one-third of their personnel were always away from their teaching or research, assigned to work with production brigades in agriculture communes or factories. This of course was a legacy of the Cultural Revolution mass mobilization of science. Not only were laboratories and field-test sites chronically understaffed, but the experimental facilities were poorly equipped and had little budget support for improvement. Active scientists were out of touch with colleagues in their fields, having inadequate means to collaborate across institutional lines, and an insufficient number of professional organizations and journals for communication. This fragmentation and isolation of the genetics research community was compounded by its lack of communication with the "international mainstream."[4]

Another set of common themes in these evaluations was the lack of "fundamental research" in genetics and the overemphasis on direct application of research to production. This was particularly true of the field of plant genetics where it was additionally observed that researchers tended to lack training in theoretical genetics, quantitative genetics, and statistics.

### *Plant Genetics*

Plant genetics, to the degree that it functioned at all, was dominated by a preoccupation with haploid breeding via pollen culture of anthers. Haploids have only one set of unpaired chromosomes in each nucleus, which gives them the potential advantage to create immediate genetic uniformity, uniform lines, and easily identified genotypes.[5] In China, this methodology was first taken seriously in 1970, by senior workers at the CAS Institute of Genetics, who claimed that the technique was suitable to Cultural Revolution populist values. They said it was easy to teach the masses and easy for the masses to use without the meddling of professional scientists; and the technique could be directly responsible for creating better varieties of plants in short order. "With the conventional method [of creating new plant varieties]," the Institute of Genetics reported, "it used to take seven to eight years before a new variety could be released to production, but varieties developed by this new method can be released to production in about three years. . . . This new method will also open up a new vista for studying the theory of genetics."[6]

By the mid-1970s, haploid breeding had become the much touted signature project of the CAS Institute of Genetics, and it had become fashionable as well in the CAS Institutes of Botany, Plant Physiology, and Agricultural Science.[7] In 1978 and again in 1984, the Institute of Genetics was able to promote itself and this approach to plant breeding in two international symposia held in Beijing, under its sponsorship. While it is not evident that these events created a "new vista for studying the theory of genetics," they were clearly useful for reestablishing the legitimacy of natural science in China and reviving the international relations of Chinese scientists.[8]

At these conferences, the Chinese were praised by like-minded foreign biologists for their pioneering work with haploidy, a technique that was supported and practiced fairly widely throughout the world.[9] American scientists who made site visits in the mid-1970s, however, remained skeptical because Chinese practitioners of this technique could only explain it with obscure genetic reasoning.[10]

The haploid technique was widely taught to masses of Chinese peasant agricultural technicians in the 1970s and encouraged for large-scale practice, probably because the larger the scale of haploid projects the greater "the chance of producing and fixing highly superior genotypes." Outside observers wondered, however, if "better results might have been obtained if the immense resources of the peasant-technician population had been devoted to making more crosses rather than to haploid breeding."[11]

*Biochemistry*

In 1977, another assessment of genetics, this time outside the field of plant studies, was even harsher and more critical. In regard to the study of nucleic acids, it acknowledged the sophistication of unpublished work, but concluded that it was "still years behind the respective fields in the West."[12] Genetics was virtually absent from biomedical research not only because of the Lysenkoist legacy in China, but also because of the lack of senior geneticists and the need to use geneticists in practical work. Microbial, radiation, and population genetics, which had been studied at Fudan University before the Cultural Revolution, still had not been revived. The CAS Institute of Microbiology was doing microbial research, however the "work was competent, but years behind related Western work."[13] In cytogenetics, the report found that in various medical schools, chromosome banding techniques had been applied to clinical material and used in experimental oncology, but that the work was "of little general interest."[14] The CAS Institute of Experimental Biology was described as doing "relatively unsophisticated work on chromosome structure," and the CAS Institute of Cell Biology in Shanghai was judged largely to be doing work of low quality on the role of nucleic acids in cell cultures and as a messenger. The quality of the work at the CAS Institute of Genetics was similarly denigrated, particularly its efforts to study the transformation by DNA of factors for sporulation in *Bacillus subtilis*.[15]

A more sanguine appraisal of genetics-related biochemistry was made in a 1978 trip report by American chemists. It found a respectable effort being made to achieve the first synthesis of transfer RNA—a cooperative effort undertaken by the CAS Institutes of Organic Chemistry, Experimental Biology, and Biochemistry in Shanghai and Biophysics in Beijing. This project was also a significant effort to reconstitute a community of scientists that was lost during the Cultural Revolution. And further hope for the future was taken in the just-completed, new, seven-story facility built for the Shanghai Institute of Biochemistry. Its staff of 200 research and technical workers, including 150 who had B.S. or higher degrees, was poised to open China's first set of laboratories dedicated to "genetic engineering."[16] This is the first concrete evidence of the 1978 national eight-year science plan which made "genetic engineering" one of a number of key areas for development.

## Institutional and Intellectual Milestones in Genetics

However accurate these evaluations may have been in the 1970s, few of them considered the immense amount of effort required of the institutions in question to overcome the inertia of the early Cultural Revolution years and the persistence of Cultural Revolution antiscience attitudes. For the genetics science community, the earliest efforts to break out of Cultural Revolution constraints were initiated by the CAS Institute of Genetics.

*CAS Institute of Genetics*

In 1971-1972, the institute started up a journal (*The Genetics Report*) and sponsored a national conference on genetics and plant selection held on Hainan Island.[17] The content of the journal and the conference was primarily concerned with hands-on breeding techniques, reminiscent of earlier Michurinist concern for simple, fast, surefire methods dominated by the institute's growing fascination with haploid techniques. Additionally, both the journal and the conference revealed that the legacies of Lysenkoism and the Cultural Revolution were not easy to leave behind. In the journal, perfunctory articles and pro forma discussions in research reports rehashed the limitations of Lysenkoism and Morganism, and saluted the 1956 "two schools" policy that required an evenhanded approach to their respective contributions. At the same time, the journal paid the required ritualistic tribute to the thought of Mao Zedong: At the beginning of each journal issue, and often in research articles, fragments of Mao's thoughts were quoted helter-skelter, in bold black type. This was a custom that persisted in many science journals until the late 1970s.

At the Hainan genetics conference, an extraordinary keynote speech was given by Zu Deming, head of the CAS Institute of Genetics. When last heard from, before the Cultural Revolution, Zu was still a supporter of Michurinism and grudgingly compliant with the "two schools" policy. At this conference, he publicly capitulated to "Morganism" and recanted his belief in all of the principles of Lysenkoism, including the belief that there is no intraspecies competition and that environment is the fundamental source of organic change and evolution. Zu further denounced Lysenko for failing to acknowledge the validity of modern gene theory, as well as the solid accomplishments in previous chromosome and gene research and in more recent work by biochemistry and biophysics.[18]

There is no earlier record of how and why Zu arrived at this position and chose to air it on this occasion. In his speech, however, he claimed that it was the thought of Mao Zedong, especially Mao's essay "On Contradiction," that brought him to his senses on these subjects. Zu's position, including his acknowledgment of Mao, was echoed throughout the conference presentations. There were even a few conference papers that made respectable surveys of current trends in various subfields of genetics. In all references to modern gene theory or to current work in molecular genetics, their dialectical materialist respectability was emphasized, as was their solid materialist foundations. If this were not sufficient grounds for the legitimacy of modern genetics, all discussions of Morganism at the conference invoked the Double Hundred policy as well.

The revival of genetics continued in 1972 with the initiation of yet another journal from the CAS Institute of Genetics. This one, *Genetics and Plant Selection*, was completely devoted to reporting on recent international developments in genetics. Its editors justified its mission—which would have been an anathema during the Cultural Revolution—with a recent pronouncement by Mao that "raising the level of China's science requires attention to foreign culture."[19] A labor of love, it was anonymously produced from handwritten, mimeographed text during its first two years of publication. The journal translated or summarized skillfully a broad range of sophisticated international materials derived equally from the United States, Great Britain, Japan, and the Soviet Union. Obviously, some major foreign science journals were again accessible in China, though they were probably circulated and used with circumspection. This publication catered to a hunger for news of the field, blacked out since 1966.

### Revival of Genetics Publications

By 1974, many CAS institutes had begun to revive their main research journal; such was the case with the Genetics Institute whose journal, *Genetics* (Yichuanxue bao), became the only national forum for genetics research until the late 1970s.[20] Geneticists from around the country and of various interests did publish here; but in the 1970s and 1980s it was nevertheless dominated by the Genetics Institute's primary concerns with plant genetics and haploidy, and its secondary concerns—stemming from its work in the mid-1960s—with environmental sources of human genetic disorders.

Just as the head of the Genetics Institute, Zu Deming, had earlier felt compelled to bring the problem of Lysenkoism to public closure at the Hainan conference, the institute's journal likewise published conspicuous articles that straightforwardly acknowledged the validity of genetics from Weismann to Morgan to DNA, and the invalidity of Lysenkoist principles. The ceremonial first article in *Genetics'* first issue, for example, was written not by a member of the institute, but by Fang Zhongxi, who had worked successfully for Lysenkoism in China, and then abandoned it to return to classical genetics. He provided an erudite, up-to-date summary of the development of gene theory, and an outline of DNA's role in the central dogma of modern genetics. Without referring to Lysenkoism, he stated the case against the inheritance of acquired characters, and against the role of extranuclear material in the hereditary process.[21]

Rather than describe such public statements as belated efforts to achieve closure with the Lysenkoist era, perhaps it would be more useful to think of them as declarations of principles for the post-Cultural Revolution revival of science; efforts to begin again with a shared understanding of genetics. And who better to deliver that message than those, like Zu Deming and Fang Zhongxi, who had once divided the community and been leading exponents of Lysenkoism?

The signal effort to reunify the genetics community occurred in 1978, with the revival of the Genetics Society of China. Unlike its predecessor, founded in 1958, this organization was not a subsidiary of the CAS, but an independent, determinedly ecumenical, professional society that provided a common organization for those doing genetics-related work, whether they be based in CAS institutes, universities, agriculture or industry, and regardless of their former relationship to Lysenkoism. The society's first president was Li Ruqi, one of China's earliest students of T. H. Morgan and now chair of Beijing University's biology department. He had been president of the 1958 organization as well. Vice presidents included Tan Jiazhen, Zu Deming, and Fang Zhongxi, who was the first editor of the society's new journal, *Hereditas*.[22] In addition to providing a national forum for genetics research in China and from around the world, the journal acted as the major source of news about the redevelopment of genetics in China and an articulate proponent of state science policies contributing to that development. The society regularly sponsored an annual national conference and, as the Four Modernizations drive gained momentum, actively promoted international conferences and the international exchange of scientists.

## Reviving the Biology Community

During these early years of its resuscitation, the biology community continued to find ways to build and express its solidarity, as well as its professional competence and authority. Dramatic examples were two large centenary commemorations, for Darwin and then for Mendel, held in the early 1980s.

### 1982 Darwin Centenary

In 1982, the CAS sponsored a conference in Beijing for the centenary of Darwin's death.[23] The convener was Chen Shixiang, the venerable entomologist who had repeatedly challenged the authority of Lysenkoism in the 1950s and 1960s. Before the Cultural Revolution, he had emerged as one of the country's leading interpreters of evolutionary theory, and at this conference his papers demonstrated that he still maintained an informed control of the literature. One of the goals of the conference was to provide a state-of-the-art picture of evolutionary theory, a subject last confronted in China over fifteen years earlier. The perspectives of scientists from many subfields of biology were expressed, but the points of view from molecular biology and contemporary genetics were the central concerns.

While the conference papers surveyed a broad range of material, there did emerge a common concern with postsynthesis evolutionary theories, especially the "neutral" (or "non-Darwinian") theory of evolution and the questions it raised about the role of natural selection. A strong consensus at the conference argued for the compatibility of the neutral theory and the synthesis, and hence for the continued place of natural selection in the heart of evolutionary theory.[24] Most of the conference participants were actively engaged in research closely involved with evolutionary theory, and all of the conference papers were notable for the depth of their familiarity with theoretical literature, especially with publications unavailable to them until the mid-1970s.

It is also notable that Lysenko's ideas on evolution were virtually ignored during the conference. Be that as it may, older scientists probably needed no references to Lysenko to be reminded that the last time such a centenary commemoration was held in China, it was for Michurin, in 1955.

*1984 Mendel Centenary*

Within a year after the Darwin conference, the centenary of Mendel's death was commemorated by the Genetics Society of China in the form of an anthology of essays, edited by Tan Jiazhen, and an encyclopedic dictionary of genetics edited by Li Ruqi.[25] The latter beautifully organized contributions from the entire life sciences community and set the new standard for genetics terminology in Chinese.

Tan Jiazhen's editorial introduction to the Mendel anthology set a tone somewhat different than that of the Darwin volume. He explained that the major fields of contemporary genetics were surveyed in the volume by their leading exponents in China. Not only was this material meant to provide readers with an introduction to modern genetics, but also to indicate the contributions specifically made by geneticists in China, particularly since the 1956 Double Hundred policy liberated them from the domination of Lysenkoism. Tan claimed that even though the level of all fields of genetics in China might not yet be up to international standards, this volume would illustrate the great strides taken since 1956.

While Tan's introduction denounces the Lysenkoist era and in the strongest terms indicates the damage it did to genetics, he does not directly mention the Cultural Revolution at all. Instead, he praises the 1977 CCP Eleventh Party Congress (the first since Mao's death and the Cultural Revolution) for its fundamental contribution to subsequent science reform policy. Tan also praises the efforts in the Soviet Union that eventually led to the end of Lysenkoism and the restoration of Soviet genetics to its place of international leadership. Thus, where the Darwin commemoration dealt with Lysenkoism by ignoring it, the one for Mendel used direct and self-assured confrontation. This latter tone resonates throughout the carefully edited collection, which virtually pulsates with exuberance about the development of modern genetics and China's participation in it. Though some of the thirty-five papers in the collection address the practical applications of contemporary genetics, most enthusiasm is displayed for its broad theoretical insights and creativity.

The main articles in the Mendel volume are noteworthy for their exhaustive, up-to-date bibliographies, obviously meant to serve a pedagogical function. Additionally, the volume contains a number of well-annotated historical essays that trace the development of Mendelism in the West and in China respectively. In spite of its overly sanguine picture of the contemporary state of genetics in China, altogether, this volume provides a unique and accessible insight into genetics in China from the time of its introduction in the 1920s.

While the Genetics Society volume is scholarly and technical, it also has about it the air of a public relations prospectus, prepared perhaps for the scrutiny of party leaders involved in determining science policy in general and the budgets for genetics research in particular. It would have been easy for such influential people to read the volume and conclude that the considerable investment in genetics made thus far during the 1980s was quite worthwhile and in the hands of very competent scientists.

## Deng Xiaoping and the Beginning of Science Reform

Tan Jiazhen's promotion of genetics occurred in the midst of a revival of science that was tied to the rise of Deng Xiaoping. Science reform, along with the Four Modernizations, was central to the political processes condemning the "ultra-leftist" ideals and actions of the Cultural Revolution,[26] establishing a successor to Mao, and legitimizing Deng's accession to political preeminence in 1980.[27]

By the early 1980s, party leaders, especially Deng Xiaoping, had dramatically condemned the Gang of Four and Cultural Revolution for their attacks on science and the scientific community. By the same token, they sought to restore the legitimacy and stature of science and scientists at least to what they had been during the best of earlier times. Not incidentally, Deng and the Party also wanted to bolster their own legitimacy by recruiting scientists into the Party.

In very public forums, like the 1978 Science Congress, Mao's populist, class-based critique of science was excoriated, though the critique was attributed to the Gang of Four. Science was declared to be part of the material substructure of society (and not an element of the superstructure where it was liable to political manipulation). People engaged in scientific and technological work, Deng Xiaoping said, were indeed workers: "the overwhelming majority of them are part of the proletariat. The difference between them and the manual workers lies only in a different role in the social division of labor. Those who labor, whether by hand or by brain, are all working people in a socialist society."[28]

So much for the Maoist (and utopian socialist) critique of the division of labor, trivialized here and dismissed with an "only." The "red and expert" litmus test for science intellectuals was neutralized by Deng with equal brevity and decisiveness: "To devote oneself to our socialist science and contribute to it is an important manifestation of being red, the integration of being red with being expert."[29]

All of these positions supporting science had been taken before, albeit piecemeal, in the policy statements of Zhou Enlai, Lu Dingyi, and others during the 1950s and 1960s. In Deng's declarations, however, they are given unusual gravity by the fact that they are aimed at the crimes of the Cultural Revolution and are all packaged together as essential elements of a larger plan to drag the Chinese economy out of an embarrassing slough of self-imposed backwardness.

Deng and his colleagues played on this tension to argue further that the Chinese scientific community would not be able to raise itself up by its own bootstraps; while it must always strive for self-reliance and independence, it must also own up to its backwardness: "independence does not mean shutting the door on the world, nor does self-reliance mean blind opposition to everything foreign"—which had been the case in the Cultural Revolution. Deng wanted science and technology to be seen as "a kind of wealth created in common by all of mankind"; and he argued that the necessity of learning from the international science community is not just a temporary expedient to overcome China's current backwardness, but that it must be a permanent feature of China's science community.[30]

Ideological redemption of the science community was accompanied by an eight-year science and technology development plan that promised a great deal. For biology, the plan's most notable decision was to make "genetic engineering" one of the eight key areas targeted for special attention and concentrated resources.[31] In the immediate future, the plan called for facilitating basic research in genetic engineering by building or improving relevant laboratory facilities. Subsequently this research was to be combined with studies in "molecular biology, molecular genetics, and cell biology." The uses of research in these areas were to be considered for the pharmaceutical industry, in the treatment and prevention of certain "difficult and baffling" diseases, and in the creation of new high-yield crop varieties capable of nitrogen fixation.[32]

Accompanying his political ascendancy, Deng launched a vigorous campaign to recruit more intellectuals, especially science and technology intellectuals, into the Party. Simultaneously, he began an extended effort to purge the Party of "leftists" who had been brought in during the Cultural Revolution, and who objected to the recruitment program. Oddly reminiscent of Mao in the 1956-1957 Double Hundred campaign, Deng was concerned about the Party's lack of technical expertise. Rather than enhance the expertise of current party members, Deng sought to recruit experts from the outside, mostly from engineers, but certainly from natural scientists as well. By the Twelfth Party Congress in 1982, there had already occurred a great turnover in the Party Central Committee and in the upper echelons of the state bureaucracy.[33]

Prominent geneticists were among those who joined the Party during these years: Tan Jiazhen in 1979 and Li Jingxiong (the radiation specialist) in 1981, for example. Other veteran biologists, and heretofore distinctly independent types (like Chen Shixiang in 1979) joined up too. For these intellectuals recruited to the Party, perhaps to technical intellectuals in general, Deng Xiaoping apparently held out the prospect of enhanced professional autonomy, greater material benefits and social status, along with the development of science and technology.[34] One need not have joined the Party to have profited from Deng's promised benefits for the scientific community; but party membership did suggest at least the possibility of more closely influencing the reform process. Joining the Party at this time might also be taken as a vote of confidence for Deng's general anti-Cultural Revolution policies and his prospects for achieving relative social order and civility while advancing China's economy.

## Constructing Memory of the Double Hundred

In spite of evident mutual support of party and scientists, the latter apparently harbored a sense of caution and ambivalence about the Party and state. This was evident in the mid-1980s when the oft-violated "Hundred Flowers, Hundred Schools" policy had its thirtieth anniversary. Since Deng Xiaoping identified himself and his science policies with the Double Hundred, the Party encouraged its commemoration. Typically, commemorative publications and events, which took place over 1985-1986, were couched in terms of the experience of genetics and Lysenkoism. That is, they emphasized the poor relations between party and intellectuals that the Double Hundred was designed to ameliorate, and which it failed to ameliorate time and time again. Between April and July 1986, for example, dozens of national newspaper and periodical articles cited the experience with Lysenkoism and the ensuing Qingdao Symposium as major factors in the formulation of that part of the Double Hundred policy concerned with science.

These articles argued that for science education and research to flourish and be productive—in the present as in the past—it must be free of government and party interference.[35] An article in *People's Daily* complained that the government and party were still meddling in the work of scientists and must stop it if the scientific community is to progress.[36] In May of 1986, the CAS and other science institutions sponsored a well-attended national symposium to commemorate the Double Hundred and (implicitly) to express appreciation for its renewal under the auspices of Deng Xiaoping's science policies; but even here dissident voices claimed that in spite of recent accomplishments on behalf of science, academic freedom still had a long way to go in China.[37] In the fall of 1986, the last formal event of the commemoration was sponsored by the Genetics Society of China and was attended by a small group of biologists who poignantly discussed the profound political implications of the 1956 Qingdao Symposium.[38]

The most sophisticated and enduring acts of commemoration for the Double Hundred policy, however, were the republication and public sale of the original proceedings of the 1956 Qingdao Genetics Symposium.[39] In addition to the complete proceedings, this publication also includes a history of Lysenkoism in China and a detailed chronicle of Lysenkoism in the Soviet Union. While this volume provided the most comprehensive view yet of China's former "genetics problem," the ground had been prepared for it by bold historical studies sponsored by an extraordinary group of Chinese intellectuals interested in the history and philosophy of science, and in the relationship of science, society, and the state. From 1979, their forum was the *Dialectics of Nature Journal*, one of the most intelligent and insightful publications on science in post-Mao China.[40] In its early issues, this journal published the first critical historical studies of Lysenko to appear in China, as well as biographical portraits of Chinese scientists who were victimized by Lysenkoism, and research on the political background to the Qingdao Genetics Symposium.[41]

In this post-Cultural Revolution/post-Mao effort to reconstruct the origins of the Double Hundred policy, there was one more particularly telling project. Throughout 1986, a group of young scholars associated with the newly established Beijing University Center for the Study of Science and Society conducted an unprecedented study of the Double Hundred policy and its consequences. They interviewed or interrogated via a voluntary questionnaire three dozen prominent scientists, party cadre, and bureaucrats who were personally involved in the formulation of the policy and/or affected by its promulgation. Even some former Lysenkoites submitted to inquiries, including questions about the effectiveness of the Double Hundred policy in the 1980s and about current constraints on academic freedom.[42] Together with the new edition of the Qingdao Symposium proceedings, this study was clearly motivated by a desire to use the experience of the genetics community with Lysenkoism as a cautionary tale about the dangers to science, if not the entire intellectual community, when politics and ideology violate its independence and integrity. Collectively, all these activities share a common, passionate message: In a time of party and state largesse to science, don't become complacent, don't compromise intellectual freedom, and don't forget the lessons of the past.

## Systemic Problems of Science in the 1980s

Before the ink had dried on the eight-year science development plan, it was apparent to all concerned that just throwing money at the plan was not going to get it done. There was general agreement among reformers and planners that fundamental systemic problems with the organization and finance of science had to be fixed. Those problems stemmed not only and perhaps not even mainly from the Cultural Revolution; they were inherent in China's organization of science according the Soviet model of the 1950s. Foreign observers as well as the Chinese themselves have provided sufficient description and analysis of the problems and their solutions. A short summary here is meant to provide appropriate background for later illustrations of how reform policy was applied specifically to the field of genetics.

## Science Personnel and Production

In its most abbreviated form, the critique of science argued that there were two fundamental problems: improper treatment and use of technical personnel, and the "administrative separation of research and production units."[43] The problems' general source was a rigid, vertically organized, and fragmented science community in which research was isolated in institutes that communicated poorly or not at all with each other, with production units, or educational institutions. In the latter, science research was generally discouraged altogether; in the institutes there was no teaching. During the 1950s, this system may well have proved expedient in the short run for achieving fast and focused industrial development controlled tightly from the top down in Beijing. It failed, however, to evolve an efficiently organized, creative science community whose research intended for application had a good opportunity to be used for productive purposes.

The science research community was completely supported by state budgets, which came to be treated as entitlements. Successful research units were not necessarily rewarded beyond their budgets; lack of success was not systematically reflected in lower budgets or loss of personnel. Indeed, the personnel system which China created within the Soviet structure all but eliminated the possibility of rationally adjusting personnel needs to changing needs of the greater science community.

## Work-Unit System

The teachers and researchers at all science institutions shared a common working experience with each other and all other urban workers in China. Known as the "work-unit system" (danwei) this had come to be a basic means for the state to control the urban labor market and to control the fundamental social, economic, and political aspects of the workers' lives. All urban workers had to be associated with an identifying unit, ordinarily their workplace. The unit was the source of the "iron rice bowl," the basic, guaranteed economic and social benefits—housing, education, health care, and lifetime employment. Since a lifetime association with one and the same unit was ordinarily assumed for each worker, the work-unit system all but eliminated labor's mobility and a labor market. A scientist whose career began at one institution was very likely to remain there, even if his/her skills could be better used elsewhere, even if he/she preferred to be working elsewhere. Research or educational institutions hoarded their workers and were loath to permit them to be transferred elsewhere for (well-founded) fear that they would not be replaced.

Because budgets were allocated to the science research work-units and not to individual scientists, work-units were often accused of developing an attitude of "ownership" of their workers, their particular areas of science, and specific research problems. Nationwide, the result was redundancy, waste, and lack of concern for the broader development of a science discipline or a cooperative approach to a common scientific problem.

Not only did science research units suffer from their insulation from each other, they also showed the ill effects of isolation from the mainstream international science community. Clearly this was not an effect of Soviet-style organization, but rather it was the combined result of Cold War isolation exacerbated by Chinese nativism and xenophobia during the Great Leap and the Cultural Revolution. Potentially valuable transfers of knowledge and technology had been obviated for years. Add this to the collapse of China's educational system during the Cultural Revolution and the result was a major shortage of the science personnel required by the Four Modernizations and the eight-year science development plan.

## Systemic Solutions

Solutions to these problems were formulated and proposed, refined and promoted again over the course of the 1980s. The following summary of the reform process isolates reform categories and their respective intentions, and suggests some of their immediate general effects. The main categories that encompass the reforms are the individual contract system and the commercialization of science, the state's new research funding agencies, and strategies to increase the size and mobility of the science personnel pool. In the next chapter, there will be concrete illustrations of how the reforms influenced specific institutions devoted to genetics research.

### Individual Contract and Commercialization

The overwhelming concerns of the reforms were coordinating science development with overall economic development, and, collaterally, finding revenue to pay for a greatly expanded and modernized science establishment. The paid contract system was seen as the foundation to a solution.

Research institutions were gradually weaned from their total reliance on state allocated budgets and required to find other sources of revenue. While the state continued in some cases to support the salaries of senior researchers and overhead costs, the costs of research and potential additional income ideally were to derive from contracts between the scientists and national or international clients in industry or agriculture.

It was argued that this would compel research units to gear themselves to economic development; help to break down boundaries between economic sectors and geographical regions; promote the rational flow of personnel (assuming that individuals had the right to leave their work-units); and stimulate initiative and creativity. The contract system was the hammer that would break "the iron rice bowl," and destroy the counterproductive notion that work-units "owned" their personnel and their special field of science or technology.[44]

New production needs were thought of as a market force for science research; and, to some degree, the research agenda would be set by this market, rather than by abstract notions of what constituted appropriate research. Production units would be able to contract out for needed research; research units would be able to compete for and bid on contracts, keeping the consequent revenues for overhead, expansion, and future research projects. Research and production units were even encouraged to form "joint ventures"—that is, longer term associations that crossed heretofore impenetrable institutional barriers.[45]

In 1985, the president of the CAS said that the commercialization of science and technology was gradually being accommodated in the CAS by transferring authority down to the institutes, making them independent and autonomous, while making them responsible for supporting themselves. By this date, some CAS institutes were already beginning to raise revenues not only through contracted research, but also from renting out the use of their research facilities, and from specialized training or postgraduate degree programs for educators and industrial or agricultural workers. The CAS was encouraging its institutes to create and operate companies that applied their scientific or technical research and then marketed the product—not just in China—but internationally as well, for much-prized foreign exchange.[46]

Even as research units began to support themselves, however, there was persistent concern that the play of market forces was still not sufficiently linking research to production needs. This was largely attributed to the "rigid verticality" of the old system which discouraged rapid and open communication, the diffusion and sharing of scientific information and technology, and freewheeling mobility for science and technology personnel.[47]

## Science and Technology Research Fund

To promote the best research in the appropriate areas, in 1982 the CAS created the Science and Technology Fund, modeled on the United States National Science Foundation. It was a response to the contract system's emphasis on applied research. The CAS was rightly concerned that a contract system, unchecked, could drive virtually all research into the applied area, narrowly conceived, and deprive the CAS of any opportunity to encourage and support creative research that had no immediate market value or application. The fund catered to non-CAS scientists and was primarily meant to support science research in the universities, from which it had largely been absent since the mid-1950s.

The fund was open to applications from individuals or from teams for project grants (including the hiring of personnel, or buying and using equipment). At the end of the fund's first three years of operation, 51 percent of the 4,770 project proposals were approved, to the tune of 120.33 million yuan. Ninety percent of the funding went to individuals associated with units outside of the CAS; almost 80 percent to researchers in higher educational institutions.[48] In an effort to support "younger" scientists, 80 percent of the funded project directors were under fifty-five years of age. By 1985, however, in order to combat a deep-rooted seniority system, the CAS had established a "young scientists fund" specifically for researchers under the age of thirty-five.[49]

Project proposals were evaluated strictly by peer reviewers who were selected nationwide, and it was claimed this process "has promoted the leading role of scientific research personnel [instead of party officials or bureaucrats] in scientific research work."[50] Consequently, the fund administration felt it had successfully begun to promote "competition and integration" through its allocation of funds according to the best projects instead of routinely budgeting and allocating funds by work-units (danwei).

The new approach of the Science and Technology Fund made it possible, it was claimed, for "topical groups" of researchers to be assembled across "unit, departmental, and regional boundaries," benefiting "the exchange and inter-flow of scholarly ideas, the growth of marginal sciences, rational circulation of personnel, and achievement of large cooperative tasks." In addition to loosening the constraints of the work-unit system, the Science and Technology Fund was also aiming to bridge the typical exclusiveness of educational and research units as well.

## National Natural Science Foundation

The CAS Science and Technology Fund was superceded in 1986 by a new agency, the National Natural Science Foundation (NSF). Responsible directly to the State Council, and independent of the CAS, the NSF began with a greatly enhanced budget allocated directly from the Ministry of Finance. Its mandate continued to be the support of basic research, in the narrow sense, but "applications research with no near-term application value" was also to be considered for support.[51]

Though it expanded the number of categories for competitive grants, more than half of its budget continued to go to academic institutions. One of the new NSF categories, "Key Projects" was established for ongoing projects of established, reputable researchers. The awards here were four times greater than the general awards and given for three-year periods. In 1987, 61 percent of the recipients were in academic institutions. Another new category was "Frontiers in High Technology," for fields like biotechnology, information science, and aerospace sciences. While this category clearly emphasized applied science, only 8 percent of the NSF budget was allocated to it in 1987, and grants here were only slightly higher than general grants. Support of research in this category was soon to be subsumed by a new large-scale initiative (the "863" program) specifically to fund priority high-technology areas.

The largest group of grant recipients in 1987 was scientists between the ages of 35-65 in academic institutions. The specific "young scientist" category continued as well. It was now made more flexible in the hope of enticing foreign-trained Ph.D.'s to return to China. To that end, applications were accepted from overseas, and in 1987, 29 percent of recipients were living abroad. [52]

## Training Personnel

Personnel problems may have been the most intractable of all. Throughout the 1980s and early 1990s, efforts generally fell short of meeting the demand for properly trained scientists. CAS institutes were gradually and with some care empowered to offer masters' degrees and doctorates, but this produced only small numbers of advanced personnel whose training may not always have been up to date and of the highest international standards. To achieve that, it was necessary to rely heavily on overseas training and academic exchanges. Between 1978 and 1985, the government had already sponsored 29,000 students for overseas higher education. Many others studied abroad on their own resources or on foreign fellowship—7,800 just in 1985. There were 56,000 Chinese students and scholars who visited the United States (by far their favorite destination) from 1978 to 1987; but by the beginning of 1988, there were already 21,000 Chinese in the United States studying or working with J-1 visas and 7,000 on F-1 student visas. Forty percent of all these overseas students studied engineering; 28 percent studied natural sciences; and 7 percent agriculture.

Remarkably, over this period most of the cost of educating Chinese students in the United States was gradually assumed by United States' universities. Apparently seeing this as a way of cutting its own expenses, Chinese government support of Chinese students in the United States severely declined from 54 percent to 17 percent between 1979 and 1985. This occurred in spite of explicit requests from the science community to continue if not expand such support. [53] Getting students to return to China, however, was a problem in itself that required the science and technology reform program to try to break through the work-unit system in order to guarantee returnees their choice of jobs. [54]

Over and above accommodating the job placement of returnees, small-scale experiments were made to address the general need for a more flexible science and technology labor market. [55] Expressing considerable frustration with the constraints of the work-unit system, some reformers unrealistically called for its abolition altogether. To begin with, they argued, the belief in egalitarianism should be overcome in order to let the technical labor market be driven by monetary incentives and rewards that would give the opportunity to scientists (in Deng's words) "to get rich first." Though a "profit-above-all" mentality was discouraged, "money still mattered" and talented research personnel should be allowed to follow the money, unconstrained by work-unit "ownership."[56]

Though lack of data make it impossible to quantify mobility changes for science research personnel, it is clear that by the end of the 1980s, new kinds of research institutions were empowered by the state to draw talent from wherever they found it.[57] Illustrations of this will be provided in the next chapter. Nevertheless, despite decided progress in commercializing science research, assessments of that progress continued to claim that its labor market was still far from free.

## Critical Assessment and Lowered Expectations

All of these mid-1980s reform policies and plans were summed up in the comprehensive 1986 "Science and Technology White Paper" compiled by the State Science and Technology Commission. It assessed the just completed eight-year science plan and offered an outline of goals for the remainder of the century.[58] Biotechnology remained a high priority for development, as it had been since 1978, but more for what it promised than what it had accomplished. The white paper's assessment of the field's development in China suggested that it continued to suffer from just about every malady that the reformers had attributed to science research in general: It cited "the insufficiency of researchers, the decentralization of strength, and the mismatch of specializations." Basic research in the field was still weak, it said, and "research and development are seriously out of line; technology is mismatched [and] it is very difficult to transform research results into productive ability. . . ."[59]

Policy statements from the mid-1980s consistently reveal a general attitude toward the development of natural science that differs substantially from policy goals expressed in the late 1970s. They no longer speak of raising China's science to the most advanced international levels by the new millennium. On the completion of the eight-year science and technology plan in 1986, it was evident that there were obstacles—like the serious dearth of well-trained personnel—that would make that schedule impossible. A new fifteen-year plan (1985-2000) resonated with the policy statements where it emphasized the achievement of economic goals, viewing science and technology secondarily, as means to that end. The head of the State Council's Research Center summed it up this way in 1986: "The basic goal of science and technology toward the year 2000 is not to match international levels, rather it is to satisfy production and economic priorities. . . ."[60] This more conservative approach to science and technology development was matched by a similarly cautious approach to production and economic development. By the end of the millennium, planners hoped to achieve levels typical of advanced industrial nations in the 1970s and 1980s.

## Conclusion

Controlling science, in the sense I have been using the idea, was not a priority in the state's program to rebuild science in the 1980s. It assumed the diminution of control, and a considerable and increasing dependency on foreigners. Though this took place gradually, it nevertheless required an emphatic reversal of the Cultural Revolution's virulent nativism and xenophobia. As the country opened its doors to international influences, there were periodic expressions of conservative resistance in the form of "antispiritual pollution" campaigns. Science, however, once again was seen as a cosmopolitan entity—Deng's "universal fund of human knowledge"—and hence drawing on that fund was no grounds for cultural embarrassment. Without apology, an endless stream of foreign scientists was invited to appraise China's science and advise on its improvement. The expertise of expatriate Chinese scientists was especially sought after, perhaps because it was perceived that there was no potential loss of face in accepting their criticism and advice.

At the same time, science reform programs became quite dependent on the acquisition of technical education abroad, especially in the United States. During the 1980s, there were probably as many Chinese studying technical subjects in the United States as all the Chinese studying all subjects there previously in the twentieth century. Yet, as in other instances, this dependency was not considered by the reformers to reflect ill on Chinese culture or character, much less on Deng Xiaoping's party and government.

China's science and technology problems were initially attributed to the contingencies and aberrations of the Cultural Revolution; but the reforms in the 1980s soon stopped addressing the Cultural Revolution and aimed instead at transforming the Soviet system which preceded it, and which the Cultural Revolution too, in its own way, wanted to transform. One of the most fundamental insights of the science reformers was that returning to the pre-Cultural Revolution status of science was not a solution. In effect, they determined to go back to the early 1950s and rethink the premises of science organization and practice. But now, unlike that time, science, as the entire economy, was burdened with ponderous realities such as state-owned enterprises and the work-unit system. There was also the reality that state and party needed to reestablish their legitimacy to the much abused science community, a community that was loath to give up the only thing that it could count on over the years—its entitlements.

If the science reformers were not deterred by the Party and state's diminished control of science, their approach to controlling scientists was equally confident. Central to the reform critique of science was its conclusion that the generally moribund state of science was in large part due to the near total reliance of scientists on the state. How vividly this proposition contrasts with the earlier Maoist concern that state and society had become too dependent on scientists. And of equal contrast was the reformers' solutions: disentitle the scientific establishment and require it to support itself, guided by a new set of values and controls—individual responsibility, competition, and market forces.

The next and final chapter provides a series of concrete illustrations of how these fundamental reform principles affected the organization and practice of genetics during the 1980s and 1990s.

# Notes

1. For discussion of early efforts to reverse the Cultural Revolution, see Merle Goldman, *China's Intellectuals: Advise and Dissent* (Cambridge, Mass.: Harvard University Press, 1981), 162-66. And see Zhou's statement in RMRB, October 10, 1972 (SCMP/5239: 18).

2. For detailed analysis, see Maurice Meisner, *The Deng Xiaoping Era: An Inquiry into the Fate of Chinese Socialism* (New York: Hill and Wang, 1996).

3. Hua Guofeng (b. 1921) had been the Party secretary for Mao's home province, Hunan. He rose to head of the CCP and premiership after Mao's death in 1976. The Gang of Four was arrested under his authority.

4. This composite assessment is based on the following reports: *Wheat in the People's Republic of China*, ed. Virgil Johnson et al. (Washington, D.C.: National Academy Press, 1977); *Plant Studies in the People's Republic of China* (Washington, D.C.: National Academy Press, 1978); *Chemistry in the People's Republic of China*, ed. John D. Baldeschwieler (Washington, D.C.: National Academy Press, 1979); Jack Harlan, "Plant Breeding and Genetics," in *Science in Contemporary China*, 295-312; H. E. Temin, "Basic Biomedical Research," in *Science in Contemporary China*, 255-68.

5. Harlan describes the procedure this way: "Pollen grains at an early stage of development are cultured on a sterile medium; calli so formed are induced to form plantlets; chromosome number of the haploid plantlets usually is doubled spontaneously, but sometimes artificially with colchicine. Theoretically, this produces completely homozygous diploids . . . [and] homozygosity and uniformity are achieved immediately." *Plant Studies*, 303.

6. Chou Pao, "Achievements of Open Door Science Research," *Scientia Sinica*, no. 2 (March-April 1976): 171-74.

7. Harlan, "Plant Breeding and Genetics," 301.

8. The 1978 Sino-Australian Symposium on Plant Tissue Culture is documented in *Plant Tissue Culture* (Boston: Pitman, 1981). In addition to Australia, this conference had fifty-five participants respectively coming from Canada, the U.K., France, West Germany, Japan, etc. The proceedings volume, *Genetic Manipulation in Crops* (Philadelphia: International Rice Research Institute, 1988), documents the much larger and more sophisticated symposium by the same name.

9. See for example the comments of W. R. Scowcroft who said the Chinese were "world pioneers and the leading exponents of the use of haploidy in plant improvement." He approvingly described the technique as "the rapid achievement of homozygosity and, in consequence, the rapid incorporation of new genes into breeding material. This is

particularly relevant to genes with major effect. For such genes, the use of haploids permits increased selection efficiency" (*Genetic Manipulation in Crops*, 13-14).

10. For example, Harlan argued that "It has not been demonstrated either in the PRC or in the West that a completely homozygous line is the most desirable and useful line. Indeed, a number of studies . . . indicate that a high degree of homozygosity is unstable even in self fertilizing species, and that there is a selective advantage for some heterozygous allelic combinations. Moreover, there are theoretical reasons for believing that a too rapid approach to homozygosity does not provide for adequate recombination and breakage of linkage. Immediate fixation can cause loss of highly desirable combinations" ("Plant Breeding and Genetics," 303).

11. Harlan, 303. And compare the survey of accomplishments in plant genetics cited by Y. C. Ting, "Genetics in the People's Republic of China," *BioScience*, no. 8 (1978): 506-10. At the time of his 1976 and 1977 trips to China, Dr. Ting was on the biology department faculty of Boston College. While not addressing the problems of the scientific community cited above in the text, Dr. Ting does provide a broader set of illustrations of contemporary research, emphasizing its practical, applied orientation. Also see China trip summary by the geneticist, James F. Crow, "Genetics in Postwar China," in *Science and Medicine in Twentieth-Century China*, 164-69.

12. Temin, "Basic Biomedical Research," 257.

13. Temin, 265.

14. Temin, 259.

15. Temin, 263, 265.

16. Baldeschwieler, *Chemistry and Chemical Engineering*, 209-10, 228.

17. The *Genetics Report* (Yichuanxue tongxun). The first issue of this quarterly appeared February 1971. Title changed to *Yichuan yu yuzhong*, January 1975. The conference, sponsored by the CAS and Ministry of Agriculture, had 226 participants from 24 provinces and 71 different work-units (danwei). Report and paper abstracts published in *Yichuan yuzhong xueshu taolunhui wenji* (Beijing, n.p., 1973).

18. Zu, conference report, 1-15.

19. *Genetics and Plant Selection: Foreign Research Trends* (Yichuan yu yuzhong: guowai yanjiu dongtai), first issue January 1972, editorial introduction. Translations, summaries, and editorial work were all anonymous—just in case. This monthly journal changed its mission to reportage on indigenous developments in plant selection in 1974, when its international mission was taken over by the *Yichuanxue bao*, the main journal of the CAS Institute of Genetics.

20. *Yichuanxue bao*, first issue, June 1974.

21. Fang Zongxi, "Chairman Mao's 'On Contradiction' and Genetic Studies" [alternate English title], *Yichuanxue bao*, no. 1 (1974): 1-14. And also see *Yichuanxue bao*, no. 4 (1976): 255-76 for series of articles similarly addressing Lysenkoism and the "two schools" problem.

22. *Hereditas*, first issue, January 1979. In 1983, Tan Jiazhen succeeded Fang Zhongxi as rotating editor.

23. Proceedings of the April 19-23 conference were published as *Jinhualun xuanji*, ed. Chen Shixiang (Beijing: Science Press, 1983). Eighty-three papers were presented.

24. And on this issue they were of course in good company, this being the position taken, for example, by Stebbins and Ayala in "The Evolution of Darwinism," *Scientific American*, no. 7 (1985): 72-82. "Non-Darwinian" evolution, especially Kimura's "neutral theory" had begun to be examined in China in 1980, in Wu Wei, "A Challenge to Modern Darwinism" [alternate English title], ZBT, no. 2 (1980): 45-48. Kimura's 1974 *Evolu-*

*tionary Genetics and Mankind's Future* was translated as *Jinhua Yichuanxue ho renlei di molai* (Beijing: Science Press, 1985).

25. *Meng-de-er shishi yibai zhounian jinian wenji*, ed. Genetics Society of China (Beijing: Science Press, 1985). The superb companion volume edited by Li Ruqi, with the collaboration of Tan Jiazhen and other Fudan University geneticists is *Yichuanxue* (Beijing: Science Press, 1983). Forty-seven biologists contributed entries.

26. See William A. Joseph, *The Critique of Ultra-Leftism in China, 1958-1981* (Stanford: Stanford University Press, 1984).

27. Deng Xiaoping (1904-1997) joined the CCP in 1920, while in a work-study program in France. He was a veteran of the Long March and Yan'an, rising to important positions in the Central Committee during the 1950s and 1960s. He was victimized during the Cultural Revolution but returned to high office in 1973, under the sponsorship of Zhou Enlai. As a result of political conflict with Cultural Revolution proponents he was jailed from mid-1976 to mid-1977. After his return, he gradually established himself as China's "paramount leader" by the end of 1980 (BDCC, 2: 819-26). The link between science reform and Deng's political legitimacy was noted early on by Richard P. Suttmeier, *Science, Technology, and China's Drive for Modernization* (Stanford, Calif.: Stanford University Press, 1980).

28. From the translation in *Peking Review* (March 24, 1978): 9-10.

29. *Peking Review* translation.

30. *Peking Review* translation.

31. The other "comprehensive scientific and technical spheres, important new techniques, and pace-setting disciplines" were "agriculture, energy, materials, electronic computers, lasers, space, and high energy physics." Fang Yi, March 18, 1978, "Outline national plan for the development of science and technology, relevant policies and measures," excerpted in *Peking Review* (April 7, 1978): 7-10.

32. Fang Yi, 10.

33. See discussion in Meisner, *Deng Xiaoping Era*, 167-69, 178-80. Also see Tony Saich, *China's Science Policy in the 1980s* (Atlantic Highlands, N.J.: Humanities Press, 1989): Saich (155) says that "specialized personnel and skilled workers of various trades and professions" constituted 8.3 percent of those recruited to the Party in 1978, 23.6 percent in 1982, and 27 percent in 1983. In the all-important Beijing municipality, this group comprised 20 percent of party recruits between 1979-1982. This was a trend that continued in other big cities up to 1985.

For further discussion of the Party's recruitment efforts, see Lynn T. Whyte III, "Thought Workers in Deng's Time," in *China's Intellectuals and the State*, ed. Merle Goldman (Cambridge: Harvard University Press, 1987), 253-74. Wang Yeufarn's study suggests that at this time of heavy recruitment there emerged "science interest groups" and "lobbies" (some with overseas Chinese scientists) supporting specific aspects of science reform or research fields. See *China's Science and Technology Policy, 1949-1989* (Brookfield, Mass.: Avebury, 1993), 87-88. Also see sensitive quantitative analysis of CCP recruitment of CAS members in Cong Cao, "Red or Expert: Membership in the Chinese Academy of Sciences," *Problems of Post-Communism*, no. 4 (1999): 42-45.

34. Meisner, *Deng Xiaoping Era*, 90, 129. The context of Meisner's observation also seems to suggest that the attractiveness of Deng's program served to co-opt intellectuals who might otherwise have participated in and expanded political activities, such as the short-lived 1978-1981 Democracy movement.

35. For example, see articles by Gong Yuzhi, *Wenhui bao*, no. 29 (1986); Lu Dingyi, GMRB, May 7, 1986; Yu Guangyuan, RMRB, May 16, 1986.

36. See Li Peishan and Huang Shun'e, RMRB, May 2, 1986. In the *New York Times*, June 16, 1986, John Burns reviews an English language version of this article which appeared in *Beijing Review* on May 26, 1986.

37. See RMRB, June 16, 1986, for a description of the event and some of the speeches.

38. This August 1986 event took place in Beijing and was sponsored by the Genetics Society of China and the Dialectics of Nature Society (publishers of the ZBT). Some of the speeches presented there are reproduced in *Yichuanxue yu baijia zhengming: 1956 Qingdao zuotanhui yanjiu* (Genetics and the Hundred Schools policy: A study of the consequences of the 1956 Qingdao Genetics Symposium) (Beijing: Beijing University Press, 1996), 183-218.

39. Li Peishan et al., eds., *Baijia zhengming—fazhan kexue di biyou zhi lu: 1956 Qingdao Yichuanxue zuotanhui jishi* (A hundred schools contending—develop the road that must be followed in science) (Beijing: Commercial Press, 1985).

40. *Dialectics of Nature Journal* (Zeran bianzhengfa tongxun [ZBT]) began publication in 1979. Its political patron was Yu Guangyuan, who had played a central role in the science division of the CCP Central Committee's Propaganda subcommittee, especially in the Qingdao Symposium and in the committee's various long-term science plans. The journal's editor and informing intelligence was Fan Dainian, a trained physicist and editor of *Kexue tongbao* before he was purged in the 1957 anti-rightist campaign. Post-Mao, Fan was politically "rehabilitated" and became a member of the Chinese Academy of Social Sciences.

41. For example see Shi Xiyuan, "Lysenko qi ren" (This man Lysenko), ZBT, no. 1 (1979): 65-81; Yu Youlin, "Bao Wenkui" (Biography of Bao), ZBT, no. 3 (1979): 85-93; Li Peishan, "Qie qie lao-ji lishi shang di jiangxian jiaoxun" (Never forget the lessons of historical experiences), ZBT, no. 1 (1981): 13-16; Li Peishan et al., "Qingdao . . . di lishi beijing ho jiben jingxian" (Background of the Qingdao symposium), ZBT, no. 4 (1985): 41-49.

42. See *Yichuanxue yu baijia zhengming*. The study was not published until 1996.

43. See discussion in Richard P. Suttmeier, "Chinese Science and Technology Reforms: Toward a Post-Socialist Knowledge System?" *China Exchange News*, no. 4 (1988): 9. And also see his "Science and Technology under Reform" in *China's Economy Looks toward the Year 2000*, ed. U.S. Congress, Joint Economic Committee (Washington, D.C.: Government Printing Office, 1986).

44. Zhao Dongwan (administrative director of the powerful Science and Technology Leading Group), "Jiefang sixiang yangyu shijian kaichuang keji tizhi gaige xin jumian" (With liberated thought and bold practice create new conditions for the S&T system), May 16, 1984, in *Keji tizhi gaige di tansuo yu shijian* (Theory and practice in science and technology reform) (Changsha: Hunan Science and Technology Press, January 1985), 3-15.

Zhao announced that already, in nineteen provinces and municipalities reporting, 87 research institutions had switched to funding research entirely on paid contracts "with the likelihood of gradually becoming economically independent." Also see general discussion of the initiation of the contract system in White, "Thought Workers," 259-62.

45. See the State Science and Technology Commission, "Outline Report on Policy Concerning Development of National Science and Technology," *Issues and Studies*, 18.5: 93. Also see Wang Yeufarn, *China's Science and Technology Policy*, 104-5.

46. Lu Jiaxi, "Quicken the Pace of Reform to Make Even Greater Contributions to Building the Four Modernizations," *Keyan guanli*, no. 3 (1985): 1-6 (JPRS/CST/85-042,

[December 24, 1985]: 5-14). Lu (b. 1895) received his 1939 Ph.D. in chemistry from University of London. He was president of Fuzhou University and director of the CAS Fujian branch before taking over the CAS presidency from Fang Yi in 1981.

47. For example, see summary of statements by Hu Yaobang in Saich, *China's Science Policy*, 24. And also see Wang Minxi (CAS Science Management Group), "Shilun Zhongguo kexueyuan di gaige qianjing" (Prospects for reforms in the CAS), *Keyan guanli* (Science research management), no. 2 (1985): 1-4, 16. And Zhao Ziyang, "Revamping China's Research System" (excerpts), *Beijing Review* (April 8, 1985): 15-20.

48. Hu Xian (CAS Fund Committee Office), "Trial Implementation of the Fund System," *Tianjin kexuexue yu kexue jishu quanli* (Science of Science and Management of science and technology Systems), no. 5 (1985): 43-45. (JPRS/CST/85-033, September 30, 1985: 17-23). Hu Xian notes that 46 percent of the funded project directors were located in institutions directly under the Ministry of Education; 42 percent were variously under ministries of industry, defense, agriculture, medicine, and under local jurisdictions; 10 percent were in CAS institutes; and 2 percent were large national cooperative projects. A 1985 *Renmin ribao* survey of the fund's first three years of work has some different overall numbers: 4,424 funded projects for a total of 172 million yuan. See RMRB (December 18, 1985): 1. (JPRS/CST/86-007, February 20, 1986: 2).

49. See GMRB (August 21, 1985): 2 (JPRS/CST/86-001, January 17, 1986: 48). For the current funding cycle, 60 projects were funded for 2.5 million yuan. The Shanghai CAS independently set up a special research fund for postgraduate scientists and graduate students. See RMRB-Overseas (August 14, 1985): 4 (JPRS/CST/86-001, January 17, 1986: 27).

50. Hu Xian, JPRS: 21-22.

51. Guo Shuyuan (member of the State Science and Technology Commission), on the new NSF, *Kexuexue yu kexue jishu quanli*, no. 5 (1986): 2-5 (JPRS/CST/86-01, December 10, 1986). Also see Wang Huijiong (Director, State Council Research Center, Beijing), "The Organization of Science and Technology in China," in *Technology Transfer in China*, ed. Lisbeth A. Levey (Selected Papers from AAAS Annual Meeting Symposium, n.p., May 1986), 20.

52. Dean H. Hamer and Shain-dow Kung, *Biotechnology in China* (Washington, D.C.: National Academy Press, 1989), 9-11.

53. See party summary of the National Science Conference, "Decision of the Central Committee of the CCP on Reform of the Science and Technology Management System," *Beijing Review* (April 8, 1985): 19-21. And also see "Overseas Study Gains Popularity," *Beijing Review* (August 26, 1985): 31; and Leo Orleans, *Chinese Students in America: Policies, Issues and Numbers* (Washington, D.C.: National Academy Press, 1988).

54. See Saich, *China's Science Policy*, 126-38 for discussion of job choice for returnees.

55. On the issue of free market versus state assignment of jobs, see discussion in White, "Thought Workers," 262-63. Also see Saich, *China's Science Policy*, 126-38, especially for "Talents Exchange and Consultation Centers" which try to put talent in the right place and to negotiate with danwei.

56. Yang Zhun, "Lizhi gaige, kaituo qianjin" (Be resolved to initiate reform), May 22, 1984, in *Keji tizhi gaige*, 24-36.

57. For an assessment of changes in the industrial labor work-unit system during this period, see Barry Naughton, "Danwei: Economic Foundations of a Unique Institution," in *Danwei: The Changing Chinese Workplace in Historical and Comparative Perspective*, ed. Xiaobo Lu and Elizabeth J. Perry (Armonk, N.Y.: M. E. Sharpe, 1997). This article

shows dramatically how entrenched and powerful the system became, especially in the large, state-owned enterprises. It also quantifies mobility changes due to limited contract employment, moves between state-owned enterprises, and moves out of them into the public sector.

58. "Zhongguo kexue jishu zhengce zhinan," no. 1 (Beijing, August 1986). (JPRS/CST/87-013, April 2, 1987, 366 pages)

59. White paper, 218.

60. Wang Huijiong, "Organization of Science and Technology in China," 22.

# 9

# Biotechnology Becomes a
# Developmental Priority, 1978-2002

How did Four Modernizations' policy and reform materially affect the quantitative and qualitative development of genetics and biotechnology? This question is addressed here variously from the perspective of the individual scientist, the research institute, the university, and hybrid institutions created specifically to meet the demands of policy and reform. Illustrations of these perspectives are largely framed by the initiation and conclusion of the "863" High Technology Program whose first fifteen-year run began in 1986. The massive funding this program put into biotechnology made it possible for the field to build up its infrastructure and personnel at a rapid pace and to be prepared, in the mid-1990s, to take advantage of breakthroughs in the genetic modification of plants and the complete sequencing of genomes.

At the beginning and end of 863's first phase, biotechnology in China was surveyed by teams of Chinese and foreign scientists. Our analysis begins with a review of their findings and then takes up a series of illustrations that are exemplary in a number of ways. Each reveals the strategies that disparate institutions use to meet the demands of policy and reform; and the efforts of conventional institutions (the CAS, universities) are juxtaposed to those newly created to fit the specifications of science policy.

Moreover, the final illustrations to be analyzed are more about individual scientists and their personal visions. These are young people who are notable not only for being accomplished scientists, but also for thinking about the fundamental issues of development outside of the constraints of conventional institutions. Some will bring to biotechnology a model of development that emulates that of Silicon Valley information technology—that is, one that begins with venture capital, proceeds to patents, and ends with a product that has commanding market share. Its proponents in China argue that you cannot get rich or powerful doing catch-up science or technology. In the 1990s, the argument continues, it was already too late for China to become a world leader in information technology, but it was just the beginning of the biotech revolution. Here is where China should get out in front and stay.

We shall see young scientists take on the unlikely role of national celebrities as a result of their scientific accomplishments combined with a competitive, entrepreneurial spirit and aggressive patriotism. They will take China to the leading edge of powerful new biotechnologies, and at the same time to international controversies that will determine how well those technologies can become the engine to a development scheme of enormous potential.

## The BIC Survey

Between 1986 and 1989 invaluable information was gathered about China's biotechnology development through the efforts of a program sponsored by the CAS and the United States National Academy of Sciences. This program began with a set of biotechnology advanced laboratory and lecture minicourses, taught in China by distinguished American scientists. Their respective topics were Nematode Molecular Genetics, Gene Expression and Amplification in Yeast, Immunotoxins and Tumor Markers, and Plant Molecular Genetics. (The original program also called for a major symposium on gene expression and regulation, but the June 1989 Tiananmen massacre required its cancellation.) Additionally, a detailed assessment of biotechnology in China was conducted by some of those same visiting scientists, based on a wide survey of published literature and extensive site visits conducted in 1988. That assessment is published as *Biotechnology in China* (BIC), an insightful state-of-the-field analysis.[1]

BIC notes that between 1984 and 1989, biotechnology experienced a twenty-five-fold increase in funding—from five million to more than one hundred million yuan annually. This put China's investment in biotechnology at a percentage of GNP comparable to many Western countries.[2] Over this same period, the CAS was virtually eliminated as a recipient of direct biotechnology funding, receiving only 1.7 percent of it by 1989. In effect, there was a "decentralization" of the process through which biotechnology research funds were granted. That is, applications for funding increasingly came directly from the scientists proposing to conduct the research, and funds were disbursed directly to them—not to their work-units. And biotechnology research strongly came to emphasize the applied over the basic.[3] In all of these trends, biotechnology was responding to science policies developed over the decade.

And what of the quality of the biotechnology research being conducted in the eighties? BIC's survey of over a dozen journals assessed the originality and accuracy of basic and applied research reports. Its evaluation concluded that work done in basic research was fairly original, sometimes very much so; but work done in applied areas was generally quite derivative and repetitious. On the other hand, both basic and applied research were generally judged to be quite accurate and successful in "proving the point."[4] As a body of scientific work, however, it was judged that "China's biotechnology research and development remains far below that in developed countries." Basic research was considered to be at the international level in only a "few scattered areas," but more importantly, the review repeatedly criticized the weakening position of basic genetics research as government policy pushed in the direction of the applied. Basic genetics, the report lamented, "is now too often bypassed in favor of applied genetics, like gene cloning." This trend was particularly evident in the "paucity of developmental biology [research], which indicates a bias against using biotechnology to advance understanding of the basic knowledge of genetic expression and regulation that governs organisms' development."[5]

Notwithstanding these serious concerns, the BIC evaluation of biotechnology does indicate a significant qualitative and quantitative improvement in all genetics-related science over the situations described a decade earlier in the trip reports of American scientists. Furthermore, BIC suggests a considerably more promising situation for biotechnology in China than the "white paper" published just two years earlier.

## The "863" High Technology Program

Among the most important reasons for BIC's positive outlook was the creation and initiation of a well-endowed "High Technology Program" in 1986-1987. Known as the "863" program (for its March 1986 approval), it was the result of intense lobbying by scientists who argued that, where important new areas of science were desired, state research funds had to be much more carefully targeted, and they had to be greatly increased.[6]

The 863 program, according to one analyst, "allows China to anticipate short-term technological trajectories and to participate in them at an early phase."[7] Instead of the all-encompassing category of biotechnology utilized for national NSF grants, three foci were designated by the 863 program: Genetic engineering for high-yield, high-quality plants and animals received about one-third of the biotechnology funding; research and development of new medicines, vaccines, and gene therapies received another third; and protein engineering for food industries and agriculture received about 8 percent. Unlike the NSF, the

863 program set aside 20 percent of its funds specifically to aid the commercialization of research outcomes.

The complex bureaucracy administering the 863 funds encouraged projects to be proposed by prospective investigators to appropriate committees of peer specialists, who were in turn supervised by a senior expert advisory committee. In spite of the mandated peer review process, grants have been "more like contracts than Western-style investigator-initiated grants."[8] Projects have been selected through a complex negotiation process involving not only peer reviewers, but no less importantly, representatives of the policymaking central State Science and Technology Commission, and the central administrators of 863 biotechnology funds (the China National Expert Committee for Biotechnology Development).

During the first phase of the 863 program, from 1987 through 2001, according to the government, it disbursed 5.7 billion yuan for all high-tech areas.[9] (By comparison, in the early 1990s the National Natural Science Foundation was operating with a budget on the order of one-half that of the 863 program.)[10] Biotechnology's share of the 863 budget was approximately 16 percent.[11] Principal investigators have been about evenly divided between those associated with higher educational institutions and those associated with the CAS.

## The "Torch" Program

While a significant portion of the 863 budget itself was devoted to commercialization, another state program was specifically mandated to facilitate that goal. From 1988, the Torch program acted as gadfly and broker for commercializing, industrializing, and internationalizing new and high technology. It did not allocate funds, but rather it provided policy guidance, incubators, and training;[12] and it "recommended projects to the banks designated to loan funds for Torch activities."[13] Further, it provided contacts for foreign companies interested in joint ventures, and "encouraged domestic and foreign research and commercial entities to take advantage of preferential policies offered to investors in the . . . [dozens] of high tech zones [and industrial parks] it established across the country."[14] While the Torch program gave relatively little help to the biotechnology sector during the early 1990s, by the end of the decade it was playing an important role in the development of the biopharmaceutical industry.

During the 1990s, as the importance of biotechnology became more apparent, and as Chinese scientists became more adept lobbyists, the state supplemented 863 support of biotechnology with other smaller, more narrowly focused programs. For example, beginning in 1991, the National Key Laboratories program created dozens of state-of-the-art facilities for higher educational institutions and research institutes all over the country. In 1997, the "973" program was initiated to support "big science" projects, giving priority to biotechnology ventures like China's participation in the International Human Genome Project. A special fund was set up for research on transgenic plants in 1999; and the following year, under the Key Science Engineering Program, a major agricultural biotechnology project was funded for work on crop germplasm and quality improvement.[13]

## The Development of Plant Biotechnology as an Indicator

No area of biotechnology has developed as dramatically as plant biotechnology, both in the size of the enterprise and the quality of its work. Thanks to a comprehensive survey of China's plant biotechnology in the year 2000 (published in the AAAS's *Science* magazine), one can see the general trends of its development since the initiation of the 863 program.[16] The resources given to plant biotechnology research, from 863 and elsewhere, have been far greater than those given to all other areas of biotechnology combined; therefore, it can act as a useful indicator of more general growth and development in the field.

During the 1986-1999 period covered by the survey, plant biotechnology's overall budget increased sixfold. Its total full-time professional research, management, and support staff had increased by 46 percent to 1,657 (at the 29 institutions addressed in the survey). For all of China, it is estimated that full-time plant biotech workers numbered almost 2,000. And the survey shows a continuous improvement in the training of these workers, resulting in 1999 with 20 percent of them holding Ph.D.'s, 28 percent with an M.S., and 35 percent with a B.S. These are educational levels far beyond those of the agricultural science community outside of plant biotechnology.

The survey is quite sanguine about what China's enormous investments in plant biotechnology have yielded so far. For example, "over 50 plant species and more than 120 functional genes" are being used in plant genetic engineering, "making China a global leader in the field."[17] Furthermore, as a result of China's aggressive program to improve food and industrial plants through genetic modification (GMO), dozens of plants have been tested and released for production; but unlike GMO trials elsewhere in the world, China's scientists have concentrated on food crops, like rice, that are central to the country's food security. And while industrialized countries invest heavily in producing GMO herbicide-resistant plants, China's plant technology field trials are virtually all devoted to insect and disease resistance. Its outstanding success with insect resistant Bt cotton is a major case in point, illustrating how savings from one plant (from reduced pesticide use) and revenue (from increased production) have been more than sufficient to pay for the cost of research.

In spite of its generally positive appraisal, the plant biotechnology survey suggests some areas that future development should address. Most general is the concern about the role of the public sector, which has been virtually the sole source of support for the field's development. The survey marvels at the public sector's commitment here, making China by far the industrializing world's leader in the field. (India's government is the next largest contributor to the field, but it does so at only 20 percent of China's budget.) In the near future, however, the survey suggests that China's public sector may have difficulties competing for the best personnel who are likely to be attracted by private-sector salaries and incentives.[18]

A second area of concern, reminiscent of the 1988 BIC survey, is the "fragmentation of research projects over many research institutions," and "unnecessary and inefficient duplication of efforts, particularly at the local level."[19] The survey suggests that more sophisticated and effective work will require coordination and consolidation, especially at the local and provincial levels.[20] What the survey does not consider is a potentially positive side to the proliferation and wide disbursement of research projects: These may well have contributed to the diffusion of the latest in biotechnology, which in turn may prove to be very useful as efforts are made to spread the application of GMO technology throughout agriculture in China.

Finally, the survey emphasizes the need to raise still higher the educational level of the plant biotechnology workforce. It acknowledges that it is considerably easier to purchase and set up the research hardware than to produce and attract Ph.D.'s. It suggests that a proper job was done in the 1990s to acquire the appropriate infrastructure, a reality which ought to make more of current funding available to do what is necessary to attract the best educated. Nevertheless, between the dual temptations of foreign employment and jobs in China's private sector, developing the "human capacity" of public plant biotechnology programs promises to be its most difficult challenge.

## The Development of Biotechnology: Variations on Institutional and Personal Themes

Against these broad overviews of biotechnology, a number of exemplary institutions and careers are going to be examined with historical specificity. This should provide a better grasp of the significance of survey generalizations and judgments.

### The CAS Institute of Developmental Biology

In 1980, well before the enactment of new science development policies and programs, the drive, imagination, and connections of two veteran embryologists—Tong Dizhou and Niu Mann-chang (M. C. Niu)—were responsible for the initiation of the CAS Institute of Developmental Biology.[21] This was among the earliest post-Cultural Revolution efforts to revive science research and to rejoin the international science community. Strictly speaking, at this time biotechnology and the use of genetic engineering techniques were only just being introduced to China; however, much of this institute's initial research was devoted to the goal of genetic modification, though it sometimes used techniques that were controversial and soon superceded. Nevertheless, because of the nature of its development and the general goals of its research, this institute anticipated many of the issues associated with biotechnology over the next two decades.

## The Institute's Founders

At a time when China was promoting the international relations of science, the institute's conception was exemplary: Niu, a Chinese American scientist, mediated its outside relations, and Tong, a well-connected member of the CAS, handled domestic ones. Its considerable seed money, which came primarily from the United Nations and the Rockefeller Foundation, gave it international visibility and domestic legitimacy. Its stated mission was to engage in both basic and applied research in the biotechnology area. With revenue from the latter it expected to become self-supporting (in the manner later promoted by the "local responsibility" policy).

Tong (1902-1979), who unfortunately did not live to see the institute open, had a long and successful career both as an experimental embryologist and academic administrator. From 1950, he was affiliated with the CAS Institute of Experimental Biology and Shandong University. As an influential administrator at the university, he was responsible for hiring Fang Zhongxi onto the faculty at a time when Fang was one of China's most prominent explicators of Michurin biology. Until the Qingdao Genetics Symposium, of which Tong was appointed chair by the CAS, he supported Michurin biology in China. He made no secret of his lack of familiarity with the latest developments in mainstream genetics and his general distrust of "Mendel-Morganism."[22] During the Cultural Revolution he was dramatically victimized, a fact that made him a widely sympathetic figure, notwithstanding his earlier relationship to Lysenkoism. Along with other prominent scientists, he joined the CCP in 1978, a year before his death.

Niu (b. 1914), a native of China, resided in the United States and was on the biology faculty of Temple University during the 1970s when he renewed his contact with China. Starting in 1972, he was among the earliest American scientists to do cooperative research with his counterparts in China, eventually spending half of his time in China, collaborating with Tong in research and in pursuit of a new CAS institute. The latter project was doubtlessly furthered when Niu arranged to have Temple University give an honorary degree to Deng Xiaoping during his grand tour of the United States in 1979.[23]

## Establishing the New Institute

The institute was formally established in 1980, initially funded by an extraordinary combination of foreign and domestic financial support.[24] The United Nations Fund for Population Activities provided a U.S.$550,000 grant to the project; the Rockefeller Foundation Population Council granted U.S.$300,000 up front and, upon renewing the grant, U.S.$100,000 a year for three subsequent years. S. J. Segal, who had collaborated on a book with Niu, was head of the Population Council when it approved of its grant for the institute. And it should be noted that, beginning with this grant to the institute, the Rockefeller Foundation played a continuous and growing role in the development of genetics and biotechnology research and education in China. While its policy and funding roles were minor compared to those it performed before 1950, BIC judges its U.S. currency grants to be well targeted and much welcomed.

All of these funds made it possible to build a first-rate, well-equipped research facility that was fully up and running in 1983. Initially, the institute had a staff of four senior researchers (including Niu) and two associates; by 1986, those numbers had increased to eight and seven, respectively. When the BIC evaluation was made in 1988, the staff had nearly doubled again to thirteen senior and fifteen associate researchers, and the evaluation found that "the institute's laboratories are the most modern, well-equipped, and best maintained" of those major facilities visited by the American scientists who contributed to the BIC evaluation.[25]

*Institute Research and Controversy*

In spite of this glowing picture of Four Modernizations science policy in action, the new institute had a serious problem. During the 1980s, strong criticism was made about the research being done by Niu and some of his senior associates, who came to constitute a discrete division of the institute, and "whose seniority gave it control over the institute's resources, positions, and power."[26]

The work of the Niu group at the institute was mainly devoted to transforming specific traits of organisms, particularly fish and sea urchins, through nuclear-cytoplasmic interactions during early development. Some of the group's experiments were meant to demonstrate that the eggs of such organisms could be transformed heritably with injected egg poly(A)RNA; that RNA injected into an egg could be efficiently reverse transcribed and incorporated into the genome. The group's various experiments included injecting carp mRNA (messenger RNA) into goldfish; and the induction of soybean protein in rice by injecting soy seedling mRNA into rice ovaries. Goldfish developing from eggs microinjected with rabbit globin mRNA were purported to express rabbit globin. Fish injected with plant chloroplast RNA were purported to synthesize chlorophyll, which was indicated when the fish turned a green color. (Chloroplasts are small bodies containing chlorophyll in the cell protoplasm, outside the nucleus.) And of potentially great commercial value, there was their cross between nuclei of the common carp and enucleated eggs of the crucian carp. This was claimed to have produced a "hybrid that has a higher growth rate, higher protein content, and lower fat content than the common carp and that it can be bred stably."[27] In the mid-1980s, the institute's deputy director and public relations spokesperson commonly sighted the work with carp as an illustration of the institute's best and most lucrative science.[28]

In its evaluation of the institute's research, BIC cited a devastating review by a prominent American scientist of the Niu group's work. It concluded "there is no experimental support for the contention that maternal poly(A)RNA can be utilized for transformation of eggs, nor, on theoretical grounds, should it be possible."[29] Generally, BIC alleged that "every aspect of the research conducted by this group is open to question. Moreover, in some cases, the experiments are so poorly controlled, and the results so unbelievable, that the question of deliberate deceit and fraud must be raised."[30]

BIC chastised the Chinese government for continuing to fund Niu even after criticism was leveled at home and abroad about his group's work; but it did not raise the obvious question about why no problems were detected with the work of Tong and Niu when they first applied for funds from the UN and the Rockefeller Foundation. After all, for years before the institute was funded, Tong and Niu—independently and then in collaboration—had been conducting research along the lines criticized so thoroughly by BIC.[31] On the other hand, BIC found the institute's less senior research group to be performing "creditable research on worthwhile topics," independent of Niu's approach.[32] Among other projects, they were successfully working on the structure and regulated expression of various genes in rice chloroplasts.

Niu did respond to some of the criticism, but never acknowledged or attempted to explain the anomalous theoretical implications of some of his conclusions.[33] Nevertheless, the institute continued to thrive, employing strategies encouraged by the new independent contract and individual responsibility policies, among others. The institute increased its revenues by contracting out specific services as well as encouraging its scientists to do outside consulting. As an example of the former, fisheries from all over the country sent technicians to the institute to learn nuclear transplant techniques for the purpose of breeding bigger, more nutritious fish.[34]

Additionally, the institute—like many in the CAS—initiated a master of science program in 1984. Up to this time, the CAS institutes had left science education and degree programs exclusively to the universities; however, they had provided valuable advanced training for younger associates who had only received an undergraduate education. Now the CAS began to encourage institutes to develop M.S. and Ph.D. programs. Students for such graduate degree programs were selected by competitive examinations administered by the CAS graduate school. By 1986, the Institute of Developmental Biology M.S. program had already graduated six students, and there were twenty students currently in the program. These students had received their undergraduate training primarily from Beijing, Fudan, or Nankai Universities. M.S. program tuition was yet another source of revenue for the institute.[35]

And finally, it should be emphasized, that while the salaries of the institute scientists and technicians were covered by the CAS, all of the institute's research was supposed to be supported eventually by competitive grants and contracts. Institute scientists were successful at supporting their projects with CAS National Science Fund grants, and later, with National Natural Science Foundation and 863 program grants.[36]

By the mid-1990s Niu had retired, and in the spring of 2001 the CAS began implementing a plan, apparently long on the drawing board, to consolidate his institute and the CAS Institute of Genetics (both in Beijing) into a new entity—the CAS Institute of Genetics and Developmental Biology. This was part of a nationwide consolidation and downsizing drive (the Knowledge Innovation Initiative), which was begun in 1997 for all CAS institutes. By the turn of the century, it had begun to affect all educational and research institutions.[37] In the case of these two institutes, this would seem to be a particular response to two expensive and sometimes controversial organizations that have been slow to answer the call to devote themselves fully to mainstream biotechnology.

For most of its existence, since its creation in 1959, the CAS Institute of Genetics seldom operated near the mainstream of genetics or biotechnology research. In the 1990s, it was drawn to the periphery mainly by becoming a major field test site for genetically modified plants. Li Jiayang, the forty-four-year-old incumbent director of the Institute of Genetics will head the new institute—which is described as a national base for basic and applied research in "high-tech farming."[38] One of China's leaders in plant genomics, Li is expected to fold its aggregated resources into the mainstream of plant biotechnology.

## Fudan University Institute of Genetics

Among the most immediately effective science policy reforms of the 1980s was the revival of research in China's major universities. The predominant flow of National NSF research funding to universities (as opposed to CAS and production ministry institutes) was initially the main engine of this transformation. Research organizations attached to universities or enterprises (known as non-independents) proliferated during the 1980s and represented more than half of all research institutes by the end of the decade.[39]

In the development of university-based biotechnology research, Fudan University was peerless.[40] The 1988 BIC survey calculates that, on a per capita basis, Fudan's Institute of Genetics was the best funded of all biotechnology research centers of its kind, whether based in a university or elsewhere. In 1988, total research support was 8 million yuan. The largest grant, 6.2 million yuan, came from the 863 program; 960,000 yuan came from special biotechnology funds built into the seventh five-year plan (1986-1990), and another 150,000 yuan came from the National NSF. From abroad, there was U.S.$100,000 from the Rockefeller Foundation and U.S.$70,000 from Interferon Sciences. Added to this was a five-million-yuan grant from the State S&T Commission for making Fudan the location of one of eleven "Key Biotechnology Laboratories" nationwide.[41]

The institute's venerable director, Tan Jiazhen, tirelessly encouraged his relatively small, highly trained, and experienced team to take advantage of every aspect of new science policy and every source of funding, domestic and foreign. The institute, it will be remembered, was established in 1961, in the wake of the Qingdao Symposium reforms. It was meant to be a device for acquiring funding from outside the university for graduate training and for faculty research. (By design, research was not encouraged in universities.)

The Cultural Revolution closed the institute down just as it began to achieve some success. After the Cultural Revolution, Tan revived the institute, relying primarily on the small coterie of talented geneticists, some of whom he had worked with since the 1940s, when they were his students.

By the late 1980s, the institute numbered about fifteen senior researchers, all of them having American doctorates and extensive research experience abroad. They worked with a support staff of over one hundred and with approximately ninety graduate students (at the M.S. and Ph.D. levels). At this time, the institute reorganized its research into five main sections, consonant with the 863 program's biotechnology main funding categories: human and medical genetics; microbial; plant cellular and molecular; insect evolutionary and behavioral; and genetic engineering. This ambitious coverage became more feasible when five senior researchers returned to the institute after extended stays (up to three years) in the United States for advanced training in their special areas.[42]

Unlike the structure of the Developmental Biology Institute, at Fudan there was no discrete senior group or laboratory that dominated funds and agenda. Indeed, Tan Jiazhen, the institute's senior figure, worked very closely with his admiring colleagues and placed his "insect evolution and behavior section" in a relatively modest position within the institute.

## Research at Fudan

A sampling of the institute's prolific research, generally praised by BIC, reveals that by the mid-1980s much of it had purposely been shifted in an applied direction—the institute claiming nonetheless to be continuing its devotion to basic research as well. Its 1984-1985 annual report pointedly noted that the institute strengthened the integration of medical genetics with environmental detection of genetic disorders and carcinogens; microbial genetics with industrial fermentation; and plant genetics with agricultural and horticultural practice.[43]

A closer look at their research shows, for example, that the section on microbial molecular genetics did manage to perform both basic and applied research. The former focused on DNA replication in *E. coli*, and the biology of thermophilic bacteria, especially *Bacillus stearothermophilus*. On the other hand, the section on human and medical genetics studied human gene therapy with the ultimate aim of curing inherited diseases by inserting a normal gene into the somatic cells of an affected individual. A treatment for hemophilia B became the model system for this technique, for which the institute pursued commercial application in the early 1990s.[44] Other groups in this section studied various aspects of human cancer. In the molecular biology section, among the main projects were the production of useful proteins by recombinant DNA techniques, and developing methods for the synthesis and overexpression of genes encoding peptides. And in the genetic engineering group, interest was developing in bacterial organisms that are used in amino acid production and have advantages over *E. coli* for fermentation purposes.[45]

The Fudan Institute of Genetics lost no time finding commercial applications for this research. Indeed, it is fair to say that, in a number of instances, a commercial contract was the inducement for the research in the first place. For example, by the mid-1980s, the institute had contracted for joint research in the fields of antibiotic and enzyme products with the Hangzhou Second Pharmaceutical Factory and the Jiangxi Agricultural Drug Factory. It also had accepted an 800,000 yuan investment from the Shanghai Tienzu Vizhen Factory for a series of amino acid products.[46]

While this pattern of applied research typified much of the institute's work, it should be noted that it also conducted some that was rather remote from such applied projects, and rather closer to the genetics earlier practiced by Tan Jiazhen. Tan personally revived Drosophila research and interest in evolutionary genetics. He established a prominent guest lecture series and a visiting scholar program for these subjects, always extending invitations to the greater biology community throughout China. In academic year 1982-1983, for example, a number of Japanese biologists accepted invitations to give lectures on insect evolution and behavior.[47] The following year, the prominent evolutionary theorist J. Ayala came from the University of California, Davis, to offer a course on evolutionary genetics.[48]

### Fudan Receives a Key Laboratory

In 1984, in the midst of all this activity, the institute successfully negotiated with the State S&T Commission to become one of the eleven national sites for a "key genetics engineering laboratory." The object of this program was to create first-class research facilities open to all qualified scientists who wished to upgrade their knowledge and skills by working with the best equipment and the best research scientists. Rather than use the five-million-yuan grant to build a new and discrete facility, Fudan opted to upgrade the existing plant and spend most of the funds for the very best equipment. The resulting refurbished laboratories became an important site for training Ph.D.'s and postdoctoral researchers, as well as an important resource for attracting visiting Chinese and foreign scientists.[49]

Establishing the key laboratory was considered by Fudan to be one of a number of major steps it had taken in its commitment to genetics and biotechnology education. In 1978, the Fudan Genetics Institute had sponsored what Tan Jiazhen considered to be China's starting point for large-scale training of qualified genetics engineering scientists and technicians. At that early date, it had sponsored a summer lecture series on the basic principles of molecular genetics and the basic experimental techniques of genetic engineering. The series was attended by two hundred veteran scientists and conducted by an American lecture team from Caltech, Tan's alma mater.[50]

Subsequently, during the 1980s, Fudan established the first and, for a few years, only undergraduate department and major in genetics and genetic engineering. Add to this its large graduate program in these same areas and the result was a very demanding teaching obligation, to which the institute's administration was dedicated; but it was also constantly concerned about holding onto qualified teaching faculty, who were regularly hired away by other Chinese universities or by American research institutions.[51] Shortage of genetics, biotechnology, and genetic engineering teaching faculty was a chronic problem, matched if not exceeded by the shortage of researchers in those fields. In 1988, Tan Jiazhen estimated that there might be one thousand qualified genetic engineering specialists in China. That same year, BIC roughly estimated that "about one to three thousand Chinese students and scientists are currently training in biotechnology and related fields in the United States." How many returned to work in China was another story.

## The Shanghai Biotechnology Center

In genetics-related science, among other fields, a challenging dilemma arose from the fact that funding and institution building might outpace the training of necessary researchers and teachers. Fudan University was perhaps among the earliest to address this serious problem. There was, however, another set of dilemmas created by the biotechnology push that took longer to recognize and required expensive strategies that appeared to be well beyond the capabilities of a single university or CAS institute, or for that matter, beyond the reach of the CAS itself.[52]

*Dilemmas of Development*

The dilemmas in question are these. If China were to pursue recklessly the lure of biotechnology, there might well be a danger of becoming merely the end link in a global biotech "putting out" system—the equivalent of assembling TVs in the 1960s, or computer components in the 1980s and 1990s. Chinese biologists would in effect become cheap technical labor, quite remote both from the creative initial stages of basic science and the final, potentially lucrative stages of application. The second dilemma of the set could arise from Chinese biotechnologists doing creative research in their laboratories, but being dependent on foreign industrial and marketing firms in the stages of application-production and commercial distribution.

The staff of the Fudan Genetics Institute, and no doubt others as well, closely watched the state's efforts to avoid these dilemmas beginning in 1984, when the State Planning Commission allocated 54 million yuan (of a budgeted one billion) toward the establishment of a national biotechnology center in Shanghai. This center was to become a giant enterprise, encompassing all aspects of biotechnology and genetic engineering, from laboratory research to production and marketing. In the words of one of its foreign investment promotional statements, the center's purpose is "to concentrate on research and development in biotechnology, to coordinate research and production and to provide new products in biotechnology and new techniques for the expansion of production by steps of applied research, development, and pilot scale production."[53] In effect, this center was a model that was incorporated into the guidelines of the 863 program and Torch. When these programs were operational, a year or two after the center was approved, they in turn contributed significantly to its research budget and helped guide it through operational complexities.

### New Departures of the Shanghai Center

While nominally a subsidiary of CAS, in fact the center dealt directly with the State S&T Commission, and was the result of a group proposal made to the commission. Following the proposal's stipulations, the center was pointedly endowed with perquisites and power that gave it the ability to avoid the organizational, personnel, and economic problems that the 1986 "white paper," for example, had attributed to the field of biotechnology.

Where scientific personnel were concerned, the center was given carte blanche to enlist whomever they needed, from any work-unit, to fill out their projected staff of 350. Its location in Shanghai gave the center access to the greatest concentration of biological scientists in the country, including those at the CAS Institutes of Biochemistry, Cell Biology, and Plant Physiology, and at Fudan University. It was the center's option to hire people away part or full time from these institutes or from any other in the country, or to engage the institutes in joint projects.

The center was organized so that its own personnel, or outsiders, could initiate a project at any point along the line between research and marketing. For example, a biologist could bring the center a new technique for using a bacterial organism in the production of amino acids for fermentation. The center then might determine the market for the technique, and pilot test the use of the technique in production. Or the center might be brought an industrial problem which it then set about solving by finding the appropriate scientists to address it.

Even while awaiting the construction of its commodious permanent home, the center had already started some short-term research and development projects, based in its interim facilities or in cooperation with local CAS institutes. Among them were those devoted to growth hormones for animal husbandry, vaccines, immunoassay kits, polynucleotides and products of immobilized enzymes. In the late 1990s, when the center was fully operative in its planned state-of-the-art facility, its ambitious goals included research and development of new products and techniques in the fields of genetic engineering, cell fusion, enzyme technology, and fermentation. At this time, the center was a significant participant in the projects that came to dominate biotechnology in China—biopharmaceuticals, and human and plant genomics.

In addition to emphasizing its "openness" to all domestic research institutions, the center solicited foreign cooperation as well, in just about any form. Joint ventures, direct investment, joint projects, technology transfer, development and marketing were all welcomed. Within a decade of the center's establishment, other and larger biotech centers (or "parks")—in Wuhan, Hangzhou, and Shenzen—were being developed for the new millennium. At ten times the size of the Shanghai Center and requiring five times the initial state investment, the Hangzhou "biotech park," for example, is an expression of the belief that "biotechnology will become the dominant industry in the next century." All of these complex organizations are explicitly meant to stake out China's share of the "world market for bio-engineering products."[54] Thus, in China at the beginning of the 1990s, it appeared that the business of genetics was business.

## Chen Zhangliang and Beijing University

In the 1990s, no practicing biologist understood the business of genetics better or influenced it more than Beijing University's *wunderkind*, Chen Zhangliang (b.1961). His stunning rise to celebrity and position in the academic and scientific world, at home and abroad, is the stuff of legend; and indeed from the facts of his own life he has constructed a model persona that melds the scientist, entrepreneur, and patriot. What he has in fact accomplished is impressive by any measure; how he frames those accomplishments and how he frames the development of biotechnology in China are every bit as interesting and important.

Chen, from the southeastern coastal province of Fujian, received his undergraduate training in plant genetics and biotechnology beginning in 1978, on the island province of Hainan, at the South China University of Tropical Agriculture.[55] In 1986 he completed his doctorate in plant biology at Washington University, in St. Louis. His Ph.D. program director was Roger N. Beachy, who had just completed the pathbreaking research project that produced the world's first genetically altered food crop—a tomato that had been engineered to resist viral infection. Beachy's methodology was a prototype for future food crop modifications, including Chen's research project which was devoted to gene regulation in transgenic soybean plants.[56]

The BIC survey suggests that Chen could easily have found permanent employment in the United States, but chose to return to China and a position at the prestigious Beijing University. Among his inducements to take that job were an associate professorship, a large and up-to-date laboratory, and opportunities to go abroad annually. In 1988, the biology department (soon to be called "life sciences") was one of the largest in the country: The staff comprised 20 full professors, 40 associates, and over 100 lecturers and teaching assistants. There were 640 undergraduate majors, 150 M.S. and 50 Ph.D. students. Additionally, the department was training 90 premedical students for Beijing Union Medical College. In 1987, the department awarded 12 doctorates, of whom five stayed on as instructors, while the rest either went abroad or found faculty positions at other Chinese universities.[57]

Over the course of the 1980s, there was a surge of funding for the department, so that by the time Chen arrived, almost 90 percent of the senior faculty were supported by research grants. The State Education Commission had provided new laboratories and a piece of a World Bank loan that made it possible to equip them "lavishly." BIC concluded that while the department was a dynamic one, it (like many universities) had a conspicuously top-heavy, inbred faculty and would require much more hiring of new, young faculty like Chen, if the quality of its research was to continue to match its prestige.[58]

At the university, within three years of his arrival, Chen had established and was operating the National Laboratory of Protein Engineering and Plant Genetic Engineering, the first of its kind in China. It was funded by the 863 and Key Laboratories programs, as well as the Rockefeller Foundation.[59] During its first six years of research, among other projects, the laboratory conducted field tests on virus- and insect-resistant transgenic tobacco, tomato, and sweet pepper in a dozen locations throughout China. Its data showed that foreign genes in transgenic plants are "relatively stably inherited for six to seven generations." Additionally, the Rockefeller funding supported a rice antifungal gene project.[60] In 1991 Chen's work was recognized with a UNESCO Javed Husain Prize for young scientists.

*The Scientist as Entrepreneur*

In an extraordinary move, obviously calculated to break the biology department's gerontocracy and push it further in the direction of advanced biotechnology, in 1991 the university appointed Chen to be dean of the life sciences (the biology department's new designation). His first prominent act was no less extraordinary and earned him still greater celebrity: On behalf of the life sciences department, Chen and a few graduate students started up a corporation, called Beijing University Weiming (that is, "no-name") Biotechnology, for the purpose of biopharmaceutical research, manufacture, and marketing. The corporation was an instant success, and by using its initial profits as venture capital was soon able to form the joint venture Kexing Biopharmaceuticals with an American venture capital firm. The main product of these firms has been alpha-interferon, a DNA recombinant drug for hepatitis B and cancer. By the end of the 1990s, the company controlled 60 percent of the China interferon market—previously controlled by an American firm—established a subsidiary in the Shenzhen economic development zone, and went public via NASDAQ.[61]

The life sciences department, not Chen, received the profits from this venture. And in his position as dean, he is reported to have used some of the first-year and subsequent profits as bonuses for the faculty, who generally were paid very modest salaries. And while Weiming was not the first or most profitable of Beijing University-based business ventures (such as Founder, which made software for commercial printing), Chen's entrepreneurship apparently touched a sensitive nerve on a campus that was concerned about holding on to its best young faculty and conflicted about the appropriate relationship of the university to the marketplace. Within a few months of Weiming's initial success, the university literally tore down the wall separating it from the nearby hi-tech commercial district and began planning to build a high-tech commercial street. This was part of a wave of reforms, called the Weiming wave, that encouraged the university to play a greater and more systematic role in the science and technology marketplace. One journalist called the phenomenon, which was spreading to other universities, the "Chen Zhangliang effect."[62]

### The Entrepreneur as Patriot

Chen's status continued to rise at Beijing University, and so did his celebrity throughout China, thanks to his willingness to be interviewed by reporters and journalists, foreign and domestic. In 1994, *Time* magazine designated him one of the world's one hundred scientists with the most potential. The following year, he was appointed vice president of the university, which took obvious pride in reporting over and again that he was the youngest person ever to hold such a lofty position in academe.

Chen made sure that his career was understood as he would have it. In 1992, he played a prominent role in a book written by an admiring journalist who interviewed him about the disturbing effects that commercialism was having on the universities. Chen used the occasion to admonish students not to take his recent business success as an incentive to leave their studies or even compromise them with part-time work. "When money-making becomes an obsession," he said, "one should really begin to worry." Using his own life as an example, he said that things were quite hard when he studied for his doctorate in the United States, but he resisted the temptation of part-time jobs. "I achieved what I did because I didn't seek enjoyment or indulge in fun and pleasure. Money has never been the first objective in my life."[63]

In later interviews, he developed these themes by juxtaposing his humble origins, in "a poor fisherman's family," with the fact that he was "among those Chinese students who studied abroad and then returned to China." That is, he resisted the temptation to stay abroad where the financial rewards might be greater and instead (according to *People's Daily*) answered the "Chinese government's call."[64] He further makes it clear that his success as an entrepreneur was not motivated by personal gain. On the contrary, he said, "my bank deposits have not increased a bit since I returned to the motherland, and even worse, I have never been to the bank. But I don't regret this because I am living in a society where you feel changes all the time, and I can contribute to these changes. I . . . hope to use my wisdom to pay back the society."[65]

When Chen talks about the success of his biopharmaceutical enterprise, he adds an aggressive nationalistic dimension to the selfless, patriotic persona. "When I was a poor professor in 1992," he told a journalist, "I set up a plant called 'Kexing' with only 47 employees. This plant now controls 60 percent of the Chinese interferon market and [through very competitive pricing] forced the [predominant] American company out of the Chinese market. We went from zero to 60 percent of the market. This is important. If we don't do this, we'll be taking the foreigner's medicine."[66]

While Chen was concerned in the first instance to rescue pharmaceuticals from foreign control, he expressed equal concern to increase domestic production in order to lower costs and thereby make the drugs available at an affordable price to a significantly larger percentage of the people who need them.[67] He obviously never expected this to happen as long as foreigners controlled China's pharmaceutical market.

By the mid-1990s, Chen had come to put a new face on ideas and policies that had been emerging for a decade. Though it was still inchoate, he was outlining a rather bold national developmental strategy that centered on biotechnology and China's need to become a superpower in its use and control. Scientists would need to seize the opportunity to use the market as a defensive weapon—first against foreign exploitation and second against domestic poverty. When the international controversy over genetically modified organisms (GMO) exploded in 1996, Chen was encouraged to formulate his positions with greater force and cogency. At the same time, he was responding to the United States-based World Watch Institute report that raised doubts about China's ability to feed itself.[68]

*The GMO Controversy*

The GMO controversy derives from two basic issues: The first is whether any firm (but especially a large multinational like Monsanto) should be able to profit from genetically altered seeds for which it holds an exclusive patent; the second is whether the GMO is safe.

The first issue is hardly a new one. There is a body of research that addresses the problems of third-world economic dependency caused by the use of hybrid seeds in the twentieth century, a dilemma with some parallels to the use of foreign GMO seeds.[69] In both instances, the farmer using these seeds might be required to purchase new seeds annually rather than freely use seeds produced on the farm from the hybrid or GMO. Seeds produced by the hybrid plant would likely be sterile or produce weak plants; those produced by the GMO might be contractually restricted from use unless directly purchased from the producer. (In the mid-1990s case of Monsanto, it was incorrectly reported that its GMO seeds would enforce the contract by containing a "terminator" gene that would prevent second generation seeds from germinating.)[70] Thus, the dependency is costly and—because farmers stop using indigenous plant varieties—it eventuates in the loss of indigenous genetic material, which in turn increases the dependency. Be that as it may, the issue has largely come to be argued from an ethical standpoint (hungry nations should be given free access to GMO seeds), or from a legal one (a GMO is patentable and its creator can claim it as intellectual property with all the attendant rights).

The question of safety has most vigorously been raised in Europe and has resulted in a ban on the local production of GMO plants of any kind as well as their importation. (A corollary issue has been the thorough labeling of GMO plants or plant products.) Producers of GMO industrial and food plants in China and the United States, while acknowledging the need to test thoroughly the safety of GMOs, have also suggested that the emotional and indiscriminate European approach is in no small way a device to restrict competition with superior agricultural products.

## Chen's Response to the GMO Controversy

As an active participant in GMO science and true believer in its great potential benefit to the "food security" of China and many other nations, Chen's position is not to find fault with the science, the patents, or the commerce of the multinationals like Monsanto. On the contrary, he openly advocates emulating them, and then beating them at their own game.

From the perspective of global development strategies, he has explained that China must seize the moment where biotechnology is concerned, because, unlike information technology, it is still early enough for China to invest here and gain a dominant position, one that will be of immense value domestically and in international trade. He encourages China to see the European ban as a superb opportunity to race ahead in GMO biotechnology, rivaled only by the United States. "China should move quickly to seize market share in this area," and make China independent of foreign biotech products, beginning with bio-pharmaceuticals and GMOs.[71]

For domestic and foreign audiences, Chen is not above putting a nationalistic spin on this otherwise businesslike analysis. In one interview on the subject, he recounted a conversation between Premier Zhu Rongji and some troubled provincial leaders who complained to him:

> The Americans have so much! Soon the seeds will all belong to them. The seeds that our people improved for so many years become the property of the Americans just by the insertion of a gene! The new improved transgenic crop that appears after six months belongs to the Americans, not to us. More tests are done in big experimental fields but the new varieties belong to the Americans, not to us.[72]

Chen's answer to this plight is not arbitrary exclusion of Monsanto seeds; but instead, he advocates more Chinese venture capital in domestic biotech research and private-sector seed companies, plus respect for intellectual property protocols for biotech research, and safety test protocols for biotech products. He said as much before the Ninth National People's Congress, in March of 1999, where he implored the government to work out quickly and comprehensively the legal details governing genetic engineering.[73]

## GMO Safety

Accompanying its development of GMOs, China has followed the United States' example by establishing two multilevel test protocols: one that must be passed before releasing a GMO for production, and another before commercial release. On the basis of these protocols and their findings, Chen is convinced that GMOs are as safe as conventional food or industrial plants that are tested by the same protocols.[74] He straightforwardly says that "the Europeans are blowing the biotechnology safety question out of all proportion," not because he believes that biotechnology products are all inherently safe, but because testing has demonstrated that it is possible to produce safe ones.[75] In 1999, Chen's mentor,

Roger Beachy, argued similarly before the United States Senate Committee on Agriculture, Forestry, and Nutrition. He took the argument a few steps further by noting that scientific colleagues using conventional and "less precise methods of cross breeding, or mutation-induced breeding, or wide species crosses" were not subject to the same type of "scrutiny, inquiry, or personal verbal attacks" that he and other GMO scientists experienced.[76]

## *Intellectual Property Rights*

China has been slowly establishing domestic frameworks for intellectual property rights and signing on to international conventions specifically for "breeders' rights." By the end of the 1990s, and notwithstanding the apparent support of scientists like Chen for free market competition, China appears to be constructing protocols whose net effect is protectionist. This results from the use of legal and safety tests that are difficult if not impossible to satisfy, and from restricting foreign investment in GMO seed development.[77] Additionally, the government has been lax in enforcing property rights for the one firm it has licensed to sell GMO seeds: Monsanto Bt cotton seeds have been regularly pirated and sold by Chinese farmers and seed companies. (International convention, however, does permit farmers to save and replant the seeds on their own farms or to exchange the seeds with other farmers.)[78]

## *The OECD 2000 Conference*

In March of 2000, the Organization for Economic Cooperation and Development held a conference in Edinburgh on sustainable agriculture and the use of GMOs. As China's representative, Chen Zhangliang was one of the four hundred attending scientists, government officials, and environmentalists. Apparently, many delegates and journalists expected the conference to be overwhelmed by the (largely European) enemies of GMO agriculture and used by them as a forum to advocate a worldwide ban. Chen's enthusiastically positive conference presentation, however, seems to have caught everyone unprepared, and has since become a reference point for much of the global debate and journalistic coverage of GMO issues.

Delegates and journalists were surprised by the basic facts that Chen cited in his description of China's involvement with GMO agriculture, such as the extent of planting and the number of commercial releases. He did not confront anti-GMO assumptions in the abstract, but rather laid out China's agricultural needs, asserted that "China simply cannot afford to reject the new technology," and predicted that in ten years at least 50 percent of some of China's major crops would be GMOs. Moreover, he made it clear that China had been using GMOs longer and more extensively than anyone else and there was no evidence, beginning with tests on laboratory animals, of any ill effects; and he offered to share published reports of tests—some of which were conducted by European scientists invited to China for that purpose. Following Chen's presentation, it is reported, the conference chair asked the delegates "if any of them knew of any published data [on GMOs] which indicated a human health risk." No one responded.[79] That, however, did not prevent the remainder of the conference from being rancorous or inhibit anti-GMO Greenpeace delegates from proffering doomsday scenarios. It did seem to contribute strongly to the conference's final recommendations—not to ban GMOs—but to persevere with GMO research, to monitor its safety very carefully, and to label GMO products.[80]

Some delegates reported that they found Chen's summary of the evidence compelling.[81] Others were encouraged by Chen's enthusiasm and confidence and began speaking out against the European position by citing Chen and elaborating his arguments. For example, in the *Christian Science Monitor*, a South African biologist delegate wrote that

> GM foods are not arguable in poor countries—they are needs based. The developing world cannot afford to let Europe's homemade problems stunt the future growth in our countries. . . . They cannot afford to limit themselves to the industrialized world's narrow interpretation of risk assessment . . . or to allow the Western debate to slow developing countries access to already existing and expected future benefits of biotechnology.[82]

In spite of Chen's success at the OECD conference, China, fearing trade sanctions, began to restrict the commercial release of some GMO crops in the spring of 2001. There is hope in China that the nation's entrance into the World Trade Organization will offer a new and more flexible forum to negotiate the future role of GMOs in world trade.[83]

## Genomics—with Chinese Characteristics

Genomics, the sequencing of an organism's genome, was the celebrity "big science" of the late 1990s. The International Human Genome Project received the most attention, but the technology it spawned was also used successfully for other projects that may be of equal importance, like describing the genomes of important food plants. Chinese biotechnology made assertive and successful efforts to participate in these projects. Notably, it sometimes did so with pronounced nationalistic overtones, well beyond those expressed by Chen Zhangliang.

*Human Genomics*

The Chinese began working independently on human genomics as early as 1994, and by 1998 were able to gain a formal place in the international project.[84] *People's Daily* proudly noted that China was the only developing country participating.[85] They were assigned specific genes to sequence, amounting to approximately one percent of the genome. Those interested in this project in China saw it as a means of gaining access to the most advanced methods and technology as well as the most sophisticated international network in the field. It was also a means of lobbying the government for a quick commitment to considerable funding for two essentially new areas of biotechnology: technology-intensive genomics itself and human genetics.[86] The latter field in China operated with a very low profile, but it was argued that it should be given renewed and special attention because it was about to be revolutionized by genomics.

Based at the Beijing Genomics Institute, the Chinese team successfully completed their "Chinese Volume" of the human genome along with the other participants in the project in early 2001. By that time, in Beijing and Shanghai, large, state-of-the-art human genome centers had been established and in use for over two years. In May of 2000, the Beijing center alone was evaluated to have more sequencing capacity than France and Germany combined, and almost as much as Japan.[87] Its director, Yang Huanming (b. 1952), built this center up in a few years from a small dysfunctional adjunct to the CAS Institute of Genetics into a giant private, nonprofit entity.[88] Though only a handful of its 500 workers are employed by the CAS, the CAS has given it institute status.[89] Yang is described as vehemently critical of "bio-piracy" (the use of patents to restrict access to genetic materials), and he remains opposed to any sort of gene patents, preferring to place all genomic information in the public realm.[90]

*The Gene Drain*[91]

Among those who disagreed with Yang's position on patents was Fudan's Tan Jiazhen. In the latter 1990s, he took a strong, impassioned position on the subject, not in response to Yang, but rather to the issue of foreign piracy of human genes in China. This problem seems to have arisen in 1995, when—without proper authorization—a California-based biotech firm collected DNA samples in China to identify a candidate asthma gene. Similar violations were reported and deplored, and the following year an emotional, nationwide reaction occurred in response to a mistranslated news bit from *Science* magazine claiming that a Harvard University scientist was going "to take blood samples from 200 million Chinese." (The original statement was that the research program in question sought "to have access to 200 million Chinese" via various Chinese medical institutions.) Not only was the project in question canceled, but the government was impelled to study the larger problem of human genetic research and to create protocols that would protect China's "genetic resources." At the annual administrative conference of the CAS "geneticists warned that China must not allow its human genome resources to drain away to foreign countries, or to be grabbed by foreigners for their own use."[92]

After many months of spiraling rhetoric and further muckraking, in July of 1997 Tan Jiazhen wrote a much publicized letter to the CCP Central Committee summarizing the problem and calling for the government to defend China decisively and commit resources necessary to bring the field of human genetics up to international standards. "A gene is a kind of wealth," he wrote, "and the war to seize the genes has already begun." He said that if China does not get its own gene patents, "then in the next century China's biotechnology industry and especially the pharmaceutical industry will be like 'the Admiral of [China's] Northern Fleet who saw all his ships capsize and sink beneath the waves [during the 1895 Sino-Japanese war]."'[93]

Tan continued the provocative allusions to imperialism by accusing scientists who collaborate with foreigners to be naïve seekers of short-term profit, and who sign contracts that are really "unequal treaties." He concluded that

> the developed countries always use the word "cooperation" for their activities but they do not allow Chinese to share in patent rights. . . . If China does not take action, all of China's gene resources will be drained away and become simply patented genes owned by foreign companies.[94]

It is speculated that the surge in public funding for genomics, beginning in 1997, and the rapid production of protocols for genetics research were due in large part to Tan's emotional alarm.[95]

## The Rice Genome[96]

If this is so, then no one profited more than Yang Huanming and his Beijing Genomics Institute, notwithstanding Yang's contrary position on gene patents. For the International Human Genome Project, the institute was a little fish in a big pond, but thanks to the efforts of Yang's rice genome project, the Chinese were able to begin the new century with a scientific coup that received international attention and admiration.

In January of 2002, the Beijing Genomics Institute, in cooperation with the smaller University of Washington Genome Center, announced that they had completed the draft sequencing of a subspecies of *indica*, one of the key varieties of rice. This subspecies is the paternal cultivar of a "superhybrid rice" developed by China during the 1980s. It yields 20-30 percent more than other rice varieties and its sequenced genome is expected to make it possible to learn more about which genes contribute to its prowess. In addition, the "extensive DNA sequence similarity between rice and other cereals will provide a shortcut to the isolation of genes of agronomic importance in cereals as well as in other crop species."[97]

Yang Huanming, and others, considered this triumph an ethical as well as technical one. His team's sequencing of *indica* was paired with the simultaneous sequencing of another major rice variety, *japonica*, by the Swiss-based firm Syngenta. On the completion of his team's work, Yang immediately placed its findings in GenBank, the public database; but Syngenta restricted access to its findings in order to protect its commercial rights.[98]

## A Concluding Word

At the beginning of the twenty-first century in China, genetics and biotechnology already have for themselves some extraordinary accomplishments and vigorous leadership. The contrasting strategies of Yang Huanming and Chen

Zhangliang cover a wide spectrum of biotechnology research and development. Yang's massive Genomics Institute, which already has a global presence, will be spewing out data for public access on a wide variety of genomes—beginning with that of the pig and moving on to traditional Chinese medical plants. The institute will not be at a loss for domestic and foreign clients who will pay handsomely to take advantage of its powerful and dependable sequencing expertise. The only issue is whether there will be a sufficient number of geneticists in China prepared to use the data and turn it with alacrity to productive and commercial ends.

Chen, on the other hand, is intent on continuing to facilitate cooperation between university and private-sector research, and to link these closely with patent-protected, private-sector production and commerce. Apparently, the government continues to admire and bank on his efforts as much as Yang's. In April 2002, Beijing University announced that the forty-one-year-old Chen had been appointed president of China Agricultural University.[99] With over a thousand faculty, a thousand graduate students, and seven thousand undergraduates, it is by far the largest agricultural science research and education institution in the country. The product of a recent merger between the former Beijing Agricultural Engineering University and Beijing Agricultural University, it should provide Chen with a base—even more powerful than his current one at Beijing University—to influence the development of biotechnology in China for years to come.

# Notes

1. Dean H. Hamer and Shain-dow Kung, eds., *Biotechnology in China* (BIC), (Washington, D.C.: National Academy Press, 1989), 96.
2. BIC, 1, 12.
3. BIC, 12-13.
4. BIC, 29-31.
5. BIC, 75.
6. This sketch of the 863 program is drawn primarily from the following: (1) BIC, 8-9; (2) Richard P. Suttmeier, "China's High Technology: Programs, Problems, and Prospects," in *China's Economic Dilemmas in the 1990s: The Problems of Reforms, Modernization, and Interdependence*, vol. 2 U.S. Congress, Joint Economic Committee (Washington, D.C.: Government Printing Office, 1991), 546-64; (3) Asian Technology Information Program (ATIP) "Biotechnology in the People's Republic of China," United States Embassy, Beijing, Albuquerque, New Mexico, January 23, 1996 (Online at http://www.atip.com); (4) International Development Research Center, *A Decade of Reform: Science and Technology Policy in China* (Ottowa, 1999), chapter 7; (5) Huang Jikun et al., "Plant Biotechnology in China," *Science* (January 25, 2002): 674-77, and online Supplementary Materials, part D, 4-5.
7. Richard P. Suttmeier, "Science and Technology Policy: Developing a Competitive Edge," *Current History* (September 1991): 278. During 1986-2001, other target areas for the 863 program were technologies related to space, information, lasers, automation, energy, and advanced materials.
8. BIC, 9.

9. "China to Spend 15 Billion Yuan in Hi-Tech Development," *People's Daily Online* (English ed.), February 14, 2001; and Li Heng, "March '86 Scheme Reaped a Benefit of 200 Billion Yuan within 15 Years," *People's Daily Online*, February 16, 2001 (http://english.peopledaily.com.cn).

10. Suttmeier, "China's High Technology," 552.

11. This percentage was arrived at by using a figure of sixty million yuan for the annual 863 budget for biotechnology. (See ATIP, "Biotechnology in the People's Republic of China," 4.)

Quantitative data on this subject are few and incomplete. In one notable instance, the figures differ with the ones I have used: An article in *Science* claims that the overall 863 program budget, 1987-2001, was 10 billion yuan and that biotechnology's share was 700 million yuan (Huang Jikun, "Plant Biotechnology in China," especially "Supplementary Material, part D," 4).

12. Suttmeier, "China's High Technology," 557.

13. ATIP, 4.

14. ATIP, 4.

15. Huang Jikun, "Plant Biotechnology," 4-5.

16. The survey: Huang Jikun, "Plant Biotechnology." It was conducted by a team of Chinese and American scientists through detailed survey research combined with on-site interviews and observation. Unlike the BIC survey in 1988, no attempt was made to evaluate the quality of published research in the field, but insightful observations are made about the kind of research done and its outcomes in agriculture. With the cooperation of a number of ministries, detailed data were gathered on twenty-nine research centers whose budgets, staff, and research output represent over 85 percent of the field. The survey instrument, tabulated data, and extended analysis can be found in the "Supplementary Material" available from Science Online (http://www.sciencemag.org). Also see these surveys of plant biotech: Chen Zhangliang and Qu Lijia, "Plant Molecular Biology in China," *Plant Molecular Biology Reporter*, no. 15 (1997): 273-77; and Zhang Qifa, "China: Agricultural Biotechnology Opportunities to Meet the Challenges of Food Production," Consultative Group on International Agricultural Research, CGIAR Online (January 2000): 45-50 (http://www.cgiar.org).

17. Zhang Qifa, "China," 45.

18. Huang Jikun, "Plant Biotechnology," Supplementary Material, part D, 10.

19. Huang Jikun, 6, 10.

20. Huang Jikun, 15.

21. This survey is based primarily on the following: (1) This writer's site tours of the institute and interviews with Shi Yingxian, Beijing 1984 and 1986. Ms. Shi (b. 1927) was a protégé of Tong and one of the institute's founding senior researchers; (2) CASS: 115-22; (3) "Fayun shengwuxue yanjiuso," *Zhongguo kexue yuan yuankan* (Bulletin of the CAS), no. 1 (1988): 72-74; (4) BIC, 43-47, 92-95.

22. For example, see Schneider, *Lysenkoism in China*, 63-67, 72-73 where Tong promotes his lifetime adherence to cytoplasmic (that is, non-Mendelian) inheritance.

23. Honorary degree: see the *Temple Times*, April 30, 1998.

24. BIC, 44 says that the "Chinese government" provided "several million yuan" to help "build and equip its research facility."

25. BIC, 44.

26. BIC, 46.

27. BIC, 44-47, 92-95.

28. Interviews with Shi Yingxian, 1984, 1986.

29. BIC, 95, citing Caltech's Eric H. Davidson.

30. BIC, 46. The authors of BIC provide specific criticisms on 42-46; and in an appendix (92-95) they reprint the devastating Eric H. Davidson evaluation of the work of Tong and Niu and of Niu's own later work.

31. Tong Dizhou's research from 1950 to the Cultural Revolution fundamentally followed the same topics and methodology that he and Niu pursued together after the Cultural Revolution. Tong's confused understanding (if not outright rejection) of Mendelian genetics is especially evident in his utterances at the Qingdao Genetics Symposium.

Niu Mann-chiang's research methods and topics of the 1980s are likewise anticipated by his work in the 1970s. For example, see the anthology he coedited with S. J. Segal, *The Role of RNA in Reproduction and Development: Proceedings of the AAAS Symposium, December 28-30, 1972* (New York: Elsevier, 1973). And Niu's research in the 1980s closely followed the Tong-Niu research collaboration of the 1970s. For example, see Davidson's review in BIC, 92-95. In particular, see Tong and Niu, "Transmission of Nucleic Acid-Induced Characters . . . ," *Scientia Sinica*, no. 2 (1975): 223-24, and "Genetic Manipulation in Higher Organisms . . . ," *Scientia Sinica*, no. 6 (1977). Also see Tong's single-authored article, summarizing their research since 1973, "Effect of Cytoplasm on Nuclear Activity and Genetic Characters in Animal Development," *Yichuanxue bao*, no. 1 (1978): 1-8.

Also see Temin's reference to this research in his 1978 trip report "Basic Biomedical Research," 259. He briefly observed that "the biochemical work was less convincing than the nuclear transplantation."

32. BIC, 47.

33. Niu's response to the criticism summarized in BIC did not appear in the 1989 hardcopy edition of the book. It was included in the 2000 online edition in appendix E, "Statement by Dr. Niu," 97-102. (Available at National Academy Press, online at http://www.nap.edu.)

34. Interviews, Shi Yingxian, 1984, 1986.

35. Interview, Shi Yingxian, 1986.

36. Interview, Shi Yingxian, 1986. And BIC, 44.

37. For examples see "CAS Reforms Its Institutional System," Bulletin of the CAS (March 2000). (Online at http://www.bulletin.ac.cn.) The CAS indicates that from 1998-2000, "large-scale institutional consolidation" has involved 60 research institutions. Among the reforms broader aims are reducing the CAS fixed positions by one-half, reserving 60 percent of CAS positions for workers under the age of forty-five, reducing the number of institutes from 120 to 80, and lowering the average number of researchers per institute to 250. (See discussion in Richard P. Suttmeier and Cong Cao, "China Faces the New Industrial Revolution," *Asian Perspective*, no. 3 [1999]: 177-78, 197-98).

38. "CAS Institute of Genetics and Developmental Biology Is Established," *Bulletin of the CAS*, July 12, 2001 (online at http://www.bulletin.ac.cn).

39. See Richard P. Suttmeier, "China: The Search for Strategies," in *Learning by Doing: Science and Technology in the Developing World*, ed. Aaron Segal (Boulder, Colo.: Westview, 1987), 159.

40. In addition to BIC, this survey of genetics at Fudan is primarily based on the following: (1) *Fudan daxue yichuanxue yanjiuso niandu baogao* (Fudan Institute of Genetics Annual Report) vol. 1, 1979 to vol. 7, January 1984-March 1985. (2) Interviews in 1984, 1986, 1994 with senior faculty including Tan Jiazhen, Liu Zedong, Xue Jinglun. (3) Unpublished draft of Tan Jiazhen speech "Zhongguo yichuan gongcheng yanjiu

jinkuang" (Recent developments in genetic engineering research in China), March 1988, 9.

41. BIC, 53.

42. BIC, 53. And *Genetics Institute Annual Report,* January 1981 to June 1982, preface. For example, during 1979-1982, Xue Jinglun (biology B.S. Fudan 1960) was a visiting scholar at the Buffalo New York Roswell Park Memorial Institute for cancer research. His collaborative work there with C. C. Huang was the equivalent of a doctoral training program. On his return, he was a coleader of the human and medical genetics group; and in 1987, he became deputy director of the institute.

43. *Annual Report,* 1984-1985, 1.

44. Interviews with Xue Jinglun, 1986, 1994.

45. BIC, 53-57 for fuller cataloging of research. Institute *Annual Reports* provide a complete accounting of all research conducted during each fiscal year.

46. *Annual Report,* 1984-1985, preface.

47. Visiting from Tokyo Metropolitan University were Professor Osamu Kitagawa in evolutionary genetics, and Emeritus Professor Toyohi Okada in systematics. *Annual Report,* 1982-1983.

48. *Annual Report,* 1984-1985, 73.

49. *Annual Report,* 2. Also see *Xinhua* domestic news service, June 10, 1986 (JPRS/CST/86.025, July 1, 1986), 5. And BIC, 16-17.

50. Tan, 1988 draft speech, 1.

51. Interview, Xue Jinglun, 1994.

52. This section is based on (1) Interview with Luo Deng, vice director of the Shanghai Biotechnology Center, June 1986. (Coconducted with Xue Jinglun, director of Fudan Genetics Institute); (2) Center promotional materials; (2) Interviews with Fudan Genetics Institute staff, 1986; (3) BIC, 15-16.

53. Promotional letter, dated February 7, 1985.

54. "Hangzhou to Be Base for Biotech Research," *China Daily, Business Weekly,* May 29-June 4, 1994, 5.

55. Hainan is the southernmost part of China, off the coast of Guangdong Province. The university was founded in 1958 and was earlier known as the South China College of Tropical Plants, a name that is sometimes still used.

56. Chen, Zhang-Liang, "Accumulation and Regulated Gene Expression of Soybean Seed Storage Proteins in Transgenic Plants," Ph.D. diss., Department of Biology and Biomedical Sciences, Washington University, 1987.

57. BIC, 51-52.

58. BIC, 51-52.

59. Chen Zhangliang and Qu Lijia, "Plant Molecular Biology"; "Young Scientist Advocates Industrialization of Hi-Tech," *People's Daily* (English), September 24, 1999 (online at: http://www.english.peopledaily.cn); BIC, 52; ATIP, 10-11.

60. Chen and Qu, 274.

61. Wei Feng, ed., *Zhongguo zhishi jie da zhendang—shang hai chenfu zhong de zhishifenzi saomiao* (Shanghai: Zhongguo shehui chubanshe, 1993). Excerpted and translated as "The Great Tremors in China's Intellectual Circles: An Overview of Intellectuals Floundering in the Sea of Commercialism," *Chinese Education and Society,* no. 6 (1996): 71-72; Choong Tet Sieu and Keith Loveard, "Magic Genes," *Asia Week,* September 9, 1997; "Peking University Drug-Making Joint Venture Poised for Listing in NASDAQ," *South China Morning Post,* May 23, 1998; Chen Zhangliang, "Unlimited Prospects for Biotechnology," *Zhishi jingji* (Knowledge and economy) (December 1999):

8. Summary and translation by United States Embassy, Beijing, January 2000; *Asia Week*, October 6, 2000. All of these sources agree that the firms had increasingly bigger sales and made significant profit; no two agree on any of the specific figures other than the market share.

62. Wei Feng, 71-72.

63. Wei Feng, 96-97.

64. *People's Daily*, "Young Scientist Advocates."

65. People's Daily, "Young Scientist Advocates."

66. Chen, "Unlimited Prospects."

67. Chen, "Unlimited Prospects."

68. *Asia Week*, January 21, 2000.

69. For example, Jack R. Kloppenburg Jr., *First the Seed: The Political Economy of Plant Biotechnology, 1492-2000* (Cambridge: Cambridge University Press, 1988).

70. See discussion, specifically related to India's ban of GMO cotton, in Robert L. Paarlberg, "Asia's Response to Genetically Modified Food," presentation to Modern Asia Series, Harvard University Asia Center, Spring 2001, 4.

71. Chen, "Unlimited Prospects," 6, 22-28.

72. Chen, "Unlimited Prospects," 3.

73. *China Daily*, March 12, 1999; *South China Morning Post*, March 13, 1999. And for a foreign observer's agreement with Chen's position, see David Murphy, "China Uncorks the Gene in a Bottle," *Far Eastern Economic Review*, March 22, 2001.

74. Chen and Qu, "Plant Molecular Biology," 275-76.

75. Chen, "Unlimited Prospects," 4, 11.

76. "Testimony of Roger N. Beachy, President, Donald Danforth Plant Science Center," October 6, 1999. (Online at http://www.senate.gov/~agriculture/hearings). Beachy also responded to the criticism of groups "established to promote protection of the environment, but are opposed to crops that require less chemical insecticide, or reduce tillage and soil erosion. Other groups are truly concerned about the safety of the foods produced by new techniques that they do not understand, but have full confidence in foods produced by classical methods that they also do not understand, or processes that purport to produce vegetables and fruits by organic methods that use questionable materials and whose safety is not guaranteed," 2.

77. Lester Ross and Walter Hutchens, "Genetic Modification in Agriculture: The Impact of China's Regulations on Foreign Trade and Investment," *China Law and Practice*, no. 15, 2001. And also see Craig Smith, "China Rushes to Adopt Genetically Modified Crops," *New York Times*, October 7, 2000; and "China: Frankenfood Finds a Home," *South China Morning Post*, December 6, 2001; and "China's Ban on Biotech Investors May Violate WTO Obligations," *DOW Jones Newswires*, May 13, 2002.

78. Robert L. Paarlberg, "Governing the GM Crop Revolution: Policy Choices for Developing Countries" (Washington, D.C.: International Food Policy Research Institute, December 2000), 23-24.

79. "GM Crops," *CropGen*, March 4, 2000.

80. Veronique Mistaen, "China Has Appetite for Gene-Altered Food," *Washington Times*, March 5, 2000.

81. "GM Crops," *CropGen*.

82. Jennifer A. Thomson, "Poor Nations Can't Afford Debate on Gene-Altered Crops," *Christian Science Monitor*, November 13, 2000.

83. Mark O'Neill, "Modified Crop Advocate Fears Lost Chances through Cultivation Ban," *South China Morning Post*, April 18, 2001.

84. Ministry of Science and Technology, "China's Human Genome Sequencing Effort Started," *China Science and Technology Newsletter* (December 20, 1999): 1-2.

85. "Genome Number 'China Volume' of Human Genome Project Fixed," *People's Daily*, August 28, 2001.

86. David Dickson, "Back on Track: The Rebirth of Human Genetics in China," *Nature* (November 26, 1998): 1-8. (Online: http://www.newclothestribune.com/Nature.)

87. Dorothy C. Wertz, "Inside the Labs of the Human Genome Project," part 2, Shanghai, part 3, Beijing. *GeneLetter*, September 1, 2000 (online at http://www.geneletter.org).

88. Yang Huanming: from Zhejiang Province; 1988 Ph.D. in genetics from University of Copenhagen, followed by stints at gene mapping on the X chromosome at the CNRS Immunology Center in Marseilles, then Harvard Medical School, and finally UCLA, before returning to China in 1994. (See Dennis Normile, "Beijing Genomics Institute: From Standing Start to Sequencing Superpower," *Science* (April 5, 2002): 36-39.

89. Murphy, "China Uncorks the Gene in a Bottle." The CAS has given it the additional designation of Genomics and Bioinformatics Institute.

90. Wertz, "Inside the Labs."

91. The "gene drain" summary is based on: "Backlash Disrupts China Exchanges," *Science* (October 17, 1997): 376-77; David Dickson, "Back on Track"; "Alarm at US Companies Draining China's Gene Pool," U.S. Beijing Embassy Report, April 1997; and "New Human Gene Rules Reflect Chinese IPR Concerns," U.S. Beijing Embassy Report, September 1998. Embassy reports reprinted by Gentech Archive, January 22, 1999 (online at http://www.gentech.free.de).

92. "Alarm at US Companies," 1.

93. "Alarm at US Companies, 4; and Dickson, "Back on Track," 3-4.

94. Dickson, "Back on Track," 5.

95. Dickson, "Back on Track," 5.

96. Summary based in part on Xue Yongbiao and Xu Zhihong, "An Introduction to the China Rice Functional Genomics Program," *Comparative Functional Genomics*, no. 3 (2002): 161-63 (online at www.interscience.wiley.com); "Beijing Genomics Institute Sequences Rice Genome: Sun Microsystems Supercomputing Technology Supports Research" (online at http://www.sun.com); Jonathan Amos, "Scientists Detail Rice Code," and "Genome Dispute Touches Rice," BBC News Online, April 4, 2002 (http://www.news.bbc.co.uk); John Whitfield, "Rice Genome Unveiled," *Nature*, April 5, 2002 (http://www.nature.com); Normile, "Beijing Genomics Institute," and J. Yu et al., "A Draft Sequence of the Rice Genome (Oryza sativa L. ssp indica)," *Nature* (April 5, 2002): 79-92.

97. Pamela Ronald and Hei Leung, "The Rice Genome: The Most Precious Things Are Not Jade and Pearls," *Science* (April 5, 2002): 58-59.

98. Jonathan Amos, "Scientists Detail Rice Code."

99. "New Center" *Peking University Online*, April 14, 2002 (http://www.cbi.pku.edu.cn).

# Conclusion:
# Biology—with Chinese Characteristics

By way of summary and conclusion, I revisit the three issues of "control" referred to throughout this study—the control of nature by science, the control of science by nation and state, the control of scientists by state and party. Hopefully this will provide additional perspective and an added degree of clarity, especially about the character of the science community and the role of the scientist in China.

## Controlling Nature

Genetics in China begins this new century and millennium in a propitious manner that warrants recollection of its situation fifty years earlier. At the start of the People's Republic, a congeries of agriculture colleges was consolidated to become Beijing Agricultural University, and Luo Tianyu was appointed by the Party to take the school, and all of biology, away from the international mainstream of science and in the direction of Lysenkoism. Today, one of China's premier geneticists, Chen Zhangliang, has been appointed to lead that school in its new embodiment as China Agricultural University and take it to new international frontiers in biotechnology. Among all the vivid contrasts between the respective biologies of Luo and Chen, none is more fundamental than their respective understanding of the transformative power of science, and the mandate to apply this power and control aggressively. These differences are largely a result of conflating social transformation and the transformation of nature; and of attempts to link science and production.

Throughout twentieth-century China, social-political philosophy and discourse have been locked in an alternating relationship of influence with the understanding and practice of natural science. Initially, and with the piecemeal influence of American Progressivism as well as anarchism, much social and political discourse was premised on a belief in its own scientific character. It was thought that, in the eye of the reformer, the scientific gaze was sufficient to erode superstition, oppressive institutions, and anachronistic values and customs. At the same time, science education and practice were thought to have morally uplifting qualities and socially ordering powers, in addition to their ability to provide knowledge and control of nature.

As China's cultural and social revolutionary discourse developed over the first quarter of the century, it in turn seemed to alter fundamentally the understanding of science—or more accurately, to generate alternative understandings of science, some of which persisted to the end of the Maoist era. A new vision of science was constructed in the image of emerging social revolutionary values. Those values emphasized the infinite malleability of social forms as well as the ability to create "new" men and women; and they optimistically posited a progressive, evolutionary trajectory to China's future, denying the existence of any historical constraints on China's advancement—such as Marxist stages of development. In effect, the exercise of human will became the motive force of social revolution and progress. In this discourse, nature and society were equated as objects of change and control: neither was considered to be informed by any permanent structures, qualities, or tendencies; both could be altered and directed from the outside, as it were, by reshaping environments. Doing science and making social revolution became equated metaphorically, and in the flow of social revolutionary discourse from Kropotkin's anarchism, to Lysenkoism, to the Cultural Revolution the metaphor came to tyrannize Chinese understanding and practice of science.

To be clear, it should be said that biologists from various countries at various times in the twentieth century have expressed transformist attitudes and values—regarding the prowess of their techniques or the malleability of organisms. They did not necessarily succumb to the extreme positions summed up in the metaphor. Importantly for China, the United States was not only one of the contributors to the construction of this metaphoric understanding of science, it was likewise a source, specifically in biology, of transformist visions that appear to have remained independent of it. For example, the post-Mendelian development of hybridization in the United States was expressly seen as the scientific empowerment for "rational intercession into nature and controlled creation of new varieties." This contrasted to the "mere manipulation" of given varieties through conventional selection and breeding techniques.[1] In the first decade of the century, American plant breeders enthused about "a new conception of varieties as *plastic groups* [that] must replace the old idea of fixed forms of chance origin which has long been a bar to progress," and it was concluded that it was "no longer necessary to wait for nature to supply [our needs] by some chance seedling."[2]

During the 1920s, when Chinese biologists were first trained in genetics in the United States, this liberating notion of organic plasticity was becoming informed by a highly disciplined and decidedly reductionist approach to research and understanding. American breeders, in their quest to improve plants and "create" new varieties, began to explore "not for useful *plants*, but for useful *genes*."[3] Genetics in China, from the Nanjing-Cornell program to the genetic engineering of biologists like Chen Zhangliang, is rooted in this approach and the transformist attitudes it encouraged.

This approach was impugned by Lysenkoism and Mao's mass science, which could not abide its focus on constituent genetic components instead of attending to environment and the whole organism—the chromosome and gene rather than the plant. In sum, the transformist tradition of Luo Tianyu's Lysenkoism and mass science was entangled and even overwhelmed by externalities to biological science: by a dogmatic belief that organisms can only be controlled definitively through environmental factors (and not through the internal factors, like genes, stipulated by reductionist biology); and by the parallel insistence that social-political and ideological factors trump scientific discipline and method as the means to understand and control nature.

There is little doubt that Chen Zhangliang's generation of biologists, like other scientists of their generation, have thought much more, and more systematically, than their predecessors about the relation of their work to production (and the market as well). Nevertheless, earlier generations, primarily schooled in the United States, were shaped by the aforementioned Mendelian-transformist tradition, and by all appearances, made it an intimate part of their understanding and practice of science. I have cited China's plant biologists as a case in point.

Lysenkoists and Maoists mischaracterized China's Westernized biology as passive, hermetic, out of touch with nature and production. In retrospect, it is especially remarkable that the more ambitious their quest for the conquest and control of nature, the more Lysenkoism and Maoist mass science chose to employ an artless trial-and-error "methodology," while denigrating the use of experimental discipline, statistics, and control groups by the Mendel-Morganists. Echoing Mao's notions of social revolution, they claimed to be activist and interventionist where nature was concerned, but claimed they could do so spontaneously, through a sheer willpower that they felt would be constrained and diminished by intellectual discipline and conventional scientific procedures. This resulted, among other things, in a shortsighted, ad hoc notion of science devoted to solving discrete problems. There was no indication here that science might be an ongoing and coherent intellectual process, much less one that is at least partly driven by the pursuit and accumulation of knowledge.

If biology in China was less involved in production than it might have been, this often had little or nothing to do with its internal workings, and everything to do with externalities—whether organizational constraints, or the interference from social and political ideologies that often found fault with science for problems created by the ideologies themselves.

# Controlling Science

No "externality" of science has been more problematic for China than the dependence on foreigners for science development. Just as scientists in China were encouraged to reject dependency on nature and not to wait for its favors, they were also encouraged to develop their science without favors from foreigners. From the time this became a subject of discourse, the meanings of dependency, agency, and independence have continuously undergone changes in response to concrete historical situations and to perennial questions about the appropriateness for China of the science transferred from abroad.

## Impossible Autarky

When China's reformers initially determined that their nation's future depended on the acquisition of science, it was considered to be a virtually independent and self-contained force that could cure all ills. "New Culture" and "enlightenment" reformers—cosmopolitan and wide open to foreign influences—simply did not see any externalities for science. They did not express any concerns about untoward cultural influences of the Western societies from which the science was taken, or potentially dangerous political or economic dependency. At the same time, reform discourse spoke of modern education (closely emulating foreign models) in very much the same fashion. In both cases it seems never to occur to the reformers that science or education, rather than shaping the state, society, and politics, would be co-opted and shaped by them, and often taken in directions not intended or desired by the reformers.[4]

In the late 1920s, political radicalism and the rise of Guomindang nationalism reframed the issue: if the immediate goal of science and modern education was to liberate China from foreign oppression, then it seemed contradictory to permit the instrument of liberation to be dominated by foreigners. I have suggested that the contradiction, by default, was to be resolved by achieving ultimate independence through a proximate dependence on foreign support. It then fell to the new Guomindang state to initiate extensive but piecemeal efforts not only to indigenize science, but also to encourage its diffusion and nurturance.[5] At the time, the Academia Sinica and a public university system filled with Chinese scientists were considered to be significant products of these efforts. Nevertheless, overseas education, and foreign foundations and schools in China continued to play central roles in their success—not simply by providing funds for growth, but perhaps more importantly, by establishing standards of education and research, and organizational models to achieve them.

The Guomindang control or appropriation of science was thus an effort packed with quite a diverse range of motives and goals, most of them concerned with the externalities of science. Many resonated with the cultural preoccupations of earlier reformers. Having one's own science institutions, educators, and researchers was taken as evidence of modernity, progress, and cultural equivalence with the West; using them well would eventuate hopefully in political and economic equivalence as well. And there were other unresolved issues, closer to the actual form and content of science: Should emphasis be placed on applied science at the cost of theoretical and experimental? Should a state industrial development plan govern the research agenda for the Academia Sinica and the universities? Should the state determine the priorities for the use of foreign support of science?

It made a difference that virtually the entire leadership of China's emerging science community was trained abroad. They experienced science as an international phenomenon. (One thinks of Tan Jiazhen's work at Caltech under the Russian expatriate Th. Dobzhansky.) And on their return to China they worked assiduously to remain an active part of an international network, through collaborative research, international professional meetings, and students sent abroad for advanced training. I am not aware of any concern on their part that this might be considered an indication of dependency; nor that science "independence" for China would eliminate any of these activities or attitudes. By the same token, the Guomindang (whose leadership contained numerous foreign-educated intellectuals) did not express concern that the foreign education of scientists was in itself a problem, or that it would lessen the ability of science to "save" China.

The experience of the Guomindang state was the first in China to demonstrate that "controlling" science had much less to do with achieving autarky than with making informed choices among available options. Its prewar experience suggested that autarky for China's science may simply not have been possible—or even desirable.

## The Inappropriateness of Borrowed Science

In Mao's China, "controlling science" takes on more complex implications and heightened urgency. In sharp distinction from the Guomindang experience, the dependency in question is directly on one state, the Soviet Union, rather than on disparate, small, and relatively innocuous agencies such as missionaries, foundations, and foreign schools. Periodically, throughout the 1950s, the CCP explicitly considered the Soviet relationship from the viewpoint of dependency and autonomy (over the issue of the atom bomb for example, or at the end of the first five-year plan and during the Double Hundred). For the most part, however, the terms of the discourse were changed: The rhetoric surrounding Lysenkoism and the Great Leap Forward speaks less of foreign dependency and control than of the appropriateness and validity of the borrowed science. This is largely because of the introduction of the social-class analysis of science and the issue of the social relations of science (or the division of labor).

At this time, China's dependency on Western science (or at least on Western biology) is condemned first and foremost because it is tainted by its capitalist, bourgeois context. And in turn, the latter is not considered capable of producing valid science. Moreover, as the extremes of the Maoist critique emerged, China's dependence on Western science is further condemned because it brings with it a counterrevolutionary division of labor in the form of specialization and professionalization.

For a while, China's reliance on Soviet science and technological aid was legitimized on the basis of its socialist source. Soon, however, especially during the Great Leap Forward, the terms of the discourse abruptly changed again: dependency on foreign nations for science and technology was condemned, as such; and in tandem, dependency on China's technical specialists was likewise condemned. Enter mass science and the illusion of nativist, populist science/technology autarky.

Though Mao's efforts to "control science" led to disastrous consequences, I do not want to dismiss out of hand all of the issues he raised. Most important is his attempt to indicate the social relations of science as valid grounds for evaluating the science one might be borrowing from abroad. It does not seem frivolous to ask how the organization of the science community affects the kind of work it does, or how specialization and professionalization affect the relations of the scientists with each other and with the general society. For some time now, it has seemed reasonable to societies around the world to raise questions about social and ethical responsibilities of science, and to impose constraints on its practice. Unfortunately, in Mao's critique of science, these kinds of useful considerations were overwhelmed by its doctrinal and coercive character.

## Returning Science to Its Cosmopolitan Status

From the outset of Deng's regime, the "open door" policy for scientific development transformed the issue of control and dependency. It implied a recognition that no developing nation could usefully consider independence where science and technology were concerned, and it explicitly called into question the very idea of science independence for any nation by characterizing science as a global, transnational asset. Clearly, when this attitude was first expressed in the late 1970s, it was aimed squarely at the xenophobia and nativism of the Great Leap Forward and the Cultural Revolution.

In time, it also became evident that China was staking out a claim on its share of those global scientific assets: scientific education abroad, information, organizational models, and science/technology transfers were all fair game. Their acquisition became patriotic duty. A steady stream of policy statements displayed concern not about dependency, but rather about the level of China's science and technology, and how to facilitate the transfer process to raise that level. For the 1980s and 1990s, I do not recall seeing one policy statement by politicians or scientists that said that, at the end of some period of time, China would no longer have to depend on American biotechnological science. Instead, they were concerned only with achieving a higher level of competence in the field by doing whatever was necessary.

The commodification of science and the mandated linkage to the market contributed significantly to this attitude and to a reformulation of China's relationship to scientifically more sophisticated donor nations. In the last chapters of my study, I have tried to illustrate how the concern to control science has been transmuted to strategies for controlling the market.

This is seen dramatically, for example, in Chen Zhangliang's personal career in biotechnology. It is also expressed in his call to the Chinese science community and the government to do what is necessary to compete with foreign science and prevail over it in the marketplace. If the right research, patents, and products are lacking, he warns, China will be "taking the foreigners' medicine," or desirable genetically modified seeds "will all belong to [the Americans]." The venerable geneticist Tan Jiazhen plays a similar jingoistic note to sound the alarm on the "human gene drain." Using the imagery of China's earlier failure to strengthen itself against imperialism, he challenges China to do what it must to get its own gene patents or watch its gene resources be "drained away and become . . . owned by foreign companies." Issues of dependence have been replaced by issues of competition.

The purpose of Chen's and Tan's heated nationalistic rhetoric was in part to promote support from state agencies for the further development of biotechnology and for accelerated state construction of patent protocols and an intellectual property regime. Tan was additionally using China's gene drain as a catalyst to develop the moribund field of human genetics. More generally, these two prominent scientists were also exhibiting the aggressive, competitive attitude toward foreign science that has come to characterize the development of biology and, I would venture, much of the rest of science in China.

## Controlling Scientists

When Mao initially intervened in the social relations of science (in Yan'an and in the early 1950s), he apparently had in mind the science community of Republican China. Its intellectual capital, education, research and administrative experience were the points of departure for science in the People's Republic. It was also the target of Mao's ever-increasing efforts to shape and control the development of science. In order to provide one last perspective on this process, I want to draw some conclusions about the social relations of science in the Republican period—especially the role of scientists and the nature of their community—as suggested by my study of the development of biology.

## The Republican China Legacy

China's first generation of modern scientists began its work during the "New Culture" reform era (ca. 1915-1925). China's new intelligentsia (replacing the scholar-officials) has been generally characterized as one that initially acquired its freewheeling autonomy at the price of political power and social influence. Though detached from both state and society, it sought to get China back on the track of social evolution and progress through a piecemeal attack on "Confucian" patriarchy, social hierarchy, and family dominance. Through its iconoclastic literature, basic New Culture values were revealed to be individualism, egalitarianism, democracy, and scientific culture. Lacking faith in political institutions and processes, "education," especially science education, was designated as the primary medium for achieving ambitious social and cultural reforms. In short order, however, New Culture efforts and ideals ran up against the hard realities of political power and violence, and conservative resistance. This led some of the new intelligentsia away from their initial positions and into formal political organization and action; it led others into a pall of despair and alienation.

In retrospect, the anecdotal materials I have drawn on for the study of biology suggest a picture of the scientist that features many New Culture values, but differs in important ways from this general sketch of the New Culture intellectuals. The latter have not typically been thought of as scientists, but rather as writers of fiction, poetry, and social criticism, or as scholars of history and philosophy. In the context of the broader society, China's scientists do not convey a sense of social disconnection or alienation. When they completed overseas training, for example, they did not exhibit any of the problems of deracination, nor did they express concern about the foreignness of their education, or the novelty of their intellectual pursuits.

There appears to have been a set of factors that readily reconnected the young scientists with society, and eventually, with the state as well. First, there was a normative atmosphere that provided at least a sense of legitimacy and social purpose for their work. New Culture values themselves had virtually iconized science, so it followed—beginning in the intellectual community—that the scientific enterprise was important in itself and for national progress. More specifically, the popular press and the rhetoric of nationalism imbued the work of scientists with an urgent, patriotic, and even revolutionary significance.

This was not always the kind of reaction that New Culture radicalism itself received. As I have indicated, there were prominent scientists who clearly embodied some New Culture values but strongly disagreed with its iconoclastic treatment of China's cultural heritage, or its strident egalitarianism, and its "negative" approach to social reform. When the Guomindang came into power in the late 1920s, it was similarly critical of New Culture values, while giving great weight to educational modernization, especially science and technical education.

The social role of science was given unprecedented definition by Guomindang institutional reforms. A preexisting public university system was expanded and in it science and technology were designated as priorities for development. The new national academy, Academia Sinica, likewise promised science research a central and prestigious role in the creation of a modern industrial economy, a major concern of the Guomindang. To plan and manage industrial development, the state in effect established a technocracy, staffed with prominent senior scientists. The state sought to enhance its local and international image and legitimacy with this commitment to science; reciprocally, the state's imprimatur gave status and national purpose to the scientific enterprise and the scientist's role.

Can we conclude that there was a science community during the Republican era? The experience of the biologists suggests that a community was developing and defining itself ever more strongly, from the 1920s through the war. Conventionally, one thinks of science communities structured by professional associations that establish research and educational standards, and provide means of communication through journals and meetings. By the end of the Republican era, such associations had begun, their stage of development usually a function of when the field was established in China and the size of its constituency. Initially, the Science Society of China acted in this role for all fields and interests; however, much of its membership were not scientists and it aimed at a larger lay audience. In time, there were modest specialty journals and associations. In the case of biology it was typical for research publications to be sponsored by a university department or a single laboratory. The nationally distributed Science Society journal (*Kexue*) served an important community role by reprinting research and occasional policy pieces from these house journals, as well as translations from foreign periodicals.

The Academia Sinica was the single most important institution in the development of a national science community. Its institute's regularly published journals were distributed to a national audience. Though good and often better science was done outside the Academia Sinica, it still provided definition for science fields and research standards against which one's own work could be measured. Thus, while formal, professional associations were still inchoate before the war, there clearly had emerged "invisible colleges" of scientists based on the significant volume of published research coming out of the institutes, as well as Chinese and missionary schools.

The Chinese science community was perhaps even more strongly developed through informal networks and associations that came to bind specialists together like an extended family. Central to this process, as I have illustrated for biology, were the academic lineages that tied teachers and students together for an entire career. I have also indicated the web of contacts and influences that came from institutional sources: Through the exchange of students and faculty, and through joint research programs, there was an interconnectedness of the various biology departments at missionary and Chinese schools. Additionally, the China and Rockefeller Foundations and the Science Society endowed pro-

grams aimed specifically at bringing scientists from different institutions together with each other and with new groups of students. This web of relations
was further extended by overseas education. Chinese students formed new
lineages with foreign scientists who often played important roles in their subsequent research careers.

During this formative era, is it possible to typify the individual "scientist"
who was supported by this web of relationships? For the field of biology, anecdotal information suggests only two strong patterns: These scientists were almost exclusively male and from urban-based families. Economically, their
families tended to be of modest means and rarely well-off. The father's occupation was usually associated, however marginally, with some aspect of the modern, commercial, treaty-port economy. Families of the biologists typically had
little education, and virtually none at all beyond middle school. While they had a
positive attitude toward education and perhaps even a career in teaching, they
had no sense of what "science" was nor how one made a living at it. Every biologist for whom I have biographical information was initially encouraged by
family to enter a career in business where there were greater opportunities to
enhance the family's welfare.

In this prewar generation of biologists, it is rare to find anyone who did not
attend a missionary school somewhere in his/her education. Those schools
regularly acted as recruiters both for more advanced education and specialization in science. Only occasionally are there indications that the families of these
students were Christians, so it is necessary to beg the interesting question of the
role of Christianity in this process. A typical biologist, then, would have had at
least three years of education, at some level, in a mission school; would have
spent at least three years abroad—likely in the United States—for an advanced
degree, and received substantial financial support along the way from at least
two of the major foundations.

Finally, there is the question of politics. Was this science community inclined to political activity or did it display any political leanings? There is no
indication that the science community as a whole or any branch of it acted as an
interest group, let alone a political faction. Only two organizations would have
been remotely positioned to do so: the leadership of the Academia Sinica and
the Science Society. The former was considered to be part of the Guomindang
government's politically neutral technocracy, not an "outside" interest group;
the latter by all means promoted the understanding and development of science,
but did not promote specific policy. While some scientists were members of the
Guomindang Party and active proponents of its political expansion, the community in general is remarkable for its lack of political association or expression.

The wartime experience of the science community reinforced many of these
emerging characteristics. While most of the community was uprooted, nevertheless it was reconstituted in ways that brought its members together in new
configurations and in closer proximity. The pressures of danger and deprivation
forged new and tighter bonds. Further, the community seems to have emerged
from wartime exile even more dedicated to its role in national welfare and more

convinced of the legitimacy of that role. There is no evidence available to assess political positions scientists may have taken during the civil war between the Communists and the Guomindang; however, it is clear that members of the scientific community did not flock to Yan'an to support the Communists.

## Rejecting the Legacy

If my conclusions about Republican-era social relations of science are at all accurate, they suggest how the science community provided Mao with grist for his antipathy toward them: Their extended and advanced education made them elitist, ipso facto; and it surely did not foster populism. Their cosmopolitanism excluded nativism, and their professionalization was taken to show more concern about their own welfare and interests than national ones. Their patriotism, nationalism, and loyalty could be called into question by their association with foreigners, or challenged because the community had worked and prospered under the Guomindang's aegis. The scientists had come to believe that simply by virtue of being scientists they were contributing to social reform, and by doing science they were contributing to national progress. Mao saw only that their commitment to science and technical issues excluded commitment to social revolution—as he defined it.

In his 1942 Yan'an lectures on the role of literature in the revolution, Mao said that New Culture literary intellectuals were wrong to believe that literature and politics should or could exist in their separate realms. All literature is political, he said, and the only question to be asked is *whose* politics are being promoted.[6] He obviously felt the same about science (which he equated with all other cultural expressions). Perhaps as early as Yan'an, but surely by the advent of Lysenkoism, Mao had determined that the science community was politically incompatible with the Chinese Communist Party because of the community's bourgeois attitudes about the division of labor and the social relations of science. From this it followed that the community's science had to be inferior if not altogether invalid.

Mao's recurrent attacks on the science community ultimately had little or nothing to do with the internal content of science and are thus best understood as the expression of a utopian quest for social equality, and additionally, a contest for sovereign authority in China. From the example of the biological sciences, one sees how the self-contained authority of a cosmopolitan science community posed a threat to the authority of Mao's Communism. Lysenkoism in China was, at least in part, a proxy for Mao in this contest with science. No matter how loyal or patriotic the science community might be, its internal workings—its methods and intellectual processes—resisted party control.

It is true that the biology community's experience with Lysenkoism was not typical of the larger science community. Though imported Soviet science came with a number of doctrinaire positions, there was nothing that compared to Lysenkoism, in the comprehensiveness of its contestations or in the thoroughness of its organization and enforcement. The Great Leap Forward and the Cultural Revolution, however, magnified the experience of biology and generalized it for the entire science community.

## Refurbishing the Legacy

In post-Mao China, the social relations of science and the social role of scientists have undergone comprehensive transformations that parallel open-ended economic development. From the experience of biology, I conclude that two principles have come to frame these recent transformations: Science along with technology research must be embedded in the commercial economy. And second, at the core of economic development the focus should not be on particular industries or sectors as such, but on acquiring the basic science or technology that inform them—not pharmaceuticals or agriculture, but biotechnology and genetic engineering.

From the Guomindang's industrial plan to the end of the century, the Holy Grail of development has been the "linkage of research and production." The post-Mao science community has been thoroughly reconfigured to achieve this goal and this is what differentiates it most importantly from its predecessors. Consequently, the scientist is proffered unprecedented status and attention, and is given recognition not only with rewards and patents, but also with membership in a CCP that has now opened its doors to "capitalists" as well. Some scientists themselves easily fit the latter category, thanks to their successful entrepreneurial efforts. And none of this, to state the obvious, contributes to socialism—in present practice or future ideal; but I am not aware that this causes any concern within the science community.[7] Nor does it appear to have tarnished the image of the scientist as one of China's new heroes.

On February 21, 2002, the *Shanghai Star* newspaper reported on an interview with Xu Gang, the fourteen-year-old boy who has become a national celebrity in China. He dubs the voice of Harry Potter in the children's films about the young sorcerer. When asked perfunctorily what he would like to be when he grows up, Xu's answer apparently surprised the reporter:

> "I want to be a scientist, if not an actor," he said, "like the man who used to live in that building"; he pointed at the building behind his home on the crowded Shaanxi Nanlu. The house used to be the residence of Tan Jiazhen, a famous biologist.

## Notes

1. Jack R. Kloppenberg, *First the Seed: The Political Economy of Plant Biotechnology, 1492-2000* (Cambridge: Cambridge University Press, 1988), 69.
2. Kloppenberg, *First the Seed*, 69.

3. Kloppenberg, *First the Seed*, 80.

4. See Suzanne Pepper, *Radicalism and Educational Reform in Twentieth-Century China: The Search for an Ideal Developmental Model* (Cambridge: Cambridge University Press, 1996).

5. The terminology follows Richard J. Samuels, *Rich Nation, Strong Army: National Security and the Technological Transformation of Japan* (Ithaca, N.Y.: Cornell University Press, 1994).

6. Mao, "Talks at the Yan'an Forum on Literature and Art," in *Selected Readings from the Works of Mao Tsetung* (Peking: Foreign Languages Press, 1971), 271.

7. See Maurice Meisner, "The Deradicalization of Chinese Socialism," in *Marxism and the Chinese Experience*, ed. Maurice Meisner and Arif Dirlik (Armonk, N.Y.: M. E. Sharpe, 1989), 341-61.

# Glossary of Chinese Names

**Bao Wenkui** (b. 1916). Plant geneticist. Ph.D. Caltech, 1950. Member of CAS Institute of Agriculture and Forestry, and faculty of Beijing Agricultural University, 1956-1980. Publicized target of Lysenkoite censure in 1956.

**Bing Zhi** (1886-1965). Entomologist. Ph.D. Cornell, 1913. Founding chair of the biology department at National Central University, 1921.

**Chen Shixiang** (b. 1905). Entomologist. Ph.D. University of Paris, 1934. Expert in Darwinian evolutionary theory. Director, CAS Institute of Entomology, 1953.

**Chen Zhangliang** (b. 1961). Plant geneticist. Ph.D. Washington University, 1986. Beijing University, 1986-2002. Founder of National Laboratory of Protein Engineering and Plant Genetic Engineering. President of National Agricultural University, 2002-.

**Chen Zhen** (1894-1957). Geneticist. M.S. Columbia, 1920. Taught China's first genetics course at National Central University, 1920-1928, and wrote the first Chinese language general biology textbook, which emphasized genetics and evolutionary theory.

**Chen Ziying** (b. 1897). Geneticist. M.S. Yanjing, 1926, Ph.D. Columbia, 1929. One of the Yanjing "trio" along with Li Ruqi and Tan Jiazhen. He did collaborative research with the Morgan group at Caltech in the 1930s.

**Fang Zhongxi** (b. 1912). Geneticist/Lysenkoite. Ph.D. University College, London, 1949. Preeminent author of Michurin biology textbooks, 1949-1956.

**Guo Bingwen** (1880-1969). Educator. Ph.D. Teachers' College, Columbia, 1914. President of National Central, 1920-1925.

**Hu Xiansu** (1894-1968). Botanist. Ph.D. Harvard, 1925. Cofounder of the biology department at National Central University.

**Li Jingxiong** (b. 1913). Plant radiation geneticist. Ph.D. Cornell, 1948. Assisted Li Xianwen at National Wheat and Rice Improvement Institute, 1937-1944. CAS Institute of Agriculture and Forestry and Beijing Agricultural University, 1956.

**Li Jingzhun** (b. 1912). Plant geneticist, biostatistician. Ph.D. in plant breeding, Cornell, 1940. Codirector, Beijing Agricultural University agronomy department, 1946-1950. Driven from China in 1950 by Lysenkoites.

**Li Ruqi** (b. 1895). Geneticist. B.S., M.S. (Animal husbandry) Purdue, 1923; Ph.D. Columbia, 1926. One of Yanjing "trio," on the biology faculty 1926-1949. Beijing University biology department, 1949-1980.

**Li Xianwen** (b. 1902). Plant geneticist. Ph.D. Cornell, 1930. Director, National Wheat and Rice Improvement Institute, 1937-1945.

**Lu Dingyi** (b. 1901). CCP cadre. Head of the CCP Central Committee's Propaganda Department. Exponent of the Double Hundred policy.

**Luo Tianyu** (1900-1984). Lysenkoite. CCP cadre. Promoter of Yan'an "rectification" program in biology, 1944-1945. Dean of Beijing Agriculture University and leading promoter of Soviet Lysenkoism in China, 1949-1952.

**Mi Jingjiu** (b. 1925). Publicist of Lysenkoism. Interpreter for Soviet Lysenkoites and of Soviet Lysenkoist literature, 1950-1956.

**Shen Zonghan** (1895-1980). Plant geneticist. Ph.D. Cornell University, 1927. Director, Cornell-Nanjing Plant Breeding program, 1926-1937.

**Tan Jiazhen** (b. 1909). Evolutionary geneticist. M.S. Yanjing University under Li Ruqi; Ph.D. Caltech under Th. Dobzhansky. Biology faculty, Zhejiang University 1937-1946, Fudan University 1946-. A leading exponent of modern genetics in China, and a leading critic of Lysenkoism.

**Tang Peisong** (b. 1903). Plant physiologist. Ph.D. Johns Hopkins University, 1930. Faculty of Qinghua University, Kunming, 1938-1945. Codirector, Beijing Agricultural University Department of Agronomy, 1949. Target of Luo Tianyu's Lysenkoism. Member of CAS Institute of Biology, 1951-1980.

**Tong Dizhou** (1902-1979). Embryologist. Sc.D. University of Brussels, 1934. Supported Michurin biology 1950-1956. Administrator, Shandong University. Cofounder of the CAS Institute of Developmental Biology.

**Wang Shou** (b. 1897). Plant geneticist. Ph.D. Cornell, 1934. On the faculty of Nanjing University Agriculture College, 1935-1950. Breeder of "Wang barley."

**Wu Zhongxian** (b. 1911). Animal geneticist/breeder, biostatistician. Ph.D. University of Edinburgh, 1937. Director, animal sciences, Beijing Agricultural University, 1950-. Target of Luo Tianyu's Lysenkoism.

**Yu Guangyuan** (b. 1915). CCP cadre. Worked under Lu Dingyi's Propaganda Department on science-related issues like the resolution of the "Genetics Question," 1956.

**Zhou Jianren** (1890-1984). Science journalist. B.S. Agriculture School, Tokyo Imperial University. Life Sciences editor at Commercial Press. Exponent of Lamarckism and, from 1949, Lysenkoism. Brother of Lu Xun and Zhou Zuoren.

**Zhu Xi** (1899-1962). Embryologist. Ph.D. Montpellier University, France, 1935. CAS Experimental Biology Institute, 1950-. Critic of the CCP's support of Lysenkoism.

**Zu Deming** (b. 1925). Lysenkoite/CCP cadre. Advanced degree in plant improvement from Tokyo Imperial University Agriculture School, ca. 1936. Succeeded Luo Tianyu as leader of Michurin biology in 1952. First director of CAS Institute of Genetics, 1959-1966.

# Interviews

Bao Wenkui,
Plant genetics, Beijing Agricultural University. Beijing, June 3, 1986.
Raisa Berg,
Russian geneticist, formerly of Leningrad University biology faculty. St. Louis, November, 1984.
Guo Tingyi,
Formerly, student and faculty member, history department, National Central University (Zhongyang). New York City, October 25, November 15, 1974.
Hu Han,
Deputy director, CAS Institute of Genetics. Beijing, August 8, 1984.
Huang Zongzhen,
Editor, Science Press, CAS. Beijing, May 26, 1986. (Conducted jointly with Dr. William Haas.)
Lee Tsung-dao,
Physics department, Columbia University. New York, October 30, 1980. (Conducted jointly with Professor John Israel.)
Li Jingjun,
Former chair, Department of Agronomy, Beijing University; retired chair, biostatistics department, University of Pittsburgh. Pittsburgh, January 9-10, 1985.
Li Ruqi,
Biology department, Beijing University (retired). Beijing, August 6, 1984.
Liu Zedong,
Institute of Genetics, Fudan University. Shanghai, September 15, 1984; June 20, 1986.
Lu Guizhen,
Historian of science. Formerly, biochemistry faculty, St. Johns University, Shanghai. Cambridge, England, May 24, 1979.
Luo Deng,
Vice director, Shanghai Center of Biotechnology, CAS. Shanghai, June 1986.

287

Mi Jingjiu,

Biology faculty, Beijing Agricultural University (retired). Beijing, June 12, 1986

Shao Qiquan,

Institute of Genetics, CAS. Beijing, August 11, 1984; May 23, 1986.

Shi Yingxian,

Institute of Developmental Biology, CAS. Beijing, August 7, 1984; May 1986.

Tan Jiazhen,

Institute of Genetics, Fudan University. Shanghai, August 16, 1984.

Tang Peisong,

Institute of Plant Physiology, CAS (retired). Beijing, May 19, 1986.

Ting Yu-cheng,

Biology department, Boston College. Beijing, September 8, 1984.

Wang Bin,

Director, CAS Genetics Engineering Laboratory. Beijing, May 1986. (Conducted jointly with Dr. William Haas.)

Wang Zhiya,

Director, Shanghai Center of Biotechnology, CAS. Shanghai, June 1986.

Wu Dayou,

Formerly, physics department, Beijing University, and Southwest Associated University; chair, physics department SUNY/Buffalo (retired). Buffalo, New York, March 22, May 8, 1978.

Wu Zhongxian,

Beijing Agricultural University (retired). Beijing, May 31, 1986.

Xue Jinglun,

Director, Institute of Genetics, Fudan University. Shanghai, 1984, 1986, 1994.

Yang Chen-ning,

Institute for Theoretical Physics, SUNY/Stony Brook. Stony Brook, New York, November 1, 1980. (Conducted jointly with Professor John Israel.)

Yu Guangyuan,

Science Division, Propaganda Department, CCP Central Committee (retired). May 20, 1986. (Conducted jointly with Dr. William Haas.)

# Bibliography

Adams, Mark B. "Genetics and the Soviet Scientific Community." Ph.D. diss., History of Science Department, Harvard University, 1972.

——. "The Soviet Nature-Nurture Debate." Pp. 94-138 in *Science and the Soviet Social Order,* edited by Loren Graham. Cambridge, Mass.: Harvard University Press, 1990.

——. "Eugenics in Russia, 1900-1940." Pp. 153-216 in *The Wellborn Science: Eugenics in Germany, France, Brazil, and Russia,* edited by Mark B. Adams. Oxford: Oxford University Press, 1990.

Allen, Garland E. *Life Science in the Twentieth Century.* New York: Wiley, 1975.

——. *Thomas Hunt Morgan: The Man and His Science.* Princeton, N.J.: Princeton University Press, 1978.

Baldeschwieler, John D., ed. *Chemistry and Chemical Engineering in the People's Republic of China.* Washington, D.C.: American Chemical Society, 1979.

*Biographic Dictionary of Chinese Communism, 1921-1965.* 2 volumes. Ed. Donald W. Klein and Ann B. Clark. Cambridge, Mass.: Harvard University Press, 1971.

*Biographical Dictionary of Republican China.* 4 volumes. Ed. Howard L. Boorman and Richard C. Howard. New York: Columbia University Press, 1970.

Blacher, L. I. *The Problem of the Inheritance of Acquired Characters.* Ed. and trans. F. B. Churchill. Washington D.C.: Smithsonian Institution and National Science Foundation, 1982.

Bullock, Mary. *An American Transplant: The Rockefeller Foundation and Peking Union Medical College.* Berkeley, Calif.: University of California Press, 1980.

Chen, Shixiang. "Guanyu wuzhong wenti" (The speciation question). *Kexue tongbao,* no. 2 (1957): 33-42.

——. "Shengwu jinhua lun" (The evolution of life debate). *Kexue tongbao,* no. 6 (1965): 667-75.

——. "Shiying yu bu shiying: Zeran xuanshi di maodun" (Adaptive and nonadaptive: Natural selection's contradiction). *Zeran bianzhengfa tongxun,* no. 3 (1981): 20-23.

Chen Zhangliang, "Unlimited Prospects for Biotechnology," *Zhishi jingji* (Knowledge and economy), December 1999. Summary and translation by United States Embassy, Beijing, January 2000.

Chen Zhangliang and Qu Lijia. "Plant Molecular Biology in China." *Plant Molecular Biology Reporter*, no. 15 (1997): 273-77.

Chen Zhen. *Putong shengwuxue* (General Biology). Shangahi: Commercial Press, 1924.

China Foundation. *Annual Report*. Peking: 1926-1940, 1947, 1948.

China Medical Board. *Annual Report*. New York: Rockefeller Foundation, 1916-1921.

*China Report: Science and Technology, White Paper, no. 1*. (A translation of *Zhongguo kexue jishu zhengce zhinan*, Beijing August 1986.) Washington, D.C.: FBIS, JPRS-CST-87-013 (April 2, 1987).

Clifford, Paul G. *The Intellectual Development of Wu Chih-hui*. Ph.D diss., London University, 1978.

Compton, Boyd, ed., trans. *Mao's China: Party Reform Documents, 1942-1944*. Seattle: University of Washington Press, 1952.

Dikotter, Frank. *The Discourse of Race in Modern China*. Stanford, Calif.: Stanford University Press, 1992.

*Directory of Chinese Biologists*. Ed. Hsu Yin-chi. Soochow: Biological Supply Service, Soochow University, 1934.

*Directory of Selected Scientific Institutions in Mainland China*. Stanford, Calif.: Surveys and Research Corporation for Hoover Institution, 1970.

Dirlik, Arif. *Anarchism in the Chinese Revolution*. Berkeley, Calif.: University of California Press, 1991.

Dobzhansky, Theodosius. *Genetics and the Origin of Species*. New York: Columbia University Press, 1937, 1980.

Dubrovina, A. V. *Darwin zhui* (Darwinism). Trans. Zu Deming et al. Shanghai: Chinese Academy of Science, 1953.

Fan Dainian, and Robert S. Cohen, eds. Trans. Kathleen Dugan and Jiang Mingshan. *Chinese Studies in the History & Philosophy of Science and Technology*. Boston, Mass.: Kluwer, 1996.

Fan, Jianzhong. *Yichuanxue* (Genetics). 2 vols. Chengdu: Ministry of Culture, Education, and Industry, 1942.

Fang, Zhongxi, ed. *Darwin zhui jichu* (Foundations of Darwinism). Shanghai: Renmin jiaoyu, 1952.

———. *Mi-chu-lin xueshuo* (Michurin theory). Beijing: Zhongguo qingnian, 1955.

———. *Putong Yichuanxue* (Introductory genetics). Beijing: Science Press, 1959, 1961, 1964, 1974.

———. *Yichuanxue di liangge xuehpai* (The two schools of genetics). Beijing: Science Press, 1962.

———. *Shengwu di jinhua* (The evolution of life). Beijing: Science Press, 1964, rev. ed. 1974.

Fleron, Frederic J., Jr., ed. *Technology and Communist Culture: The Socio-Cultural Impact of Technology under Socialism*. New York: Praeger, 1977.

Goldman, Merle. *China's Intellectuals: Advise and Dissent*. Cambridge, Mass.: Harvard University Press, 1981.

Goldman, Merle, et al., eds. *China's Intellectuals and the State: In Search of a New Relationship*. Cambridge, Mass.: Harvard University Press, 1967.

Gong Yuzi. "Zhongguo Gongzhandang di kexue zhengce di lishi fazhan" (Historical development of CCP's [pre-1949] science policy). *Zeran bianzhengfa tongxun*, no. 2.6 (1980): 6-11.

Gould, Sidney, ed. *Sciences in Communist China*. Washington, D.C.: AAAS, 1961.

Graham, Loren R. *Science and Philosophy in the Soviet Union*. New York: Knopf, 1972.

Gunn, Selskar M. "Report on a Visit to China, June 9-30, 1931." RAC:RG1/Ser.601/B.12/Fldr.129, 108 pages.

――――. "China and the Rockefeller Foundation." Shanghai, January 23, 1934. RAC:RG1/Ser.601/B.12/Fldr.130, 60 pages.

*Guoli bianyiguan gongzuo gaikuang* (General report on the work of the National Translation Bureau). Chongqing, 1940, 40 pages.

Haas, William. *China Sojourner: Gist Gee's Life in Science*. Armonk, N.Y.: M. E. Sharpe, 1996.

*Biotechnology in China*. Ed. Dean H. Hamer and Shain-dow Kung. Washington, D.C.: National Academy Press, 1989.

Huang Jikun, et al. "Plant Biotechnology in China," *Science* (January 25, 2002): 674-77.

*Jinhualun xuanji* (An evolution anthology [commemorating the centenary of Charles Darwin's death]). Ed. Chen Shixiang. Beijing: Science Press, 1983.

Johnson, Virgil, and Halsey Beemer, eds. *Wheat in the People's Republic of China*. Washington, D.C.: National Academy of Sciences, 1977.

Joravsky, David. *The Lysenko Affair*. Cambridge, Mass.: Harvard University Press, 1970.

*Kangri zhanzheng shiji jiefang qu kexue ishu fazhan shiliao* (Historical materials on the development of science and technology in the liberated areas during the anti-Japanese war [1937-1945]). 5 volumes. Ed. Wu Heng. Beijing: Zhongguo xueshu, 1983-1986.

*Keji tizhi gaige di tansuo yu shijian* (Theory and practice in science and technology reform). Hunan: Science & Technology Press, 1985.

Kimura, Motoo. "The Neutral Theory of Molecular Evolution." *Scientific American* (November 1979): 94-104.

Kloppenburg, Jack R., Jr. *First the Seed: The Political Economy of Plant Biotechnology, 1492-2000*. Cambridge: Cambridge University Press, 1988.

Kohler, Robert E. *Lords of the Fly: Drosophila Genetics and the Experimental Life*. Chicago, Ill.: University of Chicago Press, 1994.

Krebs, Edward Skinner. *Liu Ssu-fu and Chinese Anarchism, 1905-1915*. Ph.D. diss., University of Washington, Seattle, 1977.

Kropotkin, Peter. *Mutual Aid*. (1902). New York: New York University reprint, 1972.

――――. *Fields, Factories, and Workshops: Or Industry Combined with Agriculture and Brain Work with Manual Work*. New York: Greenwood reprint, 1968.

――――. *The Conquest of Bread*. New York: Putnam, 1907.

――――. "Thoughts on Evolution [1910-15]." Pp. 111-243 in *Evolution and Environment: Collected Works of Peter Kropotkin*, vol. 11, edited by George Woodcock. Montreal: Black Rose Press, 1995.

Levey, Lisbeth A., ed. *Technology Transfer in China: Selected Papers* [AAAS annual meeting symposium, May 1986] n.d., n.p.

Li Peishan, ed. *Baijia zhengming: fazhan kexue di biyu zhi lu* (Let a hundred schools contend, develop the road which science must follow [Proceedings of the August 1956 Qingdao Genetics Symposium]). Beijing: Commercial Press, 1985.

Li Peishan, et al. "The Qingdao Conference of 1956 on Genetics: The Historical Background and Fundamental Experiences." Pp. 41-54 in *Chinese Studies in the*

*History and Philosophy of Science and Technology*, edited by Fan Dainian and Robert S. Cohen. Boston, Mass.: Kluwer, 1996.

Li Ruqi. *Yichuanxue* (Genetics). Beijing, n.p., 1983.

———. *Shixian shengwuxue lunwen xianji* (Anthology of experimental biology research). Beijing: Science Press, 1985.

Luo Tianyu, comp. *Shengwu zhong nei zhong jian guanxi* (Inter- and intra-species relations). Shanghai: Xinnong, 1953.

———. *Shengwu zhexue lunping* (Debates in philosophy of biology). Shanghai: Xingping, 1953.

Love, Harry H., and John H. Reisner. *The Cornell-Nanking Story.* Ithaca, N.Y.: Cornell University Press, 1963.

Lysenko, T. D. *Shengwu kexue chuangkuang* (The situation in biology). Trans. Li Ho, ed. Zhou Jianren. Beijing: Tianxia, 1949.

MacFarquhar, Roderick. *The Hundred Flowers Campaign and the Chinese Intellectuals.* New York: Columbia University Press, 1960.

———. *The Origins of the Cultural Revolution*, 3 vols. New York: Columbia University Press, 1974-1997.

Mao Zedong. *The Secret Speeches of Chairman Mao: From the Hundred Flowers to the Great Leap Forward.* Ed. Roderick MacFarquhar et al. Cambridge, Mass.: Harvard University Press, 1989.

———. *Writings of Mao Zedong, 1949-1976.* vol. 2. Ed. John K. Leung and Michael Y. M. Kau. Armonk, N.Y.: M. E. Sharpe, 1992.

Mayr, Ernst. *The Growth of Biological Thought.* Cambridge, Mass.: Harvard University Press, 1982.

Mayr, Ernst, and William B. Provine, eds. *The Evolutionary Synthesis.* Cambridge, Mass.: Harvard University Press, 1980.

Meisner, Maurice, *Marxism, Maoism and Utopianism.* Madison, Wisc.: University of Wisconsin Press, 1982.

———. *Mao's China and After: A History of the People's Republic.* New York: Free Press, 1986, 1999.

———. *The Deng Xiaoping Era.* New York: Hill and Wang, 1996.

*Meng-de-er shishi yibai zhounian jinian wenji* (Commemorative anthology for the 100th anniversary of Mendel's death). Beijing: Genetics Society of China, Science Press, 1985.

Mi Jingjiu, ed. *Darwin zhui jiben yuanli* (Basic principles of Darwinism). Beijing, n.p., 1952.

———. *Michurin gongzuo fangfa* (Methods of Michurin work). Beijing, n.p., 1956.

———. *Darwin zhui di jige jiben wenti* (Some fundamental problems of Darwinism). Beijing, n.p., 1958.

Miller, H. Lyman. *Science and Dissent in Post-Mao China: The Politics of Knowledge.* Seattle, Wash.: University of Washington Press, 1996.

Nakamura, Teiri. "Marxism and Biology in Japan." Pp. 253-69. In *Science and Society in Modern Japan*, edited by Shigeru Nakayama. Cambridge, Mass.: MIT Press, 1974.

Needham, Joseph. *Chinese Science.* London: Pilot Press, 1945.

Needham, Joseph, and Dorothy Needham, eds. *Science Outpost: Papers of the Sino-British Science Co-operation Office (British Council Scientific Office in China) 1942-1946.* London: Pilot Press, 1948.

Ogilvie, Marilyn Bailey. "The 'New Look' Women and the Expansion of American Zoology: Nettie Maria Stevens (1861-1912) and Alice Middleton Boring (1883-

1955)." Pp. 52-79 in *The Expansion of American Biology*, edited by Keith Benson et al. New Brunswick, N.J.: Rutgers University Press, 1991.

Orleans, Leo, ed. *Science in Contemporary China*. Stanford, Calif.: Stanford University Press, 1980.

Pepper, Suzanne. *Radicalism and Educational Reform in Twentieth-Century China: The Search for an Ideal Developmental Model*. Cambridge: Cambridge University Press, 1996.

*Plant Studies in the People's Republic of China*. Washington, D.C.: National Academy of Sciences, 1975.

Provine, William B. "Origin of the *Genetics of Natural Populations* Series." Pp. 5-83 in *Dobzhansky's Genetics of Natural Populations 1-43*, edited by R. C. Lewontin, et al. New York: Columbia University Press, 1981.

Pusey, James R. *China and Charles Darwin*. Cambridge, Mass.: Harvard University Press, 1983.

Pyenson, Lewis. *Civilizing Mission: Exact Sciences and French Overseas Expansion, 1830-1940*. Baltimore, Md.: Johns Hopkins University Press, 1993.

Reardon-Anderson, James. *The Study of Change: Chemistry in China, 1840-1949*. Cambridge: Cambridge University Press, 1991.

Rockefeller Foundation. *Annual Report*. New York, 1916-1949.

Ruse, Michael. *Monad to Man: The Conception of Progress in Evolutionary Biology*. Cambridge, Mass.: Harvard University Press, 1996.

Saich, Tony. *China's Science Policy in the 1980s*. Atlantic Heights, N.J.: Humanities Press, 1989.

Samuels, Richard J. *"Rich Nation, Strong Army": National Security and the Technological Transformation of Japan*. Ithaca, N.Y.: Cornell University Press, 1994.

Schneider, Laurence. "The Rockefeller Foundation, the China Foundation and the Development of Modern Science in China," *Social Science and Medicine*, no. 16 (1982): 1217-21.

———. "Using the Rockefeller Archives for Research on Modern Chinese Science," *Chinese Science*, no. 7 (1986): 25-31.

———. "Genetics in Republican China, 1920-1949." Pp. 3-30 in *Science and Medicine in Twentieth-Century China*, edited by John Bowers and Nathan Sivin. Ann Arbor, Mich.: University of Michigan Center for Chinese Studies, 1988.

———. "Learning from Russia: Lysenkoism and the Fate of Genetics in China, 1950-1986." Pp. 45-65 in *Science and Technology in Post-Mao China*, edited by Merle Goldman and Denis Simon. Cambridge, Mass.: Harvard University Press, 1989.

———, ed. *Lysenkoism in China: Proceedings of the 1956 Qingdao Genetics Symposium*. Trans. Laurence Schneider and Qin Shizhen. Armonk, N.Y.: M. E. Sharpe, 1986.

*Shengwu xueke jiaoxue shicha baogao* (Investigative report on teaching biology in the schools). Comp. Yu Han. Chengdu: Sichuan Provincial Government, Ministry of Education, 1941, 78 pages.

*Situation in Biological Science (The): Proceedings of the Lenin Academy of Agricultural Sciences of the USSR*. Moscow: Foreign Languages Publishing House, 1949.

Stavis, Benedict. *Making Green Revolution: The Politics of Agricultural Development in China*. Ithaca, N.Y.: Rural Development Committee, Cornell University, 1974.

Stebbens, George L. "The Evolution of Darwinism." *Scientific American* (July 1985): 72-82.

Stebbins, George. L., and F. J. Ayala. "Is a New Evolutionary Synthesis Necessary?" *Science*, no. 213 (1981): 967-71.

Stross, Randall. *The Stubborn Earth: American Agriculturalists on Chinese Soil*. Berkeley, Calif.: University of California Press, 1986.

Suttmeier, Richard P. *Research and Revolution: Science Policy and Societal Change in China*. Lexington, Mass.: D. C. Heath, 1974.

———. *Science, Technology, and China's Drive for Modernization*. Stanford, Calif.: Hoover Institution, 1980.

Tan Jiazhen. See Tan Chia-chen.

Tan Chia-chen. "Inheritance of the Elytral Color Patterns of Harmonia axyridis and a New Phenomenon of Mendelian Dominance." *Chinese Journal of Experimental Biology* (Chunking), no. 2 (1942).

———. "Mosaic Dominance in the Inheritance of Color Patterns in the Lady-Bird Beetle, Harmonia Axyridis." *Genetics*, (March 1946): 195-210.

———. "Genetics of Sexual Isolation between Drosophila Pseudoobscura and Drosophila Persimilis." *Genetics*, (November 1946): 558-73.

———. "Report of the Work of the Laboratory of Genetics and Cytology, Biology Institute, National University of Chekiang, Mutan, Kweichow, 1942-44." *Acta Brevia Sinensia*, (December 1944): 5-22.

———. "Pipan wo dui Michurin shengwu kexue di zuowu kanfa" (Criticism of my mistaken attitudes toward Michurin biological science). *Kexue tongbao*, no. 8 (1952): 562-63.

———. *Tantan Morgan xuepai yichuan xueshuo* (Discussion of the Morgan school theory of inheritance). Shanghai: Kexue buji, 1958.

———. *Jiyin ho yichuan* (Genes and inheritance). Beijing, 1962.

Tisdale, W. E. "Report of a Visit to Scientific Institutions in China," September-December 1933. RAC:RG1/Ser.601D/B.40, 73 pages.

Todes, Daniel P. *Darwin without Malthus: The Struggle for Existence in Russian Evolutionary Thought*. Oxford: Oxford University Press, 1989.

Wang, Richard T. Y. *Wu Chih-hui: Intellectual and Political Biography*. Ph.D. diss., University of Virginia, 1976.

Wang Shou. *Zhongguo zuowu yuzhongxue* (Plant improvement in China). Shanghai: Commercial Press, 1936.

Wei Feng. "The Great Tremors in China's Intellectual Circles: An Overview of Intellectuals Floundering in the Sea of Commercialism" (Excerpted and translated from *Zhongguo zhishi jie da zhendang: shang hai chenfu zhong de zhishifenzi saomiao*. Beijing: Zhongguo shehui chubanshe, 1993.) *Chinese Education and Society*, November-December 1996.

Weiner, Douglas. "Community Ecology in Stalin's Russia." *Isis*, no. 75 (1984): 684-96.

———. "The Roots of 'Michurinism': Transformist Biology and Acclimatization as Currents in the Russian Life Sciences." *Annals of Science*, no. 42 (1985): 243-60.

———. *Models of Nature: Ecology, Conservation and Cultural Revolution in Soviet Russia*. Pittsburgh, Pa.: University of Pittsburgh, 1988, 2000.

———. *A Little Corner of Freedom: Russian Nature Protection from Stalin to Gorbachev*. Berkeley, Calif.: University of California Press, 1999.

Wilson, Allan. "The Molecular Basis of Evolution." *Scientific American* (October 1985): 164-73.

Wu Zhongxian. *Zong Darwin dao jiyin gongcheng* (From Darwin to genetic engineering). Beijing: Beijing Agricultural University, 1984.

Yang Ts'ui-hua. *Zhongjihui dui kexue di zanzhu* (Patronage of science: The China Foundation for the Promotion of Education and Culture). Taipei: Institute of Modern History, Academia Sinica, 1991.

*Yichuanxue wenti taolun ji* (Anthology of discussions on the genetics question). 3 vols. Ed. Institute of Genetics, Fudan University, Shanghai, 1961-1963.

*Yichuanxue yu baijia zhengming: 1956 Qingdao zuotanhui yanjiu* (Genetics and the "Let a Hundred Schools Contend" policy: A study of the consequences of the Qingdao Genetics Symposium). Beijing: Beijing University Press, 1996.

Yu Guangyuan. "Speeches at the Qingdao Genetics Conference of 1956." Translation of original in *Zeran bianzhengfa tongxun*, no. 5 (1980): 5-13. Pp. 27-40 in *Chinese Studies in the History and Philosophy of Science and Technology*, edited by Fan Dainian and Robert S. Cohen. Boston, Mass.: Kluwer, 1996.

Yuan, Tung-li. *A Guide to Doctoral Dissertations by Chinese Students in America 1905-1960*. Washington, D.C.: Sino-American Cultural Society, 1961.

Zarrow, Peter. *Anarchism and Chinese Political Culture*. New York: Columbia University Press, 1990.

*Zhongguo kexuejia cedian* (Biographical dictionary of Chinese scientists), vols. 1-3. Ed. Xu Zhichun. Shandong, n.p., 1982-1984.

*Zhongguo kexueyuan shengwuxue ge yanjiuso gaoji yanjiu jishu renyuan jianjie* (Brief introduction to the ranking research and technical personnel in each of the biological research institutes of the Chinese Academy of Sciences). Beijing: Chinese Academy of Sciences, 1981.

*Zhongguo xiandai shengwuxue jiachuan* (Biographies of contemporary Chinese biologists), vol. 1. Ed. Tan Jiazhen. Changsha: Hunan Science and Technology Press, 1986.

*Zhongguo xiandai nongxue jiachuan* (Biographies of contemporary Chinese agricultural scientists), vol. 1. Ed. Jin Shanyu. Changsha: Hunan Science and Technology Press, 1986.

*Zhongxue shengwu jiaoxue di gaizao* (The reform of middle-school biology instruction). Shanghai: Science Press, 1951.

Zhou Jianran. *Jinhualun yu shanzhongxue* (Evolutionary theory and eugenics). Shanghai: Commercial Press, 1923.

———. *Shengwu jinhualun qianshuo* (Introduction to biological evolutionary theory). Shanghai: Commercial Press, 1946.

———. *Lun yushengxue yu zhongzu qishi* (Eugenics and racial discrimination). Shanghai: Science Press, 1950. Preface dated 1948.

Zhu Xi. *Shengwu di jinhua* (Biological evolution). Shanghai: Science Press, 1958.

# Index

# About the Author

**Laurence Schneider** has taught modern Chinese history at Washington University in St. Louis for the past decade, and at the State University of New York at Buffalo for the quarter century before that. He is author of *Ku Chieh-kang and China's New History* (1971), a *Madman of Ch'u* (1980), and short pieces dealing with modern China's confrontations with its past and its efforts to accommodate modern science.